AF371631

Lecture Notes in Applied and Computational Mechanics

Volume 57

Series Editors

Prof. Dr.-Ing. Friedrich Pfeiffer
Prof. Dr.-Ing. Peter Wriggers

Lecture Notes in Applied and Computational Mechanics

Edited by F. Pfeiffer and P. Wriggers

Further volumes of this series found on our homepage: springer.com

Vol. 57 Stephan E.,
Wriggers, P.,
Modelling, Simulation and Software Concepts for
Scientific-Technological Problems
251 p. 2011 [978-3-642-20489-0]

Vol. 54: Sanchez-Palencia, E.,
Millet, O., Béchet, F.
Singular Problems in Shell Theory
265 p. 2010 [978-3-642-13814-0]

Vol. 53: Litewka, P.
Finite Element Analysis of Beam-to-Beam Contact
159 p. 2010 [978-3-642-12939-1]

Vol. 52: Pilipchuk, V.N.
Nonlinear Dynamics: Between Linear and Impact Limits
364 p. 2010 [978-3-642-12798-4]

Vol. 51: Besdo, D., Heimann, B., Klüppel, M.,
Kröger, M., Wriggers, P., Nackenhorst, U.
Elastomere Friction
249 p. 2010 [978-3-642-10656-9]

Vol. 50: Ganghoffer, J.-F., Pastrone, F. (Eds.)
Mechanics of Microstructured Solids 2
102 p. 2010 [978-3-642-05170-8]

Vol. 49: Hazra, S.B.
Large-Scale PDE-Constrained Optimization
in Applications
224 p. 2010 [978-3-642-01501-4]

Vol. 48: Su, Z.; Ye, L.
Identification of Damage Using Lamb Waves
346 p. 2009 [978-1-84882-783-7]

Vol. 47: Studer, C.
Numerics of Unilateral Contacts and Friction
191 p. 2009 [978-3-642-01099-6]

Vol. 46: Ganghoffer, J.-F., Pastrone, F. (Eds.)
Mechanics of Microstructured Solids
136 p. 2009 [978-3-642-00910-5]

Vol. 45: Shevchuk, I.V.
Convective Heat and Mass Transfer in Rotating Disk
Systems
300 p. 2009 [978-3-642-00717-0]

Vol. 44: Ibrahim R.A., Babitsky, V.I., Okuma, M. (Eds.)
Vibro-Impact Dynamics of Ocean Systems and Related
Problems
280 p. 2009 [978-3-642-00628-9]

Vol.43: Ibrahim, R.A.
Vibro-Impact Dynamics
312 p. 2009 [978-3-642-00274-8]

Vol. 42: Hashiguchi, K.
Elastoplasticity Theory
432 p. 2009 [978-3-642-00272-4]

Vol. 41: Browand, F., Ross, J., McCallen, R. (Eds.)
Aerodynamics of Heavy Vehicles II: Trucks, Buses,
and Trains
486 p. 2009 [978-3-540-85069-4]

Vol. 40: Pfeiffer, F.
Mechanical System Dynamics
578 p. 2008 [978-3-540-79435-6]

Vol. 39: Lucchesi, M., Padovani, C., Pasquinelli, G., Zani, N.
Masonry Constructions: Mechanical
Models and Numerical Applications
176 p. 2008 [978-3-540-79110-2]

Vol. 38: Marynowski, K.
Dynamics of the Axially Moving Orthotropic Web
140 p. 2008 [978-3-540-78988-8]

Vol. 37: Chaudhary, H., Saha, S.K.
Dynamics and Balancing of Multibody Systems
200 p. 2008 [978-3-540-78178-3]

Vol. 36: Leine, R.I.; van de Wouw, N.
Stability and Convergence of Mechanical Systems
with Unilateral Constraints
250 p. 2008 [978-3-540-76974-3]

Vol. 35: Acary, V.; Brogliato, B.
Numerical Methods for Nonsmooth Dynamical Systems:
Applications in Mechanics and Electronics
545 p. 2008 [978-3-540-75391-9]

Vol. 34: Flores, P.; Ambrósio, J.; Pimenta Claro, J.C.;
Lankarani Hamid M.
Kinematics and Dynamics of Multibody Systems
with Imperfect Joints: Models and Case Studies
186 p. 2008 [978-3-540-74359-0]

Vol. 33: Nies ony, A.; Macha, E.
Spectral Method in Multiaxial Random Fatigue
146 p. 2007 [978-3-540-73822-0]

Vol. 32: Bardzokas, D.I.; Filshtinsky, M.L.;
Filshtinsky, L.A. (Eds.)
Mathematical Methods in Electro-Magneto-Elasticity
530 p. 2007 [978-3-540-71030-1]

Modelling, Simulation and Software Concepts for Scientific-Technological Problems

Ernst Stephan, Peter Wriggers (Eds.)

 Springer

Prof. Dr. Ernst Stephan
Gottfried Wilhelm Leibniz
Universität Hannover
Institut für Angewandte Mathematik
Welfengarten 1
30167 Hannover
Deutschland
e-mail: stephan@ifam.uni-hannover.de

Prof. Dr.-Ing. habil. Peter Wriggers
Gottfried Wilhelm Leibniz
Universität Hannover
Institut für Kontinuumsmechanik
Appelstr. 11
30167 Hannover
e-mail: wriggers@ikm.uni-hannover.de
http://www.ikm.uni-hannover.de

ISBN: 978-3-642-20489-0   ISBN 978-3-642-20490-6 (eBook)

DOI 10.1007/ 978-3-642-20490-6

Lecture Notes in Applied and Computational Mechanics   ISSN 1613-7736

e-ISSN 1860-0816

Library of Congress Control Number: 2011926506

Typeset & Cover Design: Scientific Publishing Services Pvt. Ltd., Chennai, India.

Printed on acid-free paper

9 8 7 6 5 4 3 2 1 0

springer.com

Table of Contents

Preface

The interdisciplinary Research Training Group entitled "Interaction of Modeling, Computation Methods and Software Concepts for Scientific-Technological Problems" (GRK 615) was funded by the German Research Foundation from October 2000 till September 2010. The goal was twofold: firstly, to foster interdisciplinary education on the PhD level, suitable for the demands of modern research intensive fields in industry, etc. Secondly, joint interdisciplinary research was activated among engineering, mathematics, and computer science institutes at the Leibniz Universität Hannover. The highly successful educational program resulted in 34 PhDs, all of which stand out through the interdisciplinary character of the problems treated. GRK 615 was supported by its two international partners Chalmers University, Göteborg in Sweden and Universitat Politècnica de Cataluña, Barcelona in Spain.

Within the Research Training Group research contributions were achieved in the following areas:

A Error controlled numerical methods, efficient algorithms and software development
B Elastic and inelastic deformation processes
C Models with multiscales and multi-physics

"High Performance" adaptive numerical methods with finite elements (FEM) and boundary elements (BEM), efficient solvers for linear systems, and corresponding software components for non-linear, coupled field equations were developed with applications in various branches of mechanics, electromagnetics, and geosciences. A major aspect was the industrial importance of the newly created algorithms and software tools to analyze, e.g., metal forming processes or biomechanical problems such as bone growth. The numerical simulation (with FEM and BEM) of scanning probe and atomic force microscopes is one of many examples of the fruitful cooperation of mathematicians, computer scientists, and engineers in the Research Training Group. Another class of processes, naturally connecting projects in A, B, and C, results from heat production or chemical reactions. Understanding such coupled thermohydromechanical processes in geological systems is very important for questions concerning the storing of oil, gas, and geothermal energy.

The chapters brought together in this book are examples of the research performed within the Research Training Group and mirror the interdisciplinary theme of GRK 615 in an excellent way.

Ernst P. Stephan and Peter Wriggers
Hannover, December 2010

Martensitic Phase Transformations of Mono and Polycrystalline Shape Memory Alloys – A Theoretically and Numerically Unified Concept

Gautam Sagar and Erwin Stein

Abstract The unified setting presented here is based on phase transformation (PTs) of monocrystalline shape memory alloys (SMAs) and includes polycrystalline SMAs whose microstructure is modeled using lattice variants of RVEs consisting of equal convex isotropically elastic grains with specific texture. A pre-averaging scheme for randomly distributed polycrystalline variants of PT strains is used transforming them into fictitious phase variants of a monocrystal. Thus, the integration process in parametric time and the spatial integration algorithms of the discretized variational problems for both mono and polycrystalline PTs are implemented into a unified algorithm with bifurcation within incremental time integration before spatial integration via finite element method. Furthermore, error-controlled adaptive 3D finite element method in space is presented for PT problems using an explicit a posteriori discretization error indicator with gradient smoothing and adaptive mesh refinements by new mesh generation in each adaptive step. Examples for full PT cycles and comparisons with experiment are presented.

1 Introduction

SMAs exhibit a specific feature associated with martensitic phase transformations (PTs) which is the ability to 'remember' their initial state. They have intrinsic ability to transform between austenite (parent phase) and a number of symmetry-related martensitic variants (product phases). Martensite PT is usually considered as a diffusionless first-order transformation between 'high' temperature austenitic and 'low' temperature martensitic phases [1]. The two important behaviors of martensitic PTs

Gautam Sagar
ELAN GmbH, Karnapp 25, 21079 Hamburg, Germany; e-mail: gautam.sagar@elan-edag.com

Erwin Stein,
Institute of Mechanics and Computational Mechanics (IBNM), Leibniz Universität Hannover, Appelstr. 9A, 30167 Hannover, Germany; e-mail: stein@ibnm.uni-hannover.de

E. Stephan & P. Wriggers (Eds.): Modelling, Simulation and Software Concepts, LNACM 57, pp. 1–28.
springerlink.com   © Springer-Verlag Berlin Heidelberg 2011

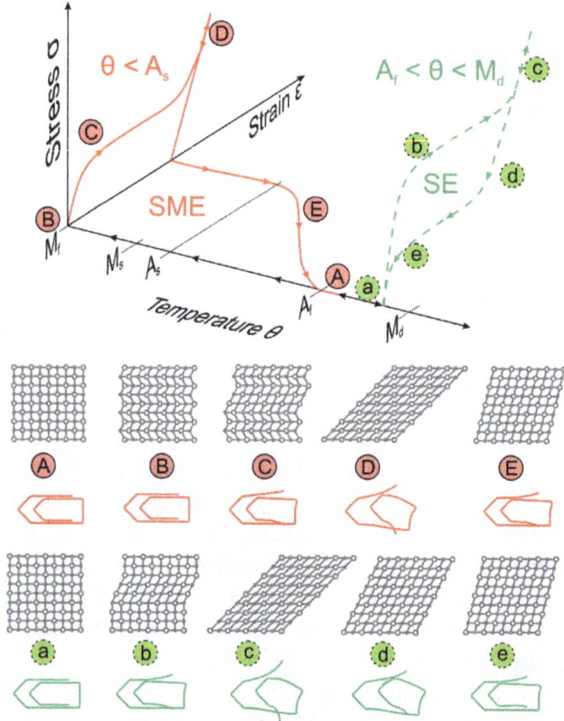

Fig. 1 Schematic illustration of the SME and SE effect by martensitic phase transformation.

are known as quasiplastic (QP) and superelastic (SE) behavior. In case of shape memory effect (SME) due to QP behavior after elastic deformation and subsequent PT (due to a critical driving force at a subcritical temperature) a SMA will only recover its old shape after unloading if a second (higher) critical temperature is reached by heating. On the other hand, to exhibit SE behavior a SMA returns immediately to its initial shape during elastic unloading if the temperature of the material has at least the second critical value from the beginning of the process (Fig. 1). Nowadays, SMAs are widely used for biomedical systems, e.g. as peripheral stents which are designed for supporting the blood vessels or orthodontic wires to correct irregularities in the position of the teeth etc. making use of superelasticity. SMAs are also employed in actuating devices for many engineering applications e.g. in robotic muscles by applying quasiplasticity. As the applications of SMAs are growing rapidly, there is a great need to develop fairly accurate and efficient models to describe this complicated material response. There are several lines of development for engineering models connected to various computational methods at different length scales of crystal properties and related phase transformation effects in mono and polycrystalline SMAs. The literature on it is very rich, hence references are restricted to the specific topics of this chapter.

Patoor et al. [2] initiated to work at macroscale using the information from microscopic scale. Therein, habit planes, Cauchy–Born hypothesis of lattice-continuum link, martensitic phase variants etc. are introduced for homogenization, allowing macroscale thermodynamics. Many sophisticated models have been developed in this line and can be found in the literature, e.g. [3–11]. However, it has been challenging to develop a constitutive theory directly based on microstructural physics and apply to solve boundary value problems at larger scales. The main challenges here has been the non-convexity of the free energy which even does not fulfill the quasi-convexity conditions. The practical advantage of partially relaxed free energy was shown by Hall and Govindjee [12].

Using Reuß assumption, a simple polycrystalline model was proposed by Mura [13] which accounts only for interaction between phase variants. Both models, proposed by Lexcellent et al. [4] and Vivet and Lexcellent [7], belong to the half-interaction group as only one variant was considered in each grain, thus the interaction between variants can not be considered. Lu and Weng [14, 15] proposed a self-consistent model where the variants are interacting within crystals. The model proposed by Jung et al. [10] considers the coupling between variants and neglects the interaction between grains. Also in the presented polycrystalline model, the interaction between the variants (of equivalent monocrystal) is considered via quasi-convex relaxation which is intimately connected to the notion of homogenization of optimal microstructure (at minimum energy). The interactions between the grains are ignored and surface energy is not considered. The model is restricted to specific texture developments.

The model used by Gall et al. [16] also considers a half-interaction because there is no interaction between variants, but grains do interact by using self-consistent method. Research work by Sun and Hwang [17–19] also belongs to this group since no variants were considered but grains are interacting.

Patoor et al. [20] proposed a model which accounts for both interactions, but those between variants are only partially considered since a simplified interaction matrix is used in single crystal case, and a self-consistent scheme is used to obtain averaged elasticity moduli over the grains. Huang et al. [21] consider the interaction between phase variants within crystal by self consistent method and elastic moduli between the grains by Eshelby's solution.

Some shortcomings among models mentioned above are that either these models are only suitable for polycrystals or for monocrystals. Modeling of both kind of materials is naturally different. The intention and goal here is to develop a combined methodology in conjunction with a unified computational algorithm to describe martensitic PTs in mono and polycrystalline SMAs for engineering applications, see also [22]. Since correspondence variants form a physically sound basis for expressing the recoverable strain under backward transformation [23], hence they are used for both models instead of habit plane variants, see e.g. [12, 24–27]. Convergence behavior of the presented error indicators and also adaptive remeshing is treated using 3D-tetrahedral elements, see also [28].

2 PTs of Monocrystalline SMAs

Several promising macro and micro-macro constitutive material models are available for quasiplastic and superelastic martensitic phase transformation cycles of mono and polycrystalline memory metal alloys. A generalized variational formulation, including quasi-convexification of energy wells for arbitrary many martensitic variants in case of monocrystals at linearized strains, was developed by Govindjee and Miehe [11] and computationally extended by Stein and Zwickert [29]. This work was further generalized by Stein and Sagar [30] for finite strain kinematics with monotonous hyperelastic stress-strain functions (Neo-Hookean model) in order to account for finite transformation strains which really take place physically. The polycrystalline PT model presented in Section 4, takes the form (after pre-averaging at mesoscale of the grains) of a fictitious monocrystalline PT model. Both mathematical models are based on the Bain's principle, e.g. [1], which states that the martensite crystal structure is built along the smallest lattice strain, and the Cauchy–Born hypothesis [31, 32]. Austenitic and martensitic crystallographic lattices and their deformations are described by Bravais lattice, using adequate linearly independent lattice vectors. According to the Cauchy–Born hypothesis, the deformation of the Bravais lattice vectors can be presented by the deformation gradient F of the C^1 point continuum, see Appendix A.

2.1 Twinning of Monocrystalline SMAs

When a shape memory alloy is cooled from austenite phase to martensite phase, the resulting microstructure shows twinned patterns of the microstructure variants. This geometric phenomenon without macroscopic deformation is known as twinning. The reason can be explained as follows. At any temperature the crystal lattice of the alloy tends to be in the minimum total potential energy state. When the temperature is greater than austenite finish temperature, only one variant (austenite phase) exists to provide the possible crystal lattice orientation satisfying minimum energy condition. However, when the temperature is between martensite finish temperature and austenite finish temperature there are many possible orientations in martensitic phases according of their symmetry orders. Thus, many different martensitic phases which have equal energy are generated simultaneously during the cooling process. Typically, the mixture of martensitic microstructure occurs at fine scales. Importantly, distinct martensitic lattices exist in such a way that rows of the atoms are kinked but unbroken across the interface (zig-zag pattern).

Under thermal loading, a minimum strain energy at the transition from austenite to martensite phases is achieved without a macroscopic deformation, and the most favorable variant outgrows in a self-accommodating manner via twinning process. During the transformation of austenitic to martensitic phases, there is a significant increase in the strain energy due to the misfit between the martensite variant and surrounding austenite. Thus the produced phase is at last characterized by twinning process, which minimizes the misfit between the evolved martensite and parent austenite matrix.

On applying mechanical loads, the internally twinned martensite may detwin. As the martensite detwins, macroscopic strains get accumulated due to growth of favorably oriented single crystals of martensite.

A geometrically nonlinear theory of martensitic transformations developed by Ball and James [24, 25], and also by Bhattacharya [1], is the main tool to model microstructure. In the following sections twinning is not considered, but a direct macro deformation from austenite to detwinned martensite is analysed.

2.2 PT Modeling at Linear and Nonlinear Kinematics

According to Ball and James [24], the macroscopic free strain energy of a SMA crystal is given by the global minimum of the energies of all possible i.e. compatible n-phase variants, $\Psi = \min_{i=1\dots n}[\psi_i^{\text{el}} + \psi_i^{\text{ch}}]$, where ψ_i^{el} is elastic energy and ψ_i^{ch} is the so-called chemical energy of the i-th phase variant. The phase variants are described through internal variables with constraints for the mass conservation of the phases.

For the defined type of problem Ψ is not quasi-convex, which implies non-existence of deformations minimizing Ψ for prescribed boundary data and hence indicates the formation of microstructure, Govindjee, Mielke and Hall [33]. To overcome the problem, quasi-convex relaxation is used connected with homogenization of the microstructure at minimum energy. The global free energy, depending on the phase fractions can be decoupled at linearized strain as [11]

$$\Psi(\boldsymbol{\varepsilon}, \boldsymbol{\xi}) = \boldsymbol{\xi} \cdot \boldsymbol{\psi}(\boldsymbol{\varepsilon}) + \Psi_{LS}^{M}(\widetilde{\boldsymbol{\varepsilon}}^{t}, \boldsymbol{\xi}), \tag{1}$$

with the vector of phase fractions $\boldsymbol{\xi} = \sum_{i=1}^{n} \xi_i \boldsymbol{e}_i$, $\boldsymbol{e}_i \cdot \boldsymbol{e}_j = \delta_{ij}$ (1 austenitic phase and $n-1$ martensitic phases), where $\widetilde{\boldsymbol{\varepsilon}}^{t}$ is the phase transformation strain tensor, transformed into global coordinates of the related specimen, and $\Psi_{LS}^{M}(\widetilde{\boldsymbol{\varepsilon}}^{t}, \boldsymbol{\xi})$ is the so-called energy of mixing as part of the convexified free energy at linear strains. At linearized strains the mixing energy for the general n-variant monocrystal was derived by Govindjee et al. [33], using Reuß bound. It reads

$$\Psi_{LS}^{M}(\widetilde{\boldsymbol{\varepsilon}}^{t}, \boldsymbol{\xi}) = -\frac{1}{2} \sum_{i=1}^{n} \xi_i \, \widetilde{\boldsymbol{\varepsilon}}_i^{t} : \mathbb{C} : \widetilde{\boldsymbol{\varepsilon}}_i^{t} + \frac{1}{2} \sum_{i=1}^{n} \sum_{j=1}^{n} \xi_i \xi_j \, \widetilde{\boldsymbol{\varepsilon}}_i^{t} : \mathbb{C} : \widetilde{\boldsymbol{\varepsilon}}_j^{t},$$

where the same elasticity tensor is assumed for all the phase variants. The mathematical model at finite strain is based on the multiplicative decomposition of the total deformation gradient $\boldsymbol{F}$ into elastic $\boldsymbol{F}^e$ and transformation part $\boldsymbol{F}^t$, and using Neo-Hookean hyperelastic isotropic material [30]. An approximated extension of the quasi-convexified free energy for phase fractions from small to finite strains, $\Psi(\boldsymbol{\varepsilon}, \boldsymbol{\xi}) \longrightarrow \Psi(\boldsymbol{b}^e, \boldsymbol{\xi})$, reads

$$\Psi(\boldsymbol{b}^e, \boldsymbol{\xi}) = \boldsymbol{\xi} \cdot \boldsymbol{\psi}(\boldsymbol{b}^e) + \Psi_{FS}^{M}(\widetilde{\boldsymbol{b}}^t, \boldsymbol{\xi}) , \tag{2}$$

$\boldsymbol{b}^e = \boldsymbol{F}\left(\boldsymbol{U}_i^{t\,2}\right)^{-1}\boldsymbol{F}^T$ is elastic left Cauchy–Green tensor, and $\widetilde{\boldsymbol{b}}^t$ is left Cauchy–Green PT tensor, transformed into the specimen coordinates. $\boldsymbol{\psi}$ is the vector of the phase energies with $\boldsymbol{\psi} = \sum_{i=1}^{n} \psi_i\, \boldsymbol{e}_i, \boldsymbol{e}_i \in \mathbb{R}^n$. $\Psi_{FS}^{M}(\widetilde{\boldsymbol{b}}^t, \boldsymbol{\xi})$ follows from a the approximated incremental extension of the linearized energy of mixing to finite strains for Neo-Hookean material (with convex stress-strain function), [30]. Its proper analytical form for finite strains is not yet known. Therefore, it is not sure whether the given incremental formulation is complete. This needs further investigation, especially the existence of local lower bounds.

The resulting free energy of mixing for a 2-phase system (1 means austenite, 2 means martensite) is given as $\Psi_{FS}^{M}(\widetilde{\boldsymbol{b}}^t, \boldsymbol{\xi}) \approx -\xi_2\ \psi_2^t(\widetilde{\boldsymbol{b}}_2^t, J_2^t) + \xi_2\ \xi_2\ \psi_2^t(\widetilde{\boldsymbol{b}}_2^t, J_2^t)$, with $\widetilde{\boldsymbol{b}}^t = \tilde{\boldsymbol{R}}\,\boldsymbol{b}^t\,\tilde{\boldsymbol{R}}^T$, which was used for the study of full PT cycle in CuAlNi [30] where only one martensitic phase was active besides the austenitic parent phase.

The mass conservation condition requires that all scaled phases have to sum up to $1 \Rightarrow \boldsymbol{e}^* \cdot \boldsymbol{\xi} - 1 = 0; \quad \boldsymbol{e}^* = \sum_{i=1}^{n} e_i^* \boldsymbol{e}_i, \ e_i^* = 1$, where the phase fractions have to be positive semi-definite $(\Rightarrow \xi_i \geq 0)$. $\boldsymbol{e}^*$ is the normal vector of the $n - 1$ dimensional convex PT polytope.

Each phase energy ψ_i results from an elastic and a temperature dependent 'chemical' part, $\psi_i = \psi_i^{\mathrm{el}} + \psi_i^{\mathrm{ch}}$, with the elastic energy, ψ_i^{el}. The elastic energy for phase i at linearized strain reads $\psi_i^{\mathrm{el}} = \frac{1}{2}\boldsymbol{\varepsilon}_i^{\mathrm{el}} : \mathbb{C}_i : \boldsymbol{\varepsilon}_i^{\mathrm{el}}$, with the elastic strain $\boldsymbol{\varepsilon}_i^{\mathrm{el}} = (\boldsymbol{\varepsilon} - \boldsymbol{\varepsilon}_i^t)$. Then ψ_i^{el} follows as $\psi_i^{\mathrm{el}} = \frac{1}{2}(\boldsymbol{\varepsilon} - \widetilde{\boldsymbol{\varepsilon}}_i^t) : \mathbb{C}_i : (\boldsymbol{\varepsilon} - \widetilde{\boldsymbol{\varepsilon}}_i^t)$, with $\widetilde{\boldsymbol{\varepsilon}}_i^t = \tilde{\boldsymbol{R}}\,\boldsymbol{\varepsilon}_i^t\,\tilde{\boldsymbol{R}}^T$, $\mathbb{C}$ the linear elasticity tensor for each phase. The approximation $\mathbb{C} = \mathbb{C}_i$ is used in the above equations. $\tilde{\boldsymbol{R}}$ is the crystal orientation matrix and $\boldsymbol{\varepsilon}_i^t$ is linearized transformation strain of phase i as $\boldsymbol{\varepsilon}_i^t = \frac{1}{2}\left(\boldsymbol{U}_i^{t\,2} - \mathbf{1}\right) \approx \boldsymbol{U}_i^t - \mathbf{1}$.

The hyperelastic elastic free energy of phase i is split into the volumetric and deviatoric terms as $\psi_i^{\mathrm{el}} = W_i(J_i^e) + \overline{W}_i\,(\overline{\boldsymbol{b}}_i^e)$, with the deviatoric part of elastic left Cauchy–Green tensor $\overline{\boldsymbol{b}}_i^e := J_i^{e-2/3}\,\boldsymbol{F}_i^e\,\boldsymbol{F}_i^{eT} \equiv J_i^{e-2/3}\,\boldsymbol{b}_i^e$, W_i is a convex function of $J_i^e := \det \boldsymbol{F}_i^e$. Herein $\boldsymbol{F}_i^e$ is elastic deformation gradient of i-th phase and, $\boldsymbol{F}_i^t$ is phase transformation gradient of i-th phase for n-phase system. The following explicit forms of a Neo-Hookean hyperelastic material are considered:

$$W_i(J_i^e) := \frac{1}{2}\kappa\left[\frac{1}{2}(J_i^{e2} - 1) - \ln J_i^e\right], \quad \overline{W}_i\,(\overline{\boldsymbol{b}}_i^e) := \frac{1}{2}\mu\,(\mathrm{tr}[\overline{\boldsymbol{b}}_i^e] - 3), \tag{3}$$

where μ and κ are the shear modulus and bulk modulus for linearized strains, respectively.

The chemical free energy ψ^{ch} for linearized strains, i.e. with mass density ρ_0 in reference state, was given in [34] for three phases and used for n-phase material in [11, 12, 29] as

$$\psi^{\mathrm{ch}} = \sum_{i=1}^{n} \psi_i^{\mathrm{ch}} \boldsymbol{e}_i = \sum_{i=1}^{n}\left[\rho_0 c_i \theta\left(1 - \log\left(\frac{\theta}{\theta_0}\right)\right) + \rho_0 l_i\left(\frac{\Delta\theta}{\theta_0}\right)\right]\boldsymbol{e}_i,$$

where $\Delta\theta = \theta - \theta_0$ is the temperature difference from the actual to the absolute temperature θ_0, the latent heat l_i of the i-th phase is exothermic for PT from austenite to martensite $(A \xrightarrow{\text{exo}} M)$, and endothermic for the reverse process, $(M \xrightarrow{\text{endo}} A)$. From crystallographic symmetry $l_i = l_j, \forall i, j$ holds. c_i is the specific heat capacity of the i-th phase where $c_i = c$ can be assumed. The heat energy related to volumetric expansion is neglected.

The presence of kinematic constraints of the material suggests the enhancement of the free energy function, equation (1), by a Lagrangian functional [6,8,11,29] for small strains which reads

$$\mathcal{L}(\boldsymbol{\varepsilon}, \boldsymbol{\xi}, \boldsymbol{\gamma}, \delta) = \Psi(\boldsymbol{\varepsilon}, \boldsymbol{\xi}) - \boldsymbol{\gamma} \cdot \boldsymbol{\xi} + \delta(\boldsymbol{e}^* \cdot \boldsymbol{\xi} - 1), \tag{4}$$

and at finite strain for incremental analysis, equation (2), given in [30] as

$$\mathcal{L}(\boldsymbol{b}^e, \boldsymbol{\xi}, \boldsymbol{\gamma}, \delta) = \Psi(\boldsymbol{b}^e, \boldsymbol{\xi}) - \boldsymbol{\gamma} \cdot \boldsymbol{\xi} + \delta(\boldsymbol{e}^* \cdot \boldsymbol{\xi} - 1), \tag{5}$$

with the vector $\boldsymbol{\gamma}$ and the scalar δ as Lagrangian parameters for n phases, which have to fulfill the Kuhn–Tucker conditions of the global saddle point problem $\gamma_i \geq 0, -\xi_i \leq 0; i = 1$ to n and $\boldsymbol{\gamma} \cdot \boldsymbol{\xi} = 0$. The extension of the free energy to a Lagrangian functional by adding the constraint conditions multiplied with Lagrangian multipliers is called generalized formulation.

2.2.1 Stress Response and Thermodynamic Forces

Using the material theory with internal variables and the Coleman and Noll argument [35,36] yields the stress tensor including phase fractions, denoted by $\boldsymbol{\sigma}^*$, as the partial derivative of $\mathcal{L}$ with respect to the elastic strain tensor. The derivation of $\mathcal{L}$ with respect to the phase fraction vector $\boldsymbol{\xi}$ yields the vector of driving thermodynamic forces $\boldsymbol{f}$ for PT.

The expressions for the complete stress tensor at linearized strains, $\boldsymbol{\sigma}^*(\boldsymbol{\varepsilon}, \boldsymbol{\xi})$, and at finite strains (FS), $\boldsymbol{\sigma}^*(\boldsymbol{b}^e, \boldsymbol{\xi})$, depending on phase fraction, $\boldsymbol{\xi}$, are as follows:

$$\text{linear: } \boldsymbol{\sigma}^*(\boldsymbol{\varepsilon}, \boldsymbol{\xi}) = \sum_{i=1}^{n} \xi_i \, \mathbb{C}_i : (\boldsymbol{\varepsilon} - \tilde{\boldsymbol{\varepsilon}}_i^t); \quad \text{FS: } \boldsymbol{\sigma}^*(\boldsymbol{b}^e, \boldsymbol{\xi}) = \sum_{i=1}^{n} \xi_i \, \boldsymbol{\sigma}_i(\boldsymbol{b}_i^e, J_i^e), \tag{6}$$

where $\boldsymbol{\sigma}_i(\boldsymbol{b}_i^e, J_i^e)$ is the Cauchy stress tensor for the elastic energy function $\psi_i^{\text{el}}(\boldsymbol{b}_i^e, J_i^e)$. Index i represents the phase number, and n is the total number of phases.

The complete Cauchy stress tensor including phase fractions for the n-phase system yields

$$\boldsymbol{\sigma}^*(\boldsymbol{b}^e, \boldsymbol{\xi}) = \sum_{i=1}^{n} \frac{1}{J_i^e} \xi_i \left[\frac{1}{2}\kappa \left(J_i^{e2} - 1 \right) \mathbf{I} + \mu \left(\overline{\boldsymbol{b}}_i^e - \frac{1}{3}\text{tr}[\overline{\boldsymbol{b}}_i^e] \mathbf{I} \right) \right].$$

The thermodynamic driving force, $f = -\partial \mathcal{L}/\partial \xi$, follows in a canonical way as

$$f_{\mathrm{LS}} = -\psi(\varepsilon) - \partial_\xi \Psi_{LS}^M(\widetilde{\varepsilon}^t, \xi) + \gamma - \delta e^*$$

$$f_{\mathrm{FS}} = -\psi(b^e) - \partial_\xi \Psi_{FS}^M(\widetilde{b}^t, \xi) + \gamma - \delta e^* \tag{7}$$

at linearized and finite strains, respectively.

From the local energy dissipation condition,

$$\mathcal{D} = f \cdot \dot{\xi} \geq 0 \,, \tag{8}$$

the phase fraction vector ξ can be determined using the local maximum dissipation principle $f \cdot \dot{\xi} \to$ Max.

2.2.2 PT Inequality and Hypothesis of Maximum Dissipation

With analogies to the theory of stable inelastic deformations in elastoplasticity it is deduced that the local 'transformation function' ϕ has to be convex (similar to the convex plastic yield function) as

$$\phi = \|f\| - f_c \leq 0 \,; \quad \text{with elastic domain} \quad \mathbb{E} = \{f \mid \phi(f) < 0\} \,, \tag{9}$$

with the important critical driving force f_c for initiating PT at the energy barrier. Transformation can only take place if the norm of the conjugate force is equal to f_c. In [11], the L_2-norm is used to determine an elastic step, which is utilized for this work too. It was also used in [29].

In order to complete the constitutive model, an evolution law for phase fractions is required. With equations (8) and (9), the additional local Lagrangian functional

$$\Pi(f, \lambda) = -\mathcal{D} + \lambda \phi \longrightarrow \text{stat} \tag{10}$$

is introduced with the Lagrangian parameter λ (the loading factor), describing a saddle point problem with the Kuhn–Tucker conditions $\lambda \geq 0$, $\phi \leq 0$, and $\lambda\phi = 0$.

The stationary condition (partial derivative of Π has to be 0) yields the desired evolution equation with the normality rule, $\dot{\xi} = \lambda\partial_f\phi$, completing the constitutive equations of the monocrystalline PT model.

3 Time Integration of Constitutive Equations

At first, the resulting equations are integrated in process time with increments Δt (parameterized by prescribed displacement increments at Dirichlet boundaries of the system), using an implicit Euler-backward finite difference method to the evolution equation. Then integration in space follows via finite elements.

The main difficulty is to know in advance which variant will become active in time interval $\Delta t = t_{n+1} - t_n$, i.e. which variant will have active constraints associated with the condition $-\xi_i \leq 0$. A robust active set strategy for locally quasi-convexified energy minimization as given in [11] is used for time-integration of the resulting four nonlinear incremental equations (25 to 28) in Section 3. Herein, time integration is presented only for finite strain formulation. For simplicity the subscript $n+1$ indicating discrete time steps is omitted in the sequel. The time-invariant driving force f reads for finite strains from equation (7),

$$f + \psi(b^e) - \gamma + \partial_\xi \Psi_{FS}^M(\widetilde{b}^t, \xi) + \delta e^* = 0 . \tag{11}$$

The Kuhn–Tucker conditions for the Lagrange multiplier λ, are

$$\Delta\lambda\phi = 0 \quad \text{and} \quad \Delta\lambda \geq 0, \tag{12}$$

with loading increment $\Delta\lambda = \Delta t \bar{\lambda}$ in the process time increment Δt, and scaling value $\bar{\lambda}$. The PT criteria reads

$$\psi - \|f\| \quad f_c \leq 0 . \tag{13}$$

The evolution equation for ξ (the growth of phase fraction at time t_n) follows from the normality rule with the incremental load factor $\Delta\lambda$ as

$$\xi - \xi_n - \Delta\lambda\partial_f\phi = 0 , \tag{14}$$

by which the PT process is controlled.

Table 1 describes the active set strategy for time integration. The further time-invariant constraints for the polytope of phase fractions are

Table 1 Active set solution strategy for every material point of the system at each time.

Step 1. Initialization of active constraint set: $B = \{i \mid \xi_{i\,n} = 0\}$
Step 2. Solve the constitutive equations (25 to 28)
Step 3. Check the constraints and update B by:

Removing constraints: $\quad B = B \setminus \{i \mid \gamma_i < 0\}$
and adding the constraints:

$$\xi_{\min} = \min_\beta \xi_\beta; \quad \text{if} \quad \xi_{\min} < 0, \quad \text{then} \quad B = B \cup \{i \mid \xi_i = \xi_{\min}\}$$

Step 4. Re-solve equations (25 to 28) and go to Step 3 until all equations are satisfied with the given error tolerance

---

$$e^* \cdot \boldsymbol{\xi} - 1 = 0 \quad \text{mass conservation condition,} \tag{15}$$

$$-\xi_i \le 0 \quad \text{positiveness of the phase fractions,} \tag{16}$$

$$\gamma_i \ge 0 \quad \text{second Kuhn–Tucker condition,} \quad \text{and} \tag{17}$$

$$-\boldsymbol{\gamma} \cdot \boldsymbol{\xi} = 0 \quad \text{third Kuhn–Tucker condition.} \tag{18}$$

Equations (11–18) have to be fulfilled in every Gaussian integration point of all finite elements in space, for which tetrahedrons and hexahedrons are used in this work. In order to detect whether a deformation increment is still elastic or already in the state of PT, the following PT inequality (at a frozen deformation state) is introduced with a trial driving force

$$\boldsymbol{f}^{tr} = -\boldsymbol{\psi}(\boldsymbol{b}^e) + \boldsymbol{\gamma} + \partial_{\boldsymbol{\xi}} \Psi^M_{FS}(\widetilde{\boldsymbol{b}}^t, \boldsymbol{\xi}) - \delta e^* \tag{19}$$

at the beginning of a (process) time increment Δt with the conditions

$$\phi(\boldsymbol{f}^{tr}) \begin{cases} < 0 \Rightarrow & \text{step } \Delta\lambda \text{ is elastic} \\ \ge 0 \Rightarrow & \text{PT takes place in this step } \Delta\lambda. \end{cases} \tag{20}$$

After elimination of δ from (19) by taking the dot product with e^*, the trial driving force results as

$$\boldsymbol{f}^{tr} = -\boldsymbol{s} + \mathcal{P}^* \boldsymbol{\gamma} \,, \tag{21}$$

with $\boldsymbol{s}$ as points (driving forces) in the f_c neighborhood (inside or outside of the admissible region of critical driving force vector) and $\mathcal{P}^*$ as the orthogonal projection tensor on the PT surface,

$$\boldsymbol{s} = \mathcal{P}^* \left[\boldsymbol{\psi}(\boldsymbol{b}^e) + \partial_{\boldsymbol{\xi}} \Psi^M_{FS}(\widetilde{\boldsymbol{b}}^t, \boldsymbol{\xi}) \right]; \quad \mathcal{P}^* = \left(\mathbb{I} - \frac{1}{n} e^* \otimes e^* \right),$$

where $\mathbb{I}$ is the rank 2 identity tensor on $\mathbb{R}^n$. It should be noted that $\boldsymbol{f}^{tr}$ is still dependent on the unknown Lagrangian parameter $\boldsymbol{\gamma}$.

Equation (21) can be geometrically interpreted that an elastic step can occur if the known point $\boldsymbol{s}$ is within a f_c neighborhood of the projection of the positive span,

$$\mathbb{K}^+ = \left\{ x \,\middle|\, x = \sum_{i \in \boldsymbol{B}} \gamma_i e_i \text{ with } \gamma_i \ge 0 \right\},$$

on the hyperplane orthogonal to e^* [11]. This is the set of points $\mathcal{P}^* \mathbb{K}^+$. e_i are the basis vectors of the canonical orthonormal basis on $\mathbb{R}^n$. The distance between point $\boldsymbol{s}$ and the projected *positive span* $\mathbb{K}^+$ is denoted by d_1.

The possibility of a non-physical elastic step has to be regarded. It can arise during the active set selection process if too many phase fraction constraints are assumed to be active. Such a situation may occur when there is a solution of equation (21) for negative components in the vector $\boldsymbol{\gamma}$. It can be expressed by introducing the total span, $\mathbb{K} = \left\{ x \mid x = \sum_{i \in \boldsymbol{B}} \gamma_i e_i \text{ with } \gamma_i \in \mathbb{R} \right\}$. A non-physical step will take place if the known point $\boldsymbol{s}$ is within a f_c neighborhood of $\mathcal{P}^* \mathbb{K}$ and not within a

f_c neighborhood of $\mathcal{P}^*\mathbb{K}^+$. The distance between point s and the projected *total span* $\mathbb{K}$ is denoted by d_2. Hence to know when an elastic step is taking place, one has to solve two nonlinear problems: (i) the constrained elastic minimization problem, $d_1 = \min_{\boldsymbol{\gamma}_B \geq 0} \|s - \boldsymbol{V}\boldsymbol{\gamma}_B\|$, where $\boldsymbol{V} = [\mathcal{P}^*\boldsymbol{e}_{B(1)} \ldots \mathcal{P}^*\boldsymbol{e}_{B(m)}]$, $\boldsymbol{\gamma}_B \in \boldsymbol{B}$, and $\boldsymbol{\gamma}_B \cdot \boldsymbol{w}_i = \boldsymbol{\gamma} \cdot \boldsymbol{e}_{B(i)}$. The $\boldsymbol{w}_i$ are the basis vectors of the canonical orthonormal basis of $\mathbb{R}^m$, m is the number of active constraints in active set $\boldsymbol{B}$, and (ii) the unconstrained elastic minimization problem, $d_2 = \min_{\boldsymbol{\gamma}_B} \|s - \boldsymbol{V}\,\boldsymbol{\gamma}_B\|$, $\boldsymbol{\gamma}_B :$ all γ in the active constrain set.

In case of L_2-norm the stated problems are classical quadratic programming problems and can be easily solved by using standard methods [37]. Based on the values of d_1 and d_2 elastic step selection is done by the following selection algorithm:

- If $d_1 \leq f_c$, then time step is elastic.
- If $d_1 > f_c$ and $d_2 \leq f_c$, then time step is non-physical elastic with some negative γ_i in vector $\boldsymbol{\gamma}$.
- If $d_1 > f_c$ and $d_2 > f_c$, then PT evolution takes place.

In case of PT evolution, additional quantities have to be computed. As outlined above the local iterative solution algorithm for the unknowns $\boldsymbol{f}, \boldsymbol{\gamma}, \Delta\lambda, \delta$ in the process time is based on an active set strategy (Table 1) for which all inactive partial phases (phase fractions that are zero) have to be stored in each spatial integration point of finite elements. The initialization of unknowns is done as following:

$$\gamma_{B(i)} = \gamma_{B_i} \quad \text{(using } \gamma_B \text{ from the computation of } d_1\text{)}, \quad \boldsymbol{f} = -s + \mathcal{P}^* \cdot \boldsymbol{\gamma} \quad (22)$$

$$\text{and} \quad \delta = \frac{1}{n} e^* \cdot [-\boldsymbol{\psi}(\boldsymbol{b}^e) + \boldsymbol{\gamma} - \partial_\xi \Psi_{FS}^M(\boldsymbol{\xi}_n + \Delta\lambda\partial_f\phi)]. \quad (23)$$

Introducing the time derivative, $\dot{\boldsymbol{f}} = \frac{\partial f}{\partial \xi} \cdot \dot{\boldsymbol{\xi}}$, where $\dot{\boldsymbol{f}}$ is non-zero when PT is taking place ($\dot{\boldsymbol{\xi}} \neq \boldsymbol{0}$), a frozen deformation state is defined as $\dot{\boldsymbol{f}} = \boldsymbol{f}^{tr}$, and then $\Delta\lambda$ can be computed by

$$\|\partial_\xi \boldsymbol{f} \cdot \dot{\boldsymbol{\xi}}\| = \|\boldsymbol{f}^{tr}\| = d_1 \Rightarrow \Delta\lambda = \frac{d_1}{\|\partial_\xi \partial_\xi \Psi_{FS}^M(\widetilde{\boldsymbol{b}}^t, \boldsymbol{\xi}) \cdot \partial_f\phi\|}. \quad (24)$$

The constitutive equations (11), (13), (14), the mass conservation condition (15), and above explained constraints (16–18) following from the stationary conditions of the Lagrangian functional yield the final coupled incremental equations for every material point of the whole system or – in spatially discretized form – in every Gaussian integration point of the finite elements; they read as follows:

$$\boldsymbol{f} + \boldsymbol{\psi}(\boldsymbol{b}^e) - \boldsymbol{\gamma} + \partial_\xi \Psi_{FS}^M(\boldsymbol{\xi}_n + \Delta\lambda\partial_f\phi) + \delta e^* = \boldsymbol{0} \quad (25)$$

$$\phi(\boldsymbol{f}) = \|\boldsymbol{f}\| - f_c = 0 \quad (26)$$

$$\xi_{in} + \Delta\lambda\partial_{f_i}\phi = 0 \quad \text{for all } i \in \text{active set} \quad (27)$$

$$e^* \cdot (\boldsymbol{\xi}_n + \Delta\lambda\partial_f\phi) - 1 = 0 \quad (28)$$

and are solved iteratively by Newton's method. For the evolution of PT, $n + m + 2$ unknowns are to be determined by $n + m + 2$ equations (25 to 28); n is the total number of phases, and m is the number of constraints for the active set. The global unknowns are stored in the vector $X := \{f, \gamma_B, \Delta\lambda, \delta\}$ and the global residuals in the vector $R := \{(25), (26), (27), (28)\}$. The iteration tangent matrix reads (see also [11]), $K = \partial R / \partial X$ from which the material tangent at finite strain follows as

$$\mathfrak{c}^{\text{algo}} = \sum_{i=1}^{n} \xi_i \mathfrak{c}_i - \sum_{i=1}^{n} \sum_{j=1}^{n} \mathfrak{c}_{ij}^{PT}, \tag{29}$$

where $\mathfrak{c}_{ij}^{PT} = D_{ij} [\sigma_i(b_i^e) \otimes \sigma_j(b_j^e)]$ is the spatial PT tangent for interacting phases i and j, with the abbreviation, $D = \delta_f \phi \otimes X + \Delta\lambda \delta_{ff}^2 \phi A = D_{ij} e_i \otimes e_j$,

Herein A is the upper left $n \times n$ block of K^{-1}, and v is the vector of the first n entries of the $(n + m + 1)$-th row of K^{-1}.

The time integration of presented model is performed within the FE-program Abaqus via UMAT interface; in case of nonlinear deformation processes this interface requires the Jaumann rate of Kirchoff stress tensor as the tangent in the current configuration in order to get quadratic convergence [38].

4 PT of Polycrystalline SMAs

4.1 Presumptions for Modeling Polycrystals

With the aim to develop a unified computational PT concept, the following consistent assumptions for specific engineering application are introduced using the RVE concept, yielding first order approximation which is fairly approved by related experiments (Section 6):

1. All grains of a polycrystalline RVE have the same topology, and they are convex without empty volume [21, 39]. A RVE consists of sufficient finite number of grains.
2. All grains of a RVE have same size (volume) [21, 40].
3. All grains are kinematically C^1 compatible.
4. All grains have the same number of phase variants.
5. The volumetric size of a particular phase variant is the same in every grain.
   Remark: In polycrystalline shape memory materials the real grains may have different topology and size. The transformation strains differ between grains, and all neighboring grains interact with each other due to incompatibility of the different transformation strains [41]. However, in the presented work it is assumed that all grains have the same topology and sizes as well as they are kinematically C^1 compatible. The response of one grain does not impact the stress or strain state in the neighboring grain. These assumptions are in line with the work of Huang

et al. [21]. Furthermore, the total number of martensitic variants depends on the crystalline structure [10], and it is reported in [42] that the active variants in each grain are the same in each load-unload cycle according to the type of alloy. This has been illustrated in the micrograph of activated variants at 2% strain in one set of grains at different cycle number for NiTi alloy [42].

6. All grain orientations are deterministic and represented by Eulerian angles [21, 40].

Remark: This is a further approximation because generally, there is a stochastic distribution of grain orientations which can be described by Young's measure theory, provided the required data are available [39]. Published calculations show that the influence of this stochastic behavior can be neglected for the computed examples in this chapter [39]. From these results it can be deduced that pre-texturing by pre-training reduces the stochastic effect. Deeper knowledge needs further research.

7. A constant transformation strain is assumed in each phase variant of a grain [21, 39, 40].

Remark: This is motivated from the micromechanical theory of an inclusion in an elastic body for which a constant strain is assumed (instead of a phase variant in this work).

8. The elastic anisotropy of martensitic and austenitic phases caused by texture is neglected [6, 8, 10–12, 21, 29, 30, 33, 40, 42, 43].

Remark: In general, elastic anisotropy in the material can arise especially due to rolling process which is considered in [39]. Here, it is assumed that this anisotropy can be neglected in the macro model.

9. The Reuß homogenization assumption (yielding a lower bound) is used which states that all grains of a RVE are subjected to a uniform stress state which is equal to the applied macroscopic stress state.

Remark: According to Daly et al. [44], the phase transformation initiates in 'favorably-oriented' grains. This is consistent with the observation that the Sachs model, which assumes that each grain deforms independently in response to the uniformly applied stress without constraints imposed by its neighbors, is a good predictor for stress at initiation. The Reuss approximation has also been used by other researchers, e.g. [40–42]. The Taylor approximation yielding an upper bound is not adequate for presented algorithm because for this the introduced averaging method for PT cannot be used. The upper bound is got upon assumptions on the type of microstructures, in particular lamination type, which is formed in shape memory alloys. This would lead to very complicated upper bounds with computational expense. These bounds rely on the information about the twinning effect. For example, cubic to monoclinic-II transformation yields 13-variants (1 austenite and 12 martensitic variants), and the best upper bound is achieved by evaluating and comparing all 955 compatible combinations of phase fraction vectors [45].

4.2 Modeling of Polycrystals

Based on the above assumptions, an averaging scheme is used for polycrystalline PT strains associated with each variant of the crystals which are assumed to be randomly distributed, in contrast to the assumption of equally distributed textured crystals (e.g. achieved by rolling and/or pre-training) as applied in [10]. In this chapter a simple habit plane-based multi-variant model has been proposed as extension of an earlier work [8], and the macroscopic Lagrangian transformation strains were computed as in monocrystal. Furthermore, each finite element of a discritized system was assumed to correspond to a crystal grain with proper texture. The material model was formulated in a large deformation material setting restricted to St. Venant elastic material i.e. allowing large rotations but only linearized strains.

In this section, the averaged PT strains for each variant, $\langle \varepsilon_i^t \rangle$, are treated similar to the transformed monocrystalline PT strains $\widetilde{\varepsilon}_i^t$ for each phase i. By pre-homogenization, the material properties of randomly oriented grains of a RVE are represented approximately at Gaussian integration points (GPs) of finite elements.

With the Reuß assumption (No. 9, Section 4.1), the average PT strain of variant i in an equivalent monocrystal of a polycrystalline RVE with N-grains follows as

$$\langle \varepsilon_i^t \rangle = \sum_{g=1}^{N} \frac{V_i^g}{V_i} \, \boldsymbol{R}^g \, \varepsilon_i^t \, \boldsymbol{R}^{gT} \,, \tag{30}$$

where $g = 1 \ldots N$ is the number of grains in the RVE, $i = 1 \ldots n$ is the number of phase variants, V_i^g is the volume of i-th variant in grain g, $V_i = \sum_{g=1}^{N} V_i^g$ is the sum of i-th variant over all N-grains of a RVE, and $\boldsymbol{R}^g$ is the orientation matrix of grain g. The orientation of g-th grain is described by the three random Euler angles $(\alpha^g, \theta^g, \eta^g)$, where α^g is a rotation with respect to (w.r.t.) axis x_3, θ^g is a rotation w.r.t. axis x_2, and η^g is a rotation w.r.t. axis x_3 again. The rotation matrices for these three angles are [21]

$$\boldsymbol{R}^3(\alpha^g) = \begin{bmatrix} \cos\alpha^g & -\sin\alpha^g & 0 \\ \sin\alpha^g & \cos\alpha^g & 0 \\ 0 & 0 & 1 \end{bmatrix}, \quad \boldsymbol{R}^2(\theta^g) = \begin{bmatrix} \cos\theta^g & 0 & \sin\theta^g \\ 0 & 1 & 0 \\ -\sin\theta^g & 0 & \cos\theta^g \end{bmatrix}$$

$$\text{and} \quad \boldsymbol{R}^3(\eta^g) = \begin{bmatrix} \cos\eta^g & -\sin\eta^g & 0 \\ \sin\eta^g & \cos\eta^g & 0 \\ 0 & 0 & 1 \end{bmatrix}.$$

The combined rotation matrix – the orientation matrix – is $\boldsymbol{R}^g = \boldsymbol{R}^3(\alpha^g)\boldsymbol{R}^2(\theta^g)\boldsymbol{R}^3(\eta^g)$. It is assumed that $V_i^1 = V_i^2 = \ldots = V_i^N$, which yields

$$\langle \varepsilon_i^t \rangle = \frac{1}{N} \sum_{g=1}^{N} \boldsymbol{R}^g \, \varepsilon_i^t \, \boldsymbol{R}^{gT} \,. \tag{31}$$

In case of an incremental finite strain model the averaging technique using the orientation matrix $\boldsymbol{R}_g$ yields the averaged left Cauchy–Green PT tensor for phase i

$$\langle \boldsymbol{b}_i^t \rangle = \frac{1}{N} \sum_{g=1}^{N} \boldsymbol{R}^g \, \boldsymbol{b}_i^t \, \boldsymbol{R}^{gT} \ . \tag{32}$$

This simple pre-averaging technique of PT strains in randomly oriented grains is appropriate for pronounced texture as achieved by rolling of specimen and by pre-training. The average (effective) PT strains for polycrystals using Young's measure, derived in [39], can be simplified to the presented one, equation (31), when all the grains are assumed to have the same size.

The pre-averaging technique transforms the polycrystalline material into an equivalent monocrystalline one. Hence, the assumption of Ball and James [24] can be used to get the following effective macroscopic strain energy of a RVE

$$\Psi = \min_{i=1\ldots n} \bar{\psi}_i(\boldsymbol{\varepsilon}), \quad \text{with} \quad \bar{\psi}_i = \bar{\psi}_i^{\text{el}} + \psi_i^{\text{ch}}, \quad \text{for} \ \Omega_{\text{RVE}}, \tag{33}$$

where the elastic energy at linearized strains with the assumption $\mathbb{C}_i = \mathbb{C}$ (not regarding different stiffness of austenitic with respect to martensitic phases) and neglecting thermal strain reads, $\bar{\psi}_i^{\text{el}} = \frac{1}{2}(\boldsymbol{\varepsilon} - \langle \boldsymbol{\varepsilon}_i^t \rangle) : \mathbb{C} : (\boldsymbol{\varepsilon} - \langle \boldsymbol{\varepsilon}_i^t \rangle)$, and ψ_i^{ch} can be obtained from Section 2.2.

Analogous to monocrystalline material the decoupled free energy for polycrystalline material is given here as $\bar{\Psi}(\boldsymbol{\varepsilon}, \boldsymbol{\xi}) = \boldsymbol{\xi} \cdot \bar{\boldsymbol{\psi}}(\boldsymbol{\varepsilon}) + \bar{\Psi}^M(\langle \boldsymbol{\varepsilon}^t \rangle, \boldsymbol{\xi})$, with the phase fraction vector $\boldsymbol{\xi} = \sum_{i=1}^{n} \xi_i \boldsymbol{e}_i$, $\boldsymbol{e}_i \cdot \boldsymbol{e}_j = \delta_{ij}$, $-\xi_i \leq 0$, and the mass conservation condition $\boldsymbol{e}^* \cdot \boldsymbol{\xi} - 1 = 0$; $\boldsymbol{e}^* = \sum_{i=1}^{n} \boldsymbol{e}_i^*$, $|\boldsymbol{e}_i^*| = 1$, is the normal vector of the convex PT polytope $\mathbb{P}^{n-1} \subset \mathbb{R}^n$. $\bar{\boldsymbol{\psi}}$ is the vector of the phase energies with $\bar{\boldsymbol{\psi}}(\boldsymbol{\varepsilon}) = \sum_{i=1}^{n} \bar{\psi}_i(\boldsymbol{\varepsilon}_i) \boldsymbol{e}_i$, $\boldsymbol{e}_i \in \mathbb{R}^n$, $\bar{\boldsymbol{\psi}} \in \mathbb{R}^n$, and $\bar{\Psi}^M(\boldsymbol{\xi})$ is the free energy of mixing which is convex and non-positive. Here the phase fraction vector $\boldsymbol{\xi}$ holds for a fictitious representative monocrystal of the given polycrystalline material. The mixing energy for general n-variant problem at small strains, proposed for monocrystalline SMAs by Govindjee et al. [33], now reads for polycrystals at small strains as

$$\bar{\Psi}_{LS}^{M}(\langle \boldsymbol{\varepsilon}^t \rangle, \boldsymbol{\xi}) = -\frac{1}{2} \sum_{i=1}^{n} \xi_i \langle \boldsymbol{\varepsilon}_i^t \rangle : \mathbb{C} : \langle \boldsymbol{\varepsilon}_i^t \rangle + \frac{1}{2} \sum_{i=1}^{n} \sum_{j=1}^{n} \xi_i \xi_j \langle \boldsymbol{\varepsilon}_i^t \rangle : \mathbb{C} : \langle \boldsymbol{\varepsilon}_j^t \rangle. \tag{34}$$

Similar to monocrystalline PT model presented in Section 2.2, the enhancement of the global free energy function by a Lagrangian functional as it was presented in [6, 8] is now presented as $\mathcal{L}(\boldsymbol{\varepsilon}, \boldsymbol{\xi}, \boldsymbol{\gamma}, \delta) = \bar{\boldsymbol{\psi}}(\boldsymbol{\varepsilon}, \boldsymbol{\xi}) - \boldsymbol{\gamma} \cdot \boldsymbol{\xi} + \delta(\boldsymbol{e}^* \cdot \boldsymbol{\xi} - 1)$, and $\mathcal{L}(\boldsymbol{b}^e, \boldsymbol{\xi}, \boldsymbol{\gamma}, \delta) = \bar{\boldsymbol{\psi}}(\boldsymbol{b}^e, \boldsymbol{\xi}) - \boldsymbol{\gamma} \cdot \boldsymbol{\xi} + \delta(\boldsymbol{e}^* \cdot \boldsymbol{\xi} - 1)$ in Ω at linear and finite strain kinematics, respectively. The stress response reads $\boldsymbol{\sigma}^*(\boldsymbol{\varepsilon}, \boldsymbol{\xi}) = \partial_{\boldsymbol{\varepsilon}} \mathcal{L} = \sum_{i=1}^{n} \xi_i \, \mathbb{C} : (\boldsymbol{\varepsilon} - \langle \boldsymbol{\varepsilon}_i^t \rangle)$, $\forall \boldsymbol{x} \in \Omega$, where $\boldsymbol{\sigma}_i^*(\boldsymbol{\varepsilon})$ is the contribution from phase i. The local energy dissipation condition for the driving force (thermodynamical conjugate force) at linearized strains,

$$f = -\partial_\xi \mathcal{L} \Rightarrow f + \bar{\psi}(\varepsilon) - \gamma + \partial_\xi \bar{\Psi}^M(\langle\varepsilon^t\rangle, \xi) + \delta e^* = 0, \qquad (35)$$

is $\mathcal{D} = f \cdot \dot{\xi} \geq 0$. The phase fraction vector ξ can be determined using the local maximum dissipation principle $f \cdot \dot{\xi} \rightarrow$ Max. The evolution equation and phase transformation function follow from Section 2.2.2 which completes the polycrystalline PT model.

The time integration of resulting equations of polycrystalline PT model is done in a corresponding way as presented in Section 3 for monocrystalline model and performed within Abaqus via UMAT.

5 Error Controlled Adaptive Mesh Refinement in Space via Abaqus

Here, error-controlled spatial mesh adaptivity is realized within the commercial finite element (FE) package Abaqus/CAE. Adaptive time steps coupled with adaptive FE-meshes in space are not analyzed in this chapter. The necessary time step sizes for given error tolerances are determined by pre-computations for each example in Section 5.2.

Abaqus version 6.6 onward provides an explicit gradient-smoothing [46] based a posteriori error measure associated to von Mises Stress σ_M [47] for error control of equilibrium depending from the primal discrete displacement variable u^h which reads

$$e_{\sigma_M}(u, u^h)_\Omega = \sum_{e=1}^{\text{Nel}} \int_{\Omega_e} e_{\sigma_M}(u, u^h) \mathrm{d}\Omega_e = \sum_{e=1}^{\text{Nel}} e_{\sigma_M}(u, u^h)_{\Omega_e}. \qquad (36)$$

Additionally, for phase transformation problems the error indicator of the L_2 norm of driving force vector f as a quantity of interest is introduced as

$$e_f(u, u^h)_\Omega = \sum_{e=1}^{\text{Nel}} \int_{\Omega_e} e_f(u, u^h) \mathrm{d}\Omega_e = \sum_{e=1}^{\text{Nel}} e_f(u, u^h)_{\Omega_e} \qquad (37)$$

and implemented indirectly via UMAT subroutine, but the latter indicator does not follow from the applied primal FE method and thus does not improve the converging order.

5.1 Averaging Type Error Estimator and Effectivity Index

Abaqus provides an automated process to remesh a discretized 2D or 3D system using complete new mesh generation for each adaptive step (refining or/and coarsening) [47]. It uses gradient averaging technique to obtain a quasi optimal mesh for a prescribed error distribution and tolerance.

The error of von Mises stress reads $e_{\sigma_M}^*(\boldsymbol{u}, \boldsymbol{u}^h) := \sigma_M^*(\boldsymbol{u}^h) - \sigma_M(\boldsymbol{u}^h)$. The smoothened $\sigma_M^*(\boldsymbol{u})$ is generally obtained by nodal averaging or projection of appropriate nodal values $\hat{\boldsymbol{\sigma}}^*$. The projection for the von Mises stress is obtained by taking the same shape functions $\boldsymbol{N}_e$ on element level as for the C^0-continuous displacement field [46], i.e. $\sigma_M^*(\boldsymbol{u}^h) = \boldsymbol{N}_e \, \hat{\boldsymbol{\sigma}}_M^*$ in Ω_e. Minimizing the global error in Ω with respect to $\hat{\boldsymbol{\sigma}}_{M,g}^*$ using the Least-Square method, yields

$$\bigcup_{e=1}^{\text{Nel}} \int_{\Omega_e} \left[\sigma_{M,e}^*(\boldsymbol{u}^h) - \sigma_{M,e}(\boldsymbol{u}^h)\right]^2 d\Omega_e \to \min_{\hat{\boldsymbol{\sigma}}_{M,g}^*}$$

where indices e and g represent the quantities at element level Ω_e and global level Ω, respectively. The algebraic stationarity conditions are

$$\underbrace{\delta\hat{\boldsymbol{\sigma}}_{M,e}^{*T}}_{\neq \boldsymbol{0}} \left(\underbrace{\int_{\bigcup\Omega_e} \boldsymbol{N}_e^T \boldsymbol{N}_e \, d\Omega_e \, \hat{\boldsymbol{\sigma}}_{M,e}^*}_{A_e} - \underbrace{\int_{\bigcup\Omega_e} \boldsymbol{N}_e^T \, \sigma_{M,e}(\boldsymbol{u}^h) \, d\Omega_e}_{\hat{p}_{h,e}} \right) = 0. \quad (38)$$

The global stress vector $\hat{\boldsymbol{\sigma}}_{M,g}^*$ for C^0 continuous smooth stress via Boolean matrices $\boldsymbol{a}_e$ reads, $\hat{\boldsymbol{\sigma}}_{M,e}^* = \boldsymbol{a}_e \, \hat{\boldsymbol{\sigma}}_{M,g}^*$; $\delta\hat{\boldsymbol{\sigma}}_{M,e}^* = \boldsymbol{a}_e \, \delta\hat{\boldsymbol{\sigma}}_{M,g}^*$, and by using them in (38) yields

$$\underbrace{\delta\hat{\boldsymbol{\sigma}}_{M,g}^{*T}}_{\neq \boldsymbol{0}} \left(\underbrace{\sum_{e=1}^{ne}(\boldsymbol{a}_e^T \, \boldsymbol{A}_e \, \boldsymbol{a}_e)\, \hat{\boldsymbol{\sigma}}_{M,e}^*}_{A_g} - \underbrace{\sum_{e=1}^{ne} \boldsymbol{a}_e \, \hat{\boldsymbol{p}}_{h,e}}_{\hat{P}_{h,g}} \right) = 0, \quad (39)$$

where $\boldsymbol{A}_g \cong \boldsymbol{A}_{g,\text{diag}}$. From equation (39), the smoothened nodal values of von Mises equivalent stress is obtained as $\hat{\boldsymbol{\sigma}}_{M,g}^* = \boldsymbol{A}_{g,\text{diag}}^{-1} \, \hat{\boldsymbol{P}}_{h,g}$.

Similarly, the driving force can be additionally used for error indication, and the smoothened nodal values of the L_2 norm of driving force vector f can be obtained using also averaging technique. The $\sigma_M(\boldsymbol{u}^h)$ and $f(\boldsymbol{u}^h)$ follow from finite element solution and are called 'base variable solutions' in Abaqus.

5.2 Adaptive Remeshing with Abaqus

The adaptive refinement or coarsing depends on the distribution of scaled element-wise error indicators. Abaqus uses the normalized percentage or relative error percentage which defines global or local error targets in the prescribed region. The normalized error associated with an error indicator for equilibrium, derived from gradient averaging for von Mises stress σ_M, is available as [47]

$$\eta_{\sigma_{M\Omega}} = \frac{\sum_{e=1}^{\text{Nel}} \left[\sigma^*_{M_{\Omega_e}}(\boldsymbol{u}^h) - \sigma_{M_{\Omega_e}}(\boldsymbol{u}^h) \right]}{\sum_{e=1}^{\text{Nel}} \sigma_{M_{\Omega_e}}(\boldsymbol{u}^h)} \times 100. \tag{40}$$

Additionally, the normalized error of the driving force which is responsible for phase transformation is controlled by

$$\eta_{f\Omega} = \frac{\sum_{e=1}^{\text{Nel}} \left[f^*(\boldsymbol{u}^h)_{\Omega_e} - f(\boldsymbol{u}^h)_{\Omega_e} \right]}{\sum_{e=1}^{\text{Nel}} f(\boldsymbol{u}^h)_{\Omega_e}} \times 100, \tag{41}$$

where $\sigma^*_{M_{\Omega_e}}(\boldsymbol{u}^h)$ and $f^*(\boldsymbol{u}^h)_{\Omega_e}$ are computed through a recovery technique by gradient smoothing of Zienkiewicz and Zhu [46].

The sizing method calculates new element sizes during the adaptive remeshing process. Abaqus/CAE applies the sizing method to a field of error indicators. The output of a sizing method (controlling size and shape of elements) is a set of scalar quantities located at the nodes in the region defined by the remeshing rule.

Based on the normalized error two different strategies of Abaqus/CAE are analyzed in this chapter:

1. *Uniform error distribution:* The uniform error distribution method provides a single error indicator target $\bar{\eta}$, for controlling the sizing. Abaqus/CAE applies a sizing method that attempts to meet this target in every element in the remeshing desired region, i.e. if $\eta_e > \bar{\eta}$ refine; otherwise coarse $\forall\, \Omega_e \in \Omega$; $e = 1 \ldots$ Nel.
2. *Multiple error indicators:* One can use multiple error indicators in the same region in order to obtain new elements sizes. In such case sizing methods will be applied independently to each error indicator variable with the resulting local element size based on the smallest size calculated from each sizing algorithm, i.e. $\eta_{\text{total}\Omega_e} = \max(\eta_{\sigma_{M_{\Omega_e}}}, \eta_{f_{\Omega_e}})$, for the von Mises stress and the L_2-norm of the driving force vector.

6 Numerical Examples

Three numerical examples are presented in this section. First, the strain-controlled PT computation under tensile load is analyzed for superelastic polycrystalline NiTi alloy, and comparisons with experimental data [48] are presented. Secondly, the comparison of computational results of a strain-controlled SE tension test for both, mono and polycrystalline CuAlNi alloys are investigated. Next, dovetail shaped and dog bone shaped specimens under tensile loads are computed using adaptive remeshing for phase transformation problems by error indicators for equilibrium, derived from gradient averaging of von Mises stress and for driving forces. Ad-

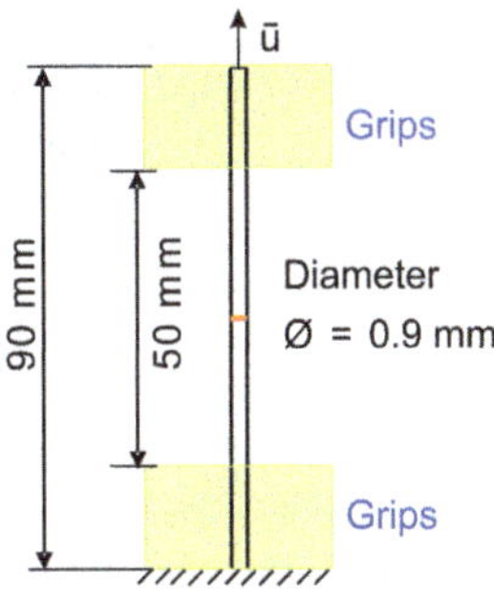

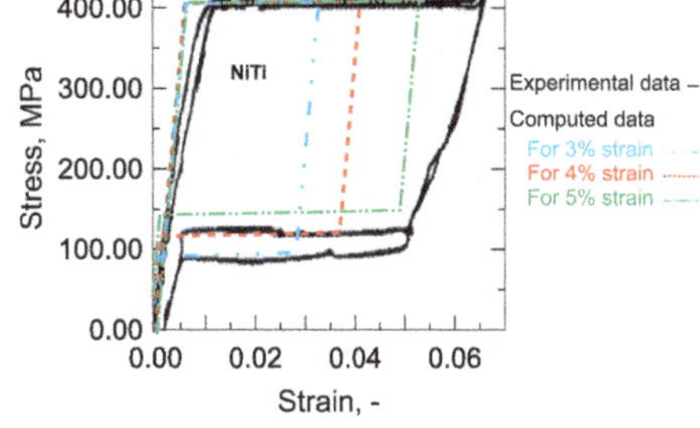

(a) Specimen geometry and BCs for FE computations

(b) Comparison of experimental data and computed stress strain for different total strains

Fig. 2 Cyclic PT of a stretched bar made from superelastic polycrystalline $NiTi$ and comparison with experimental data.

aptivity of the process time (expressed by prescribed strain increments) is realized by a priori accuracy tests for the investigated examples, see also [28].

The full PT cycle computations, using Abaqus, are strain-controlled and performed in two steps with prescribed total axial displacements in the first step and the related load reduction to zero in the second step. The temperature is kept constant during the whole PT cycles.

6.1 Superelastic Polycrystalline NiTi

Numerical results obtained for a strain controlled uniaxial tensile test are compared with experimental results carried out in [48]. The tested wire material is a nearly equiatomic NiTi polycrystalline alloy which shows superelastic behavior at room temperature. The experimental setup given in [48] is not presented here. The spatial dimensions of the wire having circular cross section and the boundary conditions (BCs) for uniaxial tension test are shown in Fig. 2a. For finite element computation the discretization was carried out with 432 linear B-bar hexahedral element (of type C3D8 in Abaqus). The material parameters used are: Young's modulus, specific heat, mass density [49], critical driving force, Poisson's ratio [10], and latent heat [50]. The remaining relevant material parameters are determined to be $\theta_0 = 293$ K and $\theta = 296$ K. All parameters are listed in Table 2. The Bain matrices (transformation stretch matrices) of the cubic to monoclinic-I transformation are given in Appendix A.

For numerical computation the RVE is chosen to consist of 3^3, 7^3 and 10^3 grains, and randomly oriented with respect to the loading axis. The orientations of grains are

Table 2 Material data of polycrystalline NiTi SE specimen.

Young's modulus E	68,200 N/mm^2	Latent heat l	167,000 J/kg
Poisson ratio ν	0.3	Energy barrier f_c	7.5 N/mm^2
Mass density ρ	6.45 kg/m^3	Equilibrium temperature θ_0	293 K
Specific heat c	460 J/(kg·K)	Ambient temperature θ	296 K

described by eulerian angles which were obtained from all possible combinations of assumed angles (10°, 20°, 30°), (10°, 12°, 15°, 20°, 22°, 25°, 30°) and (10°, 12°, 15°, 17°, 18°, 20°, 22°, 25°, 27°, 30°). From the studies on the influence of number of grains it is obtained that 3^3 randomly-oriented grains are sufficient for the RVE. The averaged PT strains of an equivalent monocrystal of N-grain polycrystalline RVE are obtained from equation (31).

During a strain-controlled PT cycle the ambient temperature is kept constant at 296 K. The loading and unloading processes are discretized by 50 displacement increments in order to realize a rather smooth development of phases in the process time. The smaller the strain increments are chosen, the better the phase development is approximated. A combined space-(parametric) time adaptivity cannot be realized in Abaqus.

The numerical results are compared with the strain-controlled SE tension experiment for polycrystalline NiTi, Fig. 2b. The computed stress and strain data are obtained from averaging the values over all the Gaussian integration points in the cross section in the middle of the prismatic specimen. Three numerical tests for 3, 4 and 5% axial tensile strain are presented. Comparison of the experimentally measured stress-strain function with the numerically obtained data at linearized strains show fairly good agreement. The specimen shows a reverse transformation from martensite to austenite during the second step because martensite is not stable at room temperature and thus transforms back at unloading. This transformation at unloading is combined with vanishing PT strains.

The reduction of experimental stiffness during elastic loading compared with computed data is evident due to the presence of a R-phase. NiTi transforms from the cubic to trigonal or rhombohedral R-phase before transforming to martensite. Thus, two transformation matrices describe the total transformation. The first describes the deformation from austenite to the R-phase and the second from austenite to martensite phase. Since the presence of R-phase is not included in the presented PT model, the computed elastic loading curve has deviation from the experimental one. Furthermore, the deviation of the computed results with same elastic modulus from the experimental ones during elastic unloading is due to higher elastic modules of austenite with respect to martensitic phases, i.e. $E_A > E_M$, for this example.

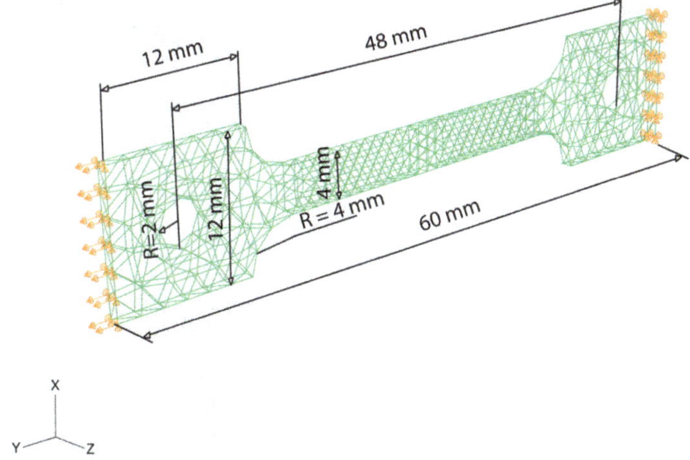

Fig. 3 Geometry, BCs and FE discretization with tetrahedrons of a SE CuAlNi specimen.

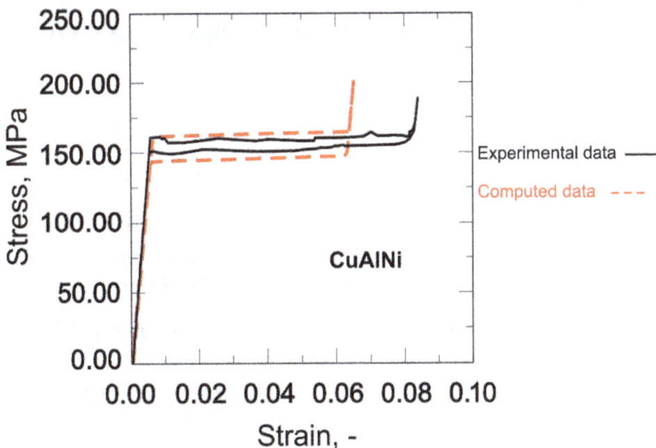

Fig. 4 Computed and experimental stress-strain data for monocrystalline CuAlNi SE specimen.

6.2 Superelastic Mono and Polycrystalline CuAlNi

In this section, results of two computations are presented. First, the calculated stress-strain data for the discretized specimen (Fig. 3) are compared with the experimental ones in Fig. 4 [51] for strain-controlled SE monocrystalline CuAlNi tension specimen. Next, comparisons of computed stress-strain data for SE mono and polycrystalline CuAlNi material are presented in Fig. 5. The computation shows the effect of pre-homogenization, equation (31), on PT strains.

The dimensions of the specimen are taken from the experiment carried out on superelastic monocrystalline CuAlNi material in [51]. The shape of the specimen with flat rectangular cross-section and 3D finite element meshing with tri-linear

Table 3 Material data of mono and polycrystalline CuAlNi SE specimen.

Young's modulus E	25,960 N/mm^2	Latent heat l	4,550 J/kg
Poisson ratio ν	0.35	Energy barrier f_c	0.001 N/mm^2
Mass density ρ	8.0 kg/m^3	Equilibrium temperature θ_0	236 K
Specific heat c	400 J/(kg·K)	Ambient temperature θ	296 K

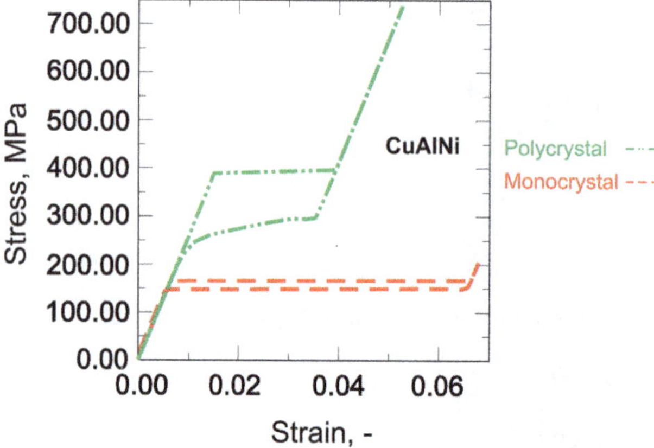

Fig. 5 Computed stress-strain curves of mono and polycrystalline SE CuAlNi.

tetrahedral elements are depicted in Fig. 3. The discretization is carried out with 1,350 linear tetrahedral elements with adaptive convergence studies. Material parameters used in computation, given in Table 3, are taken from [29]. The latent heat l is updated to 4,550 J/kg. The required transformation matrices of the cubic to orthorhombic transformation of CuAlNi crystals is given in Appendix A and the spatial orientation matrix $\widetilde{\boldsymbol{R}}$ of monocrystalline specimen [51] reads

$$\widetilde{\boldsymbol{R}} = \begin{pmatrix} 0.2019 & -0.777 & -0.596 \\ -0.0756 & 0.5934 & -0.8017 \\ 0.9767 & 0.2062 & 0.0597 \end{pmatrix}. \tag{42}$$

Numerical simulation of superelastic cyclic PT was carried out in two consecutive steps. In the first step displacement controlled load is applied, and in the second step removal of load takes place. Both steps were discretized with 50 time increments. Figure 4 shows the experimentally and numerically gained stress-strain curves for the monocrystalline CuAlNi specimen with rather good agreement, except for the difference that the experimental transformation strain is nearly 8% whereas the corresponding numerical value is about 6%. This difference can be explained by the fact that in addition to the $\beta_1 \rightarrow \gamma_1'$-transformation also a $\gamma_1' \rightarrow \beta_1''$-transformation takes place which is not considered in the presented PT model. The denotations γ_1' and β_1'' for different crystal structures are also know as 2H and 18R(2) [29]. Here, the presented stress and strain data are obtained from averaging the values over all the Gauss points in the cross section located at the middle of specimen.

Further, a finite element computation for polycrystalline CuAlNi is done in order to see the effect of pre-homogenization, equation (31), on PT strains. A fictitious polycrystalline SE CuAlNi is considered for this purpose. The RVE of this material consists of 27 different grains which are described in similar fashion as for polycrystalline NiTi in previous subsection by the same eulerian angles. The averaged PT strains of an equivalent monocrystal of N-grain polycrystalline RVE is obtained from equation (31). For comparing the computed stress-strain curve of mono and polycrystalline material, the same finite element discretization (Fig. 4) as well as same material parameters listed in Table 3 are used. Load conditions are also kept the same.

Figure 5 shows the computed stress-strain functions of mono and polycrystalline SE CuAlNi for the loading direction. One can see that the transformation strain in monocrystal is nearly 6% whereas in polycrystal the value is about 2.5% which correctly captures the experimental observation [1]. Hence the presented pre-averaging technique is a reasonably good engineering approach to model polycrystalline SMAs under the condition that the used assumptions are fairly fulfilled.

6.3 Dovetail Shaped Specimen

The verification of developed mono and polycrystalline material models is realized using two different strategies to control the a posteriori discretization error in space, see also [28]. The two applied strategies are 'uniform error distribution' and 'multiple error indicators' competing with each other, cf. Section 5.2.

The material properties for the computed examples in this section are given in Tables 2 and 3. Bain matrices are presented in Appendix A. Geometry of the dovetail problem, boundary conditions and prescribed displacement at the right boundary are shown in Fig. 6.

Computations are presented for SE monocrystalline CuAlNi and polycrystalline NiTi at linearized strains, respectively. For all the computations the same initial mesh is used which consists of 158 linear tetrahedral elements. The error tolerance is kept 5%, and the maximum allowable adaptive iterations are kept 5.

The convergence behavior of the presented error indicators can be seen in Fig. 7, where the percentage error is plotted versus the number of elements. It can also be seen from the left of Fig. 7 that in case of monocrystals with $\eta_{\sigma_{M_{\Omega_e}}}$ indicator, the convergence is better for the first adaptive iteration process, whereas for polycrystals both indicators yield in the same convergence rate in first adaptive step.

As mentioned above, when one looks for complete adaptive process, as expected the equilibrium based $\eta_{\sigma_{M_{\Omega_e}}}$ indicator yields better convergence than $\max(\eta_{\sigma_{M_{\Omega_e}}}, \eta_{f_{\Omega_e}})$ for the presented examples. This is caused by the fact that the error of f is not treated by a variational side condition requiring a mixed finite element method.

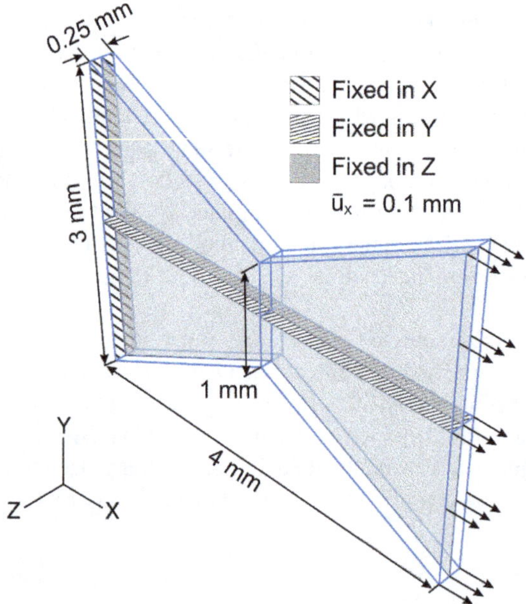

Fig. 6 Geometry and applied BCs on SE dovetail specimen.

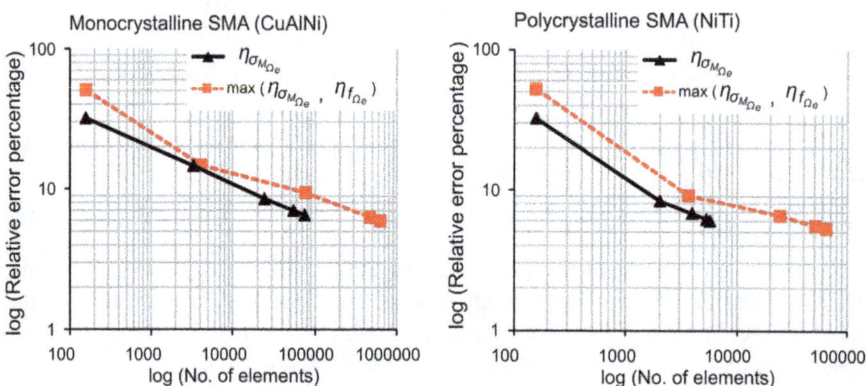

Fig. 7 Global percentage relative error for linear tetrahedral elements, left for monocrystalline SMA of CuAlNi, right for polycrystalline SMA of NiTi.

The right way to control directly the driving force as a quantity of interest is the additional treatment of the related dual problem yielding a dual error estimator next to the primal error estimator [52–55].

7 Conclusions

The unified variational concept for monocrystalline is extended to polycrystalline SMAs with pre-homogenized PT strains in a RVE of randomly distributed single crystal grains characterized by Eulerian angles. For both, mono and polycrystalline PT models, a unified algorithmic structure is obtained. The new polycrystalline PT model successfully captures basic features such as various stress-strain behaviors and phase transformations which is approved by comparisons with experimental data. Pre-trained and pre-textured SMAs best fulfill some specific assumptions. Pre-averaging technique also shows that monocrystalline SMAs usually have higher transformation strains than polycrystalline ones, which correctly captures the experimental observation.

Adaptive finite element remeshing in space is applied based on a posteriori error indicators with gradient smoothing, available in Abaqus. To control the spatial discretization error within a primal finite element method firstly equilibrium has to be controlled, which is performed for the von Mises equivalent stress with the error indicator η_{σ_M}. Additionally, the error of the driving force, η_f, is implemented. The comparison of these two indicators evidently shows that η_{σ_M} provides a better convergence rate than $\max(\eta_{\sigma_M}, \eta_f)$.

Appendix A

1. The transformation stretch matrices of the cubic to orthorhombic transformation of CuAlNi crystals are given as

$$
U_1 = \begin{pmatrix} \frac{\alpha+\gamma}{2} & 0 & \frac{\alpha-\gamma}{2} \\ 0 & \beta & 0 \\ \frac{\alpha-\gamma}{2} & 0 & \frac{\alpha+\gamma}{2} \end{pmatrix}, \quad
U_2 = \begin{pmatrix} \frac{\alpha+\gamma}{2} & 0 & \frac{\gamma-\alpha}{2} \\ 0 & \beta & 0 \\ \frac{\gamma-\alpha}{2} & 0 & \frac{\alpha+\gamma}{2} \end{pmatrix}, \quad
U_3 = \begin{pmatrix} \frac{\alpha+\gamma}{2} & \frac{\alpha-\gamma}{2} & 0 \\ \frac{\alpha-\gamma}{2} & \frac{\alpha+\gamma}{2} & 0 \\ 0 & 0 & \beta \end{pmatrix},
$$

$$
U_4 = \begin{pmatrix} \frac{\alpha+\gamma}{2} & \frac{\gamma-\alpha}{2} & 0 \\ \frac{\gamma-\alpha}{2} & \frac{\alpha+\gamma}{2} & 0 \\ 0 & 0 & \beta \end{pmatrix}, \quad
U_5 = \begin{pmatrix} \beta & 0 & 0 \\ 0 & \frac{\alpha+\gamma}{2} & \frac{\alpha-\gamma}{2} \\ 0 & \frac{\alpha-\gamma}{2} & \frac{\alpha+\gamma}{2} \end{pmatrix}, \quad
U_6 = \begin{pmatrix} \beta & 0 & 0 \\ 0 & \frac{\alpha+\gamma}{2} & \frac{\gamma-\alpha}{2} \\ 0 & \frac{\gamma-\alpha}{2} & \frac{\alpha+\gamma}{2} \end{pmatrix},
$$

where α, β and γ are the transformation stretches determined from the lattice parameters of the two phases. It holds: $\alpha = \sqrt{2}a/a_0$, $\beta = b/a_0$ and $\gamma = \sqrt{2}c/a_0$, where a_0 is the lattice parameter of the cubic parent phase, and a, b, and c are the lattice parameters of the orthorhombic product phase. For a particular CuAlNi alloy, Otsuka and Shimizu [56] found that the lattice parameter of the parent phase is $a_0 = 5.836$ Å, and the lattice parameters of the product phase are $a = 4.382$ Å, $b = 5.356$ Å and $c = 4.222$ Å [1], with the transformation stretches $\alpha = 1.0619$, $\beta = 0.9178$ and $\gamma = 1.023$.

2. The Bain matrix or transformation stretch matrices of the cubic to monoclinic-I transformation of NiTi crystals are given as

$$U_1 = \begin{pmatrix} \gamma & \epsilon & \epsilon \\ \epsilon & \alpha & \delta \\ \epsilon & \delta & \alpha \end{pmatrix}, U_2 = \begin{pmatrix} \gamma & -\epsilon & -\epsilon \\ -\epsilon & \alpha & \delta \\ -\epsilon & \delta & \alpha \end{pmatrix}, U_3 = \begin{pmatrix} \gamma & -\epsilon & \epsilon \\ -\epsilon & \alpha & -\delta \\ \epsilon & -\delta & \alpha \end{pmatrix}, U_4 = \begin{pmatrix} \gamma & \epsilon & -\epsilon \\ \epsilon & \alpha & -\delta \\ -\epsilon & -\delta & \alpha \end{pmatrix},$$

$$U_5 = \begin{pmatrix} \alpha & \epsilon & \delta \\ \epsilon & \gamma & \epsilon \\ \delta & \epsilon & \alpha \end{pmatrix}, U_6 = \begin{pmatrix} \alpha & -\epsilon & \delta \\ -\epsilon & \gamma & -\epsilon \\ \delta & -\epsilon & \alpha \end{pmatrix}, U_7 = \begin{pmatrix} \alpha & -\epsilon & -\delta \\ -\epsilon & \gamma & \epsilon \\ -\delta & \epsilon & \alpha \end{pmatrix}, U_8 = \begin{pmatrix} \alpha & \epsilon & -\delta \\ \epsilon & \gamma & -\epsilon \\ -\delta & -\epsilon & \alpha \end{pmatrix},$$

$$U_9 = \begin{pmatrix} \alpha & \delta & -\epsilon \\ \delta & \alpha & \epsilon \\ \epsilon & \epsilon & \gamma \end{pmatrix}, U_{10} = \begin{pmatrix} \alpha & \delta & -\epsilon \\ \delta & \alpha & -\epsilon \\ -\epsilon & -\epsilon & \gamma \end{pmatrix}, U_{11} = \begin{pmatrix} \alpha & -\delta & \epsilon \\ -\delta & \alpha & -\epsilon \\ \epsilon & -\epsilon & \gamma \end{pmatrix}, U_{12} = \begin{pmatrix} \alpha & -\delta & -\epsilon \\ -\delta & \alpha & \epsilon \\ -\epsilon & \epsilon & \gamma \end{pmatrix},$$

where $\alpha = 1.0243$, $\gamma = 0.9563$, $\delta = 0.058$ and $\epsilon = -0.0427$, [1] are the transformation stretches determined from the lattice parameters of the parent phase (austenite) and product phase (martensite).

Acknowledgement

Support for this work was provided by the German Science Foundation (DFG) under project No. Ste 238/51-1&2, which is gratefully acknowledged.

References

1. K. Bhattacharya. *Microstuctures of Martensite*. Oxford Series on Materials Modelling, Oxford University Press, 2003.
2. E. Patoor, A. Eberhardt, and M. Berveiller. Thermomechanical behavior of shape memory alloys. *Arch. Mech.*, 40:755–794, 1988.
3. D. Entemeyer, E. Patoor, A. Eberhardt, and M. Berveiller. Micromechanical modelling of the superelasticity behavior of materials undergoing thermoelastic phase transition. *J. Physique*, IV(C8–5):233–238, 1995.
4. C. Lexcellent, B. Goo, Q. Sun, and J. Bernardini. Charecterization, thermomechanical behavior and micromechanical-based constitutive model of shape memory Cu-Zn-Al single crystal. *Acta Mater.*, 44:3773–3780, 1996.
5. B. Goo and C. Lexcellent. Micromechanics-based modeling of two-way memory effect of a single crystalline shape-memory alloys. *Acta Mater.*, 45:727–737, 1997.
6. M. Huang and L.C. Brinson. A multivariant model for single crystal shape memory alloys behavior. *J. Mech. Phys. Solids*, 46:1379–1409, 1998.
7. A. Vivet and C. Lexcellent. Micromechanical modeling for tension-compression pseudoelastic behavior of AuCd single crystal. *Eur. Phys. J.: Appl. Phys.*, 4:125–132, 1998.
8. N. Siredey, E. Patoor, M. Berveiller, and A. Eberhardt. Constitutive equations for polycrystalline thermoelastic shape memory alloys. Part I: Intergranular interactions and behavior of the grain. *Int. J. Solids Structures*, 36:4289–4315, 1999.
9. D.C. Lagoudas (Ed.). *Shape Memory Alloys, Modeling and Engineering Application*. Springer, 2008.
10. Y. Jung, P. Papadopoulos, and R.O. Ritchie. Constitutive modelling and numerical simulation of multivariant phase transformation in superelastic shape-memory alloys. *Int. J. Num. Methods Engrg.*, 60:429–460, 2004.

11. S. Govindjee and C. Miehe. A multi-variant martensitic phase transformation model: Formulation and numerical implementation. *Comput. Methods Appl. Mech. Engrg.*, 191:215–238, 2001.

12. G.J. Hall and S. Govindjee. Application of the relaxed free energy of mixing to problems in shape memory alloy simulation. *J. Intell. Mater. Systems Structures*, 13:773–782, 2002.

13. T. Mura. *Micromechanics of Defects in Solids.* Kluwer Academic Publishers, Boston, 1987.

14. Z.K. Lu and G.J. Weng. Martensitic transformation and stress-strain relations of shape-memoryalloys. *J. Mech. Phys. Solids*, 45(11/12):1905–1928, 1997.

15. Z.K. Lu and G.J. Weng. A self-consistent model for the stress-strain behavior of shape-memory alloy polycrystals. *Acta Metall.*, 46(15):5423–5433, 1998.

16. K. Gall, H. Sehitoglu, H.J. Maier, and K. Jacobus. Stress-induced martensitic phase transfromations in polycrystalline cuznal shape memory alloys under different stress states. *Metall. Trans., A,* 29A:765–773, 1998.

17. Q.P. Sun and K.C. Hwang. A micromechanics model of transfromation plasticity with shear and dilatation effect. *J. Mech. Phys. Solids*, 39:507–524, 1991.

18. Q.P. Sun and K.C. Hwang. Micromechanics modelling for the constitutive behavior of polycrystalline shpae memory alloys. I. Derivation of general relations. *J. Mech. Phys. Solids*, 41(1):1–17, 1993.

19. Q.P. Sun and K.C. Hwang. Micromechanics modelling for the constitutive behavior of polycrystalline shape memory alloys. II. Study of individual phenomena. *J. Mech. Phys. Solids*, 41(1):19–33, 1993.

20. E. Patoor, A. Eberhardt, and M. Berveiller. Micromechanical modelling of the shape memory behavior. In: *Proceedings of ASME International Congress and Exposition,* Chicago, AMD-Vol. 189, pp. 23–27, 1994.

21. M. Huang, X. Gao, and L.C. Brinson. A multivariant micromechanical model for SMAS. Part 2. Polycrystal model. *Int. J. Plasticity*, 16:1371–1390, 2000.

22. E. Stein and G. Sagar. A unified variational setting and algorithmic framework for mono- and polycrystalline martensitic phase transformations. In: K. Hackl (Ed.), *IUTAM Symposium on Variational Concepts with Applications to the Mechanics of Materials,* pp. 245–259, Springer, 2010.

23. K. Bhattacharya and R. Kohn. Symmetry, texture, and the recoverable strain of shape memory alloys. *Acta Metall.*, 44:529–542, 1996.

24. J. Ball and R. James. Fine phase mixtures and minimizers of energy. *Arch. Rat. Mech. Anal.*, 100:13–52, 1987.

25. J. Ball and R. James. Proposed experimental tests of a theory of the fine microstructure and the two-well problem. *Philos. Trans. Roy. Soc. London*, A338:389–450, 1992.

26. T. Saburi and S. Nenno. The shap ememory effect and related phenomena. In: *Proceedings of an International Conference on Solid Solid Phase Transformations,* pp. 1455–1479, 1981.

27. D.A. Porter and K.E. Easterling. *Phase Transformations in Metal and Alloys,* 2nd edition. Chapman and Hall, London, 1992.

28. G. Sagar. Theory and computation of mono- and poly-crystalline cyclic martensitic phase transfomations. PhD Thesis, Institute of Mechanics and Computational Mechanics, Leibniz Universität Hannover, Germany, 2009.

29. E. Stein and O. Zwickert. Theory and finite element computations of a unified cyclic phase transformation model for monocrystalline materials at small strains. *Comput. Mech.*, 40:429–445, 2007.

30. E. Stein and G. Sagar. Theory and finite element computation of cyclic martensitic phase transformation at finite strain. *Int. J. Numer. Methods Engrg.*, 74(1):1–31, April 2008.

31. M. Born. *Dynamik der Krystallgitter.* Teubner, Leipzig/Berlin, 1915.

32. M. Born and K. Huang. *Dynamical Theory of Crystal Lattices.* Clarendon Press, Oxford, 1954.

33. S. Govindjee, A. Mielke, and G.J. Hall. The free-energy of mixing for n-variant martensitic phase transformations using qausi-convex analysis. *J. Mech. Phys. Solids*, 52:I–XXVI, 2003.

34. R. Abeyaratne and S. Kim. Cyclic effects in shape-memory alloys: A one-dimensional continum model. *Int. J. Solids Structures*, 34:2229–2294, 1997.

35. B. Coleman and W. Noll. On the thermostatics of continuous media. *Archive of Rational Mechanics and Analysis*, 4:97–128, 1959.

36. B.D. Colemen and M.E. Gurtin. Thermodynamics with internal state variables. *J. Chem. Phys.*, 47:597–613, 1967.

37. D.G. Luenberger. *Linear and Nonlinear Programming*. Addison-Wesley, Reading, MA, 1984.

38. E. Stein and G. Sagar. Convergence behavior of 3-D finite elements for neo-hookean material models using Abaqus-UMAT. *Int. J. Computer-Aided Engrg. Software*, 25(3):220–232, 2008.

39. K. Hackl and R. Heinen. A micromechanical model for pretextured polycrystalline shape-memory alloys including elastic anisotropy. *Continuum Mech. Thermodynam.*, 19:499–510, 2008.

40. F.A. Nae, Y. Matsuzaki, and T. Ikeda. Micromechanical modeling of polycrystalline shape-memory alloys including thermo-mechanical coupling. *Smart Mater. Structures*, 12:6–17, 2003.

41. Xiujie Gao and L.C. Brinson. A simplified multivariant sma model based on invariant plane nature of martensitic transformation. *J. Intell. Mater. System Structure*, 13:795–810, 2002.

42. L.C. Brinson, R. Lammering, and I. Schmidt. Stress induced transformation behaviour of a polycrystalline niti shape memory alloy: Micro and macromechanical investigations via in situ optical microscopy. *J. Mech. Phys. Solids*, 52:1549–1572, 2004.

43. R. Abeyaratne, S. Kim, and J. Knowels. A one-dimensional continum model for shape memory alloys. *Int. J. Solids Structures*, 31:2229–2249, 1994.

44. S. Daly, G. Ravichandran, and K. Bhattacharya. Stress-induced martensitic phase transformation in thin sheet of nitinol. *Acta Mater.*, 55:3593–3600, 2007.

45. S. Govindjee, K. Hackl, and R. Heinen. An upper bound to the free energy of mixing by twin-compatible lamination for n-variant martensitic phase transformations. *Continuum Mech. Thermodyn.*, 18:443–453, 2007.

46. O.C. Zienkiewicz and J.Z. Zhu. A simple error estimator and adaptive procedure for practical engineering analysis. *Int. J. Numer. Methods Engrg.*, 24:337–357, 1987.

47. Abaqus. *User's Manual*, Hibbit, Karlson & Sorensen, Inc.

48. R. Lammering and A. Vishnevsky. Investigation of local strain and temperature behaviour of superelastic niti wires. In: *Proceedings of the First Seminar on the Mechanics of Multifunctional Materials*, Report No. 5, pp. 77–82, 2007.

49. A. Vishnevsky, R. Lammering, and I. Schmidt. Zur Modellierung des geschwindigkeits-abhängigen Verhaltens von superelastischen Formgedchtnislegierungen. *Techn. Mech.*, 24(2):125–136, 2004.

50. K.B. Gilleo. Photon-conducting media alignment using a thermokinetic material. US Patent 6863447, http://www.patentstorm.us/patents/6863447/fulltext.html, 2005.

51. Z. Xiangyang, S. Qingping, and Y. Shouwen. A non-invariant plane model for the interface in cualni single crystal shapememory alloys. *J. Mech. Phys. Solids*, 48:2163–2182, 2000.

52. E. Stein and S. Ohnimus. Coupled model- and solution-adaptivity in the finite-element method. *Comput. Methods Appl. Mech. Engrg.*, 150:327–350, 1997.

53. E. Stein and S. Ohnimus. Anisotropic discretization- and model-error estimation in solid mechanics by local Neumann problems. *Comput. Methods Appl. Mech. Engrg.*, 176:363–385, 1999.

54. E. Stein, M. Rüter, and S. Ohnimus. Adaptive finite element analysis and modeling of solids and structures. *Int. J. Numer. Methods Engrg.*, 60:103–138, 2004.

55. Erwin Stein, Marcus Rüter, and Stephan Ohnimus. Error-controlled adaptive goal-oriented modeling and finite element approximations in elasticity. *Comput. Methods Appl. Mech. Engrg.*, 196:3598–3613, 2007.

56. K. Otsuka and K. Shimizu. Morphology and crystallography of thermoelastic cu-al-ni martensite analyzed by the phenomenological theory. *Trans. Jap. Inst. Metals*, 15:103–108, 1974.

Thermo-Hydro-Mechanical Modeling of Coupled Processes in Clay Materials

Jobst Maßmann, Gesa Ziefle, Martin Kohlmeier and Werner Zielke

Abstract Geotechnical problems are characterized by the existence of a great variety of coupled processes and non-linear effects. To focus on problems arising in the field of radioactive waste disposal, the time-dependent behavior in underground excavations in low permeable materials is investigated. For this purpose model approaches for thermo-hydro-mechanical interactions in partially saturated porous media including thermal and moisture content dependent expansion have been developed and implemented into a finite element code. As coupling phenomena the Terzaghi's effective stress concept and the mass conservation of the liquid phase in a deformable porous media are considered. The resulting numerical model is verified with analytical solutions and validated with experimental data. An extension of this model is concerned with the non-linear structural behavior of low permeable material. For this, a purely mechanical model for various kinds of material is presented,

Jobst Maßmann
Federal Institute for Geosciences and Natural Resources (BGR), Stilleweg 2, 30655 Hannover, Germany; e-mail: mail@jobst-massmann.de
Formerly at Institute of Fluid Mechanics and Environmental Physics in Civil Engineering, Leibniz Universität Hannover, Appelstraße 9A, 30167 Hannover, Germany

Gesa Ziefle
Erdwärme-Messtechnik GmbH, Überseestadt – Newport, Konsul-Smidt-Straße 8L,
28217 Bremen, Germany; e-mail: gesa@ungruh.com
Formerly at Institute of Fluid Mechanics and Environmental Physics in Civil Engineering, Leibniz Universität Hannover, Appelstraße 9A, 30167 Hannover, Germany

Martin Kohlmeier
Institute of Structural Analysis, Leibniz Universität Hannover, Appelstraße 9A, 30167 Hannover, Germany; e-mail: martin.kohlmeier@arcor.de
Formerly at Institute of Fluid Mechanics and Environmental Physics in Civil Engineering, Leibniz Universität Hannover, Appelstraße 9A, 30167 Hannover, Germany

Werner Zielke
Institute of Fluid Mechanics and Environmental Physics in Civil Engineering,
Leibniz Universität Hannover, Appelstraße 9A, 30167 Hannover, Germany;
e-mail: zielke@hydromech.uni-hannover.de

E. Stephan & P. Wriggers (Eds.): Modelling, Simulation and Software Concepts, LNACM 57, pp. 29–74.
springerlink.com   © Springer-Verlag Berlin Heidelberg 2011

which incorporates initial stresses. Finally, the long term behavior of an excavation is analyzed in detail and compared to the corresponding long term measurements of a mine-by experiment conducted in argillite at the Tournemire test site in France. Therefore damage, drying induced shrinkage and anisotropic deformation dependent permeability are considered.

1 Introduction

Geotechnical problems are often characterized by manifold interacting processes namely mechanical, hydraulic, thermal, chemical and biological ones. The research presented in the work at hand focuses on coupled thermal, hydraulic, and mechanical (THM) problems. A great variety of geotechnical applications in this field can be treated with the presented code, e.g. the geothermal energy, carbon dioxide capture and storage (CCS) or the storage of radioactive waste, which will be in the focus of the following. The International Atomic Energy Agency proclaimed that "Radioactive waste presents a potential hazard to human health and the environment and it must be managed so as to reduce any associated risks to acceptable levels" [22]. Furthermore, it recommends the storage in deep geological formations with a multi-barrier-system, consisting of a host rock as a natural geological barrier and an engineered barrier system (EBS). The most important criteria for a potential host rock are very low permeability, high thermal conductivity, no tectonic or volcanic activity, no natural resources in the neighborhood, possibility for sealing, high strength, plastic/viscous behavior, lithostatic isotropic in-situ stresses, high sorption potential, high temperature reliability, low content of water and low resolution behavior [5, 6, 22]. In addition to rock salt, cristalline formations and indurated clays are internationally discussed and investigated as a potential host rock for the storage of nuclear waste [6]. The host rock is affected by the excavation and, as the EBS, the heat production of the waste. THM processes have to be expected. These processes could change the properties of the host rock and EBS significantly. Furthermore, the complex interaction makes it nearly impossible to treat the processes separately. Hudson et al. [21] emphasize the importance of coupled THM issues related to a radioactive waste repository. In order to evaluate the impact, coupled numerical modeling of the processes can be a great support for safety assessment. Thus, numerical codes have to be developed, verified and validated.

The finite element code RockFlow was originally developed for the numerical simulation of flow and transport processes in fractured-porous media at the Institute of Fluid Mechanics and Environmental Physics in Civil Engineering (ISU) [1, 19, 20, 24, 29–31, 45, 55, 58, 65]. During the last decade further investigations have been done in order to apply RockFlow to several problems concerning the storage of radioactive waste. Basic requirement for the research was the development of a coupled thermo-hydro-mechanical finite element code as well as its belonging validation. Based on this, it was possible to focus on specific problems arising in the field of waste disposal in clay materials. As a consequence various extensions of the model were made and tested. Kohlmeier in 2006 [28] developed, implemented and validated the THM model and applied it to a case study in cristalline rock. Ziefle in 2008 [67] enhanced the mechanical part by a non-linear compressibility model and plasticity to simulate clayey EBS accurately. Finally, Maßmann in 2009 [38] investigated and simulated excavation induced processes in claystone.

This contribution gives an insight into the possibilities of numerical modeling and its underlying theory for geotechnical applications but it also emphasizes arising problems and possible further research.

One focus point is the non-linear behavior of low permeable materials due to compression. Within this context, various publications consider the effect of pore space and pore water pressure on the compressibility of the material. General remarks are given in [13]. Finally, there exist two common approaches to comprise this physical effect. On the one side, there exist purely mechanical models with a non-linear elastic compressible approach (presented in, e.g., [16]). These models are used for geometric non-linear problems with large deformations. On the other side, a common approach is the application of a bulk modulus depending on the pore space, the pore water pressure and the original bulk modulus of the material (given in [41] and others). This approach is used in various models and indicates an additional coupling of the hydraulic and the mechanical process. To assure the clearness of the coupled model and to relate numerical coupling effects directly to physical processes, the work at hand treats the problem in a purely geometrical, mechanical way.

For the French safety authority, the Institute of Radioprotection and Nuclear Safety (IRSN), France, has selected the argillaceous Tournemire site, located in the South of France, in order to study the time-dependent influence of excavation on the rock. In order to explain and model the observed delayed failure mechanism around the excavation [10, 48], a new modeling approach is needed. In contrast to former research (e.g. [49, 53]), the focus point of the work at hand is the hydro-mechanical coupling considering orthotropic non-linear shrinkage, damage, and orthotropically deformation dependent permeability. The mechanical process of stress rearrangement combined with the hydraulic process of desaturation due to the contact of the rock with dry air is simulated. The approach is evaluated by the comparison with observations as well as in-situ measurements of pore water pressures, saturations and deformations during and after the excavation.

2 Theoretical Background

Geomaterials like soils or rocks as well as artificial material like concrete or buffer materials for technical applications consist of granular and brittle materials. They have a porous skeleton. The pores are filled by a single or by multiple fluids. The behavior of the aggregate body is defined by the properties of its solid and fluid constituents. The structure of both the solid skeleton and the boundary layers of the fluids is usually not known. Thus, an averaging process is necessary to build up a continuum model. A macroscopic approach is the Theory of Porous Media (TPM) [9, 14, 15] based on the classical mixture theory of superposed continua [60]. The microscopic composition of the mixture is described by a structural quantity, the volume fraction.

In this work a geometrically linear three-phase formulation of a deformable porous medium is derived. The governing equations of the resulting thermo-hydro-mechanically coupled problem are summarized in the ensuing Section 2.1. The used multi-phase flow formulation is an approximation which assumes that the gaseous phase remains at atmospheric pressure. The transport of heat or solute matter are incorporated in the implementation but not addressed in here in detail. Thermal effects are restricted to the isothermal case. In the presented formulation non–linear behavior of solid and liquid phase are considered.

Based on the thermo-hydro-mechanical model, extensions concerning anisotropy (Section 2.2), shrinkage (Section 2.3), damage (Section 2.4), and non-linear compressibility (Section 2.5) have been investigated.

2.1 Partially Saturated Flow in Poro-Thermo-Elastic Media

In this section the governing equations of the thermo-hydro-mechanically (THM) coupled problem are summarized for a partially saturated porous medium. The solid is considered to behave as a thermoelastic material. The multiphase flow problem is treated in the framework of partially saturated porous media in a one-phase formulation, see for example Lewis and Schrefler [34]. We assume an incompressible liquid in a moving porous solid and negligible gas pressure gradients (Richards' approximation). That means that in the partially saturated zone the gaseous phase flows without resistance. Consequently, the gaseous phase remains at atmospheric pressure, which is taken as reference pressure. It is straightforward to distinguish between saturated and unsaturated zone: in the saturated zone we have positive pore pressures, whereas the pressure is negative in the unsaturated zone. The two zones are separated by the free surface which can be obtained by the isobar of zero pore pressure. Effects due to vapor transport are neglected in the formulation. Its balance equations are summarized in the following. In order to distinguish the solid and the liquid phase we use the indices s and l respectively.

1. Conservation of linear momentum of the solid phase

$$\nabla \cdot \left(\sigma - \alpha \chi p^l \, 1 \right) + \rho_b \mathbf{g} = \mathbf{0} \tag{1}$$

where σ is the effective stress tensor in the solid, α is the Biot coefficient, χ is the effective stress coefficient and the liquid pressure is denoted by p^l. According to Terzaghi's principle [63] that the total stress σ^{tot} is the sum of the effective stress and the (negative) pore water pressure, the total stress within the multiphase problem is defined by

$$\sigma^{\text{tot}} = \sigma - \alpha \chi p^l \, 1 \tag{2}$$

where χ usually is assumed to equal the liquid saturation S^l. The Biot's coefficient is defined by

$$\alpha = 1 - \frac{K_T}{K_s} \tag{3}$$

where K_T and K_s are the bulk modulus of the solid skeleton and the solid grains respectively. The acceleration due to gravity is denoted by $\mathbf{g}$ and the density of the mixture ρ_b is defined by

$$\rho_b = n\, S^l \rho^l + (1-n)\,\rho^s. \tag{4}$$

where n is the porosity which is the proportion of the non-solid volume to the total volume. The density of the liquid and the density of the solid are denoted by ρ^l and ρ^s respectively.

2. Conservation of mass

$$\left(S^{l2}\frac{\alpha-n}{K_s}+nS^l\frac{1}{K^l}\right)\frac{\partial p^l}{\partial t}+\left(\frac{\alpha-n}{K_s}\,p^l S^l+n\right)\frac{\partial S^l}{\partial t}$$

$$-\left(S^l\,\beta_T^s\,(\alpha-n)+nS^l\beta_T^l\right)\frac{\partial T}{\partial t}$$

$$+\frac{1}{\rho^l}\nabla\cdot\mathbf{J}^{ls}+S^l\,\alpha\,\nabla\cdot\frac{\partial\mathbf{u}}{\partial t}=0 \tag{5}$$

where K^l is the bulk modulus of the fluid. The corresponding term vanishes in case of an incompressible fluid. The volumetric thermal expansion coefficients of the solid and the liquid are denoted by β_T^s and β_T^l respectively, and T is the temperature. The volume averaged mass flux density of the fluid has to be defined with respect to the solid displacements $\mathbf{u}$ and is denoted by $\mathbf{J}^{ls}$.

3. Conservation of heat energy

$$\nabla\cdot\mathbf{J}_t+c\rho\frac{\partial T}{\partial t}+c^l\rho^l\mathbf{q}\nabla T=0 \tag{6}$$

where $\mathbf{J}_t$ is the volume averaged conductive thermal energy flux density and $c\rho$ is the heat capacity of the mixture given by

$$c\rho = n\,c^l\,S^l\rho^l+(1-n)\,c^s\rho^s. \tag{7}$$

The constitutive equation for the effective stresses is the stress-strain relation of linear elastic materials known as the generalized form of Hooke's law

$$\sigma = \mathbb{C}:\varepsilon^{\text{el}} = \lambda\,\text{tr}\,\varepsilon^{\text{el}}\,1+2G\varepsilon^{\text{el}} \tag{8}$$

where λ and G are the Lamé constants, 1 is the second-order unit tensor and $\mathbb{C}$ the fourth-order material tensor. In case of a thermally expanding solid material the elastic strain tensor $\varepsilon^{\mathrm{el}}$ is as follows:

$$\varepsilon^{\mathrm{el}} = \varepsilon - \varepsilon^{\mathrm{t}} \qquad \varepsilon^{\mathrm{t}} = \alpha_T (T - T_{\mathrm{init}})\,\mathbf{1}$$

where ε and ε^{t} are the total and the thermal strain tensors. The linear thermal expansion coefficient is denoted by α_T and T_{init} is the initial temperature.

The equation for the fluid flux, derived from the conservation of linear momentum in the liquid phase, is the modified form of Darcy's law

$$\mathbf{J}^{\mathrm{ls}} = n S^{\mathrm{l}} \rho^{\mathrm{l}} (\mathbf{v}^{\mathrm{l}} - \mathbf{v}^{\mathrm{s}}) = \rho^{\mathrm{l}} \frac{k^{\mathrm{l}}_{\mathrm{rel}} \mathbf{k}}{\mu^{\mathrm{l}}} (-\nabla p^{\mathrm{l}} + \rho^{\mathrm{l}} \mathbf{g}) \tag{9}$$

where $\mathbf{v}^{\mathrm{l}}$ and $\mathbf{v}^{\mathrm{s}}$ denote the velocity vector of the liquid and solid phase respectively, $\mathbf{k}$ is the permeability tensor, the scalar multiplier $k^{\mathrm{l}}_{\mathrm{rel}}$ is the relative permeability and μ^{l} the viscosity of the liquid phase.

The time derivative of the saturation $\partial S^{\mathrm{l}}/\partial t$ appearing in the mass balance equation (5) is usually replaced by $\frac{\partial S^{\mathrm{l}}}{\partial p^{\mathrm{l}}} \frac{\partial p^{\mathrm{l}}}{\partial t}$. Then, the required derivative $\partial S^{\mathrm{l}}/\partial p^{\mathrm{l}}$ is obtained from the capillary pressure-saturation relation.

The heat conduction for the multi-phase system needed in balance equation (6) is described by Fourier's law using the heat diffusion tensor of the mixture D and the temperature gradient ∇T

$$\mathbf{J}_{\mathrm{t}} = -\mathbf{D}\,\nabla T, \qquad \mathbf{D} = \lambda^{\mathrm{b}}\,\mathbf{1} \tag{10}$$

where the thermal conductivity λ^{b} of the mixture is composed by the contributions of its solid, liquid and gaseous (index g) constituents as follows:

$$\lambda^{\mathrm{b}} = (1 - n)\,\lambda^{\mathrm{s}} + n S^{\mathrm{l}} \lambda^{\mathrm{l}} + n S^{\mathrm{g}} \lambda^{\mathrm{g}}. \tag{11}$$

Remark: For the sake of simplicity, the liquid saturation S^{l} will be replaced in the following by the water saturation S, and the liquid pressure p^{l} by the pore water pressure p, because water is in fact the only liquid of interest in the following.

2.2 Anisotropic Material Modeling

The grain structure of natural rocks is often characterized by anisotropy, related to its genesis. Sedimentary rock, as claystone, has been formed by deposition and consolidation. Consequently, the micro structure is characterized by the appearance of multiple layers. This composition has a significant influence on the macroscopic behavior of the material. In practical engineering it is a common approach to replace the structured medium by an equivalent continuum, in the sense of the representative elementary volume [4]. Within this phenomenological concept, the rock is still described as a homogeneous medium and its anisotropic material properties are described by additional tensorial measures. The type of anisotropy can be specified by symmetry groups [18]. In the case of a layered structure, the material properties in one plane are isotropic. This type of symmetry is called transverse isotropy and can be seen as a special case of orthotropic material. In the hydro-mechanical model the most important material parameters are the second-order tensor of permeability k and the fourth-order material tensor $\mathbb{C}$.

In the coordinate system of anisotropy with the axes ξ, η, and ζ, the principal directions coincide with the coordinate axes and the permeability tensor $\hat{k}$ becomes:

$$\hat{k} = \begin{bmatrix} k_{\xi\xi} & 0 & 0 \\ & k_{\eta\eta} & 0 \\ \text{sym.} & & k_{\zeta\zeta} \end{bmatrix} \tag{12}$$

and the material matrix $\hat{\mathbf{C}}$ in Voigt notation for an orthotropic material [64]:

$$\hat{\mathbf{C}} = \begin{bmatrix} E_\xi \frac{1-nv_{\eta\zeta}^2}{(1+v_{\xi\eta})m} & E_\xi \frac{v_{\xi\eta}+nv_{\eta\zeta}^2}{(1+v_{\xi\eta})m} & E_\xi \frac{v_{\eta\zeta}}{m} & 0 & 0 & 0 \\ & E_\xi \frac{1-nv_{\eta\zeta}^2}{(1+v_{\xi\eta})m} & E_\xi \frac{v_{\eta\zeta}}{m} & 0 & 0 & 0 \\ & & E_\xi \frac{1-v_{\xi\eta}}{m} & 0 & 0 & 0 \\ & & & \frac{E_\xi}{2(1+v_{\xi\eta})} & 0 & 0 \\ & & & & G_{\xi\zeta} & 0 \\ \text{sym.} & & & & & G_{\xi\zeta} \end{bmatrix} \tag{13}$$

with the relations: $n = E_\xi/E_\zeta$ and $m = 1 - v_{\xi\eta} - 2nv_{\eta\zeta}^2$.

Supposed that the plane of isotropy coincides with the $\xi\eta$-plane it is common to use the Young's modulus $E_\xi = E_\eta$ in bedding plane and E_ζ perpendicular to it, the shear modulus acting in the bedding plane $G_{\xi\zeta}$, and the two independent Poisson's ratios $v_{\xi\eta}$, and $v_{\eta\zeta}$ in order to define the material matrix $\hat{\mathbf{C}}$ in the coordinate system of anisotropy. For the description of the orthotropic permeability only two values are needed.

2.3 Non-linear Shrinkage Model

A shrinkage model simulates the contraction and expansion of material due to a change of water content. In general, clayey materials show distinctive swelling/shrinkage strain.

A phenomenological model will be used to model the shrinkage/swelling behavior. The model is directly related to experimental data. It is based on the decomposition of the strain tensor ε into an elastic ε^{el} part and the swelling strain ε^{sw} ($\varepsilon = \varepsilon^{\text{el}} + \varepsilon^{\text{sw}}$). This means for the stress-strain relation (cf. Eq. 8)

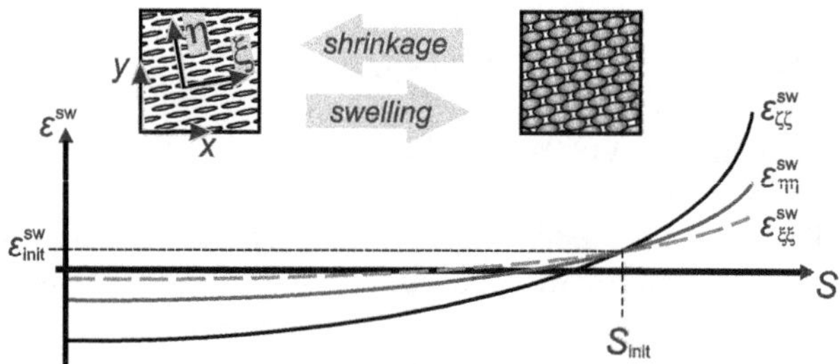

Fig. 1 Schematic behavior of an orthotropically swelling material.

$$\sigma = \mathbb{C} : \varepsilon^{\mathrm{el}} = \mathbb{C} : (\varepsilon - \varepsilon^{\mathrm{sw}}). \tag{14}$$

Following an approach from [3], the swelling strain is calculated as a function of saturation. An isotropic approach is presented in [67]:

$$\varepsilon^{\mathrm{sw}} = \beta^{\mathrm{sw}}(S - S_{\mathrm{init}})\,\mathbf{1}, \tag{15}$$

where β^{sw} is the isotropic swelling coefficient and S_{init} the initial saturation. Experiments on the shrinkage behavior of clay [62] have indicated a distinctive anisotropy in the swelling strain. In general, the swelling strain tensor consists of nine coefficients. The swelling strain tensor can be reduced to the diagonal values in its principal coordinate system by a transformation matrix $\mathbf{T}_{\mathrm{sw}}$:

$$\varepsilon^{\mathrm{sw}} = \begin{bmatrix} \varepsilon^{\mathrm{sw}}_{xx} & \varepsilon^{\mathrm{sw}}_{xy} & \varepsilon^{\mathrm{sw}}_{xz} \\ & \varepsilon^{\mathrm{sw}}_{yy} & \varepsilon^{\mathrm{sw}}_{yz} \\ \mathrm{sym.} & & \varepsilon^{\mathrm{sw}}_{zz} \end{bmatrix} = \mathbf{T}_{\mathrm{sw}}{}^{\mathrm{T}}\,\hat{\varepsilon}^{\mathrm{sw}}\,\mathbf{T}_{\mathrm{sw}} = \mathbf{T}_{\mathrm{sw}}{}^{\mathrm{T}} \begin{bmatrix} \varepsilon^{\mathrm{sw}}_{\xi\xi} & 0 & 0 \\ 0 & \varepsilon^{\mathrm{sw}}_{\eta\eta} & 0 \\ 0 & 0 & \varepsilon^{\mathrm{sw}}_{\zeta\zeta} \end{bmatrix} \mathbf{T}_{\mathrm{sw}}. \tag{16}$$

A power-law is supposed for the calculation of the diagonal values

$$\varepsilon^{\mathrm{sw}}_{\xi\xi} = \varepsilon^{\mathrm{sw}}_{\mathrm{init}} + \beta^{\mathrm{sw}}_{\xi}(S^{\gamma^{\mathrm{sw}}_{\xi}} - S^{\gamma^{\mathrm{sw}}_{\xi}}_{\mathrm{init}}), \tag{17a}$$

$$\varepsilon^{\mathrm{sw}}_{\eta\eta} = \varepsilon^{\mathrm{sw}}_{\mathrm{init}} + \beta^{\mathrm{sw}}_{\eta}(S^{\gamma^{\mathrm{sw}}_{\eta}} - S^{\gamma^{\mathrm{sw}}_{\eta}}_{\mathrm{init}}), \tag{17b}$$

$$\varepsilon^{\mathrm{sw}}_{\zeta\zeta} = \varepsilon^{\mathrm{sw}}_{\mathrm{init}} + \beta^{\mathrm{sw}}_{\zeta}(S^{\gamma^{\mathrm{sw}}_{\zeta}} - S^{\gamma^{\mathrm{sw}}_{\zeta}}_{\mathrm{init}}), \tag{17c}$$

whereby six material parameters have to be determined ($\beta^{\mathrm{sw}}_{\xi}$, $\beta^{\mathrm{sw}}_{\eta}$, $\beta^{\mathrm{sw}}_{\zeta}$, $\gamma^{\mathrm{sw}}_{\xi}$, $\gamma^{\mathrm{sw}}_{\eta}$, $\gamma^{\mathrm{sw}}_{\zeta}$), two parameters to define the initial state ($\varepsilon^{\mathrm{sw}}_{\mathrm{init}}$, S_{init}) and the corresponding coordinate system. In the case of transverse isotropy, the swelling behavior equals in two directions and the number of independent material parameters is reduced to four. In Fig. 1 a schematic sketch of an orthotropically swelling material is shown.

2.4 Continuum Damage Model

Damage is defined as irreversible degradation of mechanical properties. A continuum damage model can be used in order to simulate non-linear stress-strain relation due to microcracks. The applied model is based on the principle of strain equivalence [32]. The state of damage in the material is defined by a damage criterion in the following functional form:

$$g(\overline{\tau}_t, r_t) = \overline{\tau}_t - r_t \leq 0. \tag{18}$$

The index t refers to the value at current time, $\overline{\tau}$ is the equivalent strain, here defined as [23]:

$$\overline{\tau} = \psi_{\text{init}}^{\text{el}}(\varepsilon^{\text{el}+}) = \frac{1}{2}\varepsilon^{\text{el}+} : \mathbb{C}_{\text{init}} : \varepsilon^{\text{el}+}, \tag{19}$$

r defines the damage threshold (energy barrier), and $\psi_{\text{init}}^{\text{el}}$ is the initial elastic energy. In general it is observed that in brittle geomaterials the damage evolution is directly related to the tensile strains [11]. Thus, the use of the positive part of the strains seems to be meaningful. The positive (expansion) part of the strains $\varepsilon^{\text{el}+}$ is determined by the fourth-order positive projection tensor $\mathbb{P}^+$:

$$\varepsilon^{\text{el}+} = \mathbb{P}^+ : \varepsilon^{\text{el}}. \tag{20}$$

The positive projection tensor eliminates the negative parts of the strain tensor by setting the negative eigenvalues to zero. The first definition of the projection tensor origins from [44], further developments have been done by [23], amongst others. A detailed discussion can be found in [59]; therein, the following definition is supposed, which is used in the work at hand

$$\mathbb{P}^+{}_{ijkl} = \frac{1}{2}\sum_{A=1}^{3}\sum_{B=1}^{3}\hat{H}(\hat{\varepsilon}^A)\hat{H}(\hat{\varepsilon}^B)(n_i^A n_j^B n_k^A n_l^B + n_i^B n_j^A n_k^A n_l^B) \tag{21}$$

with the Heaviside step function $\hat{H}$ and the ith principal strain $\hat{\varepsilon}^i$ corresponding to the unit principal direction $\mathbf{n}^i$.

For the definition of the damage evolution the simple but general evolution law given by Marigo [33] has been used.

2.5 Non-linear Compressibility Model

2.5.1 Motivation

Within the simulation of coupled geotechnical problems, the use of the well-known linear elastic material model, called Hooke's law (8), is widely spread. Nevertheless, some additional non-linear effects have to be incorporated for various applications, e.g. the investigation of migration problems in mechanically loaded rocks with very low porosities and permeabilities. Incorporating the non-linearity due to the compressive behavior of the material influences the mechanical as well as the hydraulic sub problem. Particularly, the high compression of materials like dense smectite clays leads to a significant influence of the non-linear compression behavior. This kind of materials is used for example for backfills or plugs of drifts, tunnels and shafts or for plugs in boreholes as well as in various fields of waste isolation (e.g. waste landfills). Because of the low initial porosities, also the compression behavior of host rocks like claystone used for high radioactive waste disposal is significantly influenced by the non-linear compressibility.

Generally, one can state, that highly compressed materials as well as materials with a very low initial porosity have to be analyzed with an extended elastic model. Consequently, the work at hand presents an extension of the Hooke's law to a non-linear elastic compressibility model.

Assuming a linear elastic compressible model there is no limitation of the possible compaction of the material. Generally, the solid grains are presumed to be incompressible and the deformation only leads to a change of the porosity of the material. In spite of that, in many applications the incompressible state is not reached as most of the materials have relatively large porosities while the deformations are small. Applying such a model to the prescribed materials with very low porosities, even small volumetric deformations lead to a porosity near the compression point. At that point there is nearly no more pore space available and the material becomes incompressible.

To represent this situation with the theoretical model, a physically non-linear elastic compressibility model which incorporates a compression point and restricts the porosity to the valid range has to be applied. Implementing this effect in the numerical model, an extension of the classical Hooke's law by an additional term is proposed as it is done by Eipper [16] for the geometric non-linear case. The modification for the geometric linear case, incorporating the initial state is described in the following. Due to the usage of a strain dependent permeability as it is presented in [67] this effect also influences the hydraulic sub-problem.

2.5.2 Physical Background and Relating Definitions

Generally, in the Theory of Porous Media, the material is composed of air, liquid and solid grains. While the air and the liquid are stored in the pore space of the body, the solid grains provide the material matrix. Assuming the solid grains to be incompressible, deformations only lead to a change of the pore space in the body. If the so called compression point is reached, there is no more pore space available and the material becomes incompressible. As a matter of fact, the investigation of materials with very small porosities may lead to incompressible material behavior already for deformations in the range of geometric linear material behavior. Consequently, a non-linear elastic compressible material in the range of geometric linearity has to be defined. The material model has to fulfill the following requirements:

1. The valid range should be the range of geometric linear deformations.
2. The low stress case should imply a material behavior similar to the behavior of Hooke's material.
3. The compression should be limited to the compression point. Convergence to this point should lead to a significant increase of the stresses.

An additional term of the strain energy function is given here to represent the mentioned material behavior. This term should modify the linear elastic case in the required way. As the behavior of the material significantly depends on the remaining pore space of the body, the function should depend on the porosity of the body. Therefore the porosity should be treated as a time-dependent material property. This is done by the proposed strain dependent porosity, which is given by

$$n = n_{\text{init}} + \text{tr}\,\varepsilon - \text{tr}\,\varepsilon^{\text{sw}} \tag{22}$$

with the initial porosity n_{init}. As the material behavior is non-linear, there must be a definition of an initial material behavior. This initial state should depend on the initial porosity of the material. To get a relation between the initial porosity and the initial stress conditions in the body, a difference between the initial and the stress-free porosity is made.

The stress-free porosity is the porosity of the material which indicates the beginning of elastic material behavior. It results from an unconstrained storage with no (sand-like materials) or only marginal (clay materials) compaction. It yields

$$n_{\text{SF}} = n_{\sigma=0}. \tag{23}$$

The initial porosity is defined by

$$n_{\text{init}} = n_{t=0}. \tag{24}$$

The difference of both is given by

$$(\Delta n)_0 \equiv \Delta n = n_{\text{init}} - n_{\text{SF}} \tag{25}$$

and presented in Fig. 2 for a preconsolidated problem.

The strain field in the body is given by the elastic deformations due to the loads applied during the simulation time and the difference of stress-free and initial porosity Δn. The trace of the total strains results to

$$\text{tr}\,\varepsilon^{\text{tot}} = \text{tr}\,\varepsilon^{\text{el}} + \text{tr}\,\varepsilon_{\text{init}} \tag{26}$$

$$= \text{tr}\,\varepsilon^{\text{el}} + \Delta n. \tag{27}$$

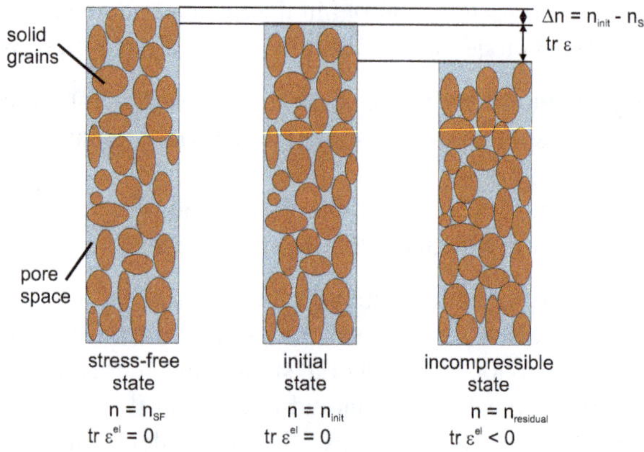

Fig. 2 Physical model for the compression of porous media with preconsolidation.

2.5.3 Theoretical Background

A strain energy function for the geometric non-linear case is given by Eipper [16]. He proposed to extend the classical Hooke approach by the additional term W_{nlc} due to the non-linear compressibility of the material

$$W = W_{\mathrm{Hooke}} + W_{\mathrm{nlc}} \tag{28}$$

with

$$W_{\mathrm{nlc}} = \frac{\lambda_{\mathrm{c}}}{\gamma\left(\gamma - 1 + \frac{1}{n_{\mathrm{SF}}^2}\right)} \left(J^{\gamma} - 1 - \gamma\ln\frac{J - (1 - n_{\mathrm{SF}})}{n_{\mathrm{SF}}} + \gamma(1 - n_{\mathrm{SF}})\frac{J - 1}{n_{\mathrm{SF}}} \right). \tag{29}$$

With the Jacobian J and γ being a control parameter for the volumetric behavior. For the given applications, this parameter is set to $\gamma = 1$. Within the framework of the compressibility model, the common Lamé parameter λ is replaced by the compression parameter λ_{c} which is defined by

$$\lambda_{\mathrm{c}} = \frac{1}{2}\lambda. \tag{30}$$

The additional term of the strain energy function remains

$$W_{\mathrm{nlc}} = \frac{1}{2}\lambda n_{\mathrm{SF}}^2 \left(J - 1 - \ln\frac{(J - 1 + n_{\mathrm{SF}})}{n_{\mathrm{SF}}} + (1 - n_{\mathrm{SF}})\frac{J - 1}{n_{\mathrm{SF}}} \right). \tag{31}$$

Incorporating this additional term to the classical linear approach, the total energy function remains

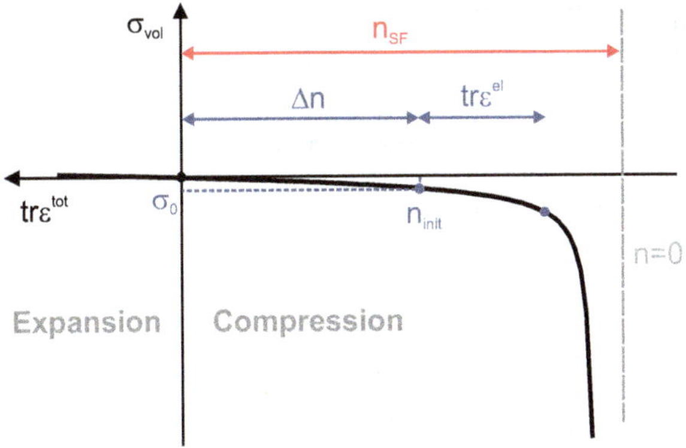

Fig. 3 Non-linear elastic compressible material model.

$$W = \frac{1}{2}\lambda\left[\frac{1}{2}(\mathrm{tr}\,\varepsilon^{el})^2 + n_{SF}^2\left(\mathrm{tr}\,\varepsilon^{el} - \ln\frac{(\mathrm{tr}\,\varepsilon^{el} + n_{SF})}{n_{SF}} + (1 - n_{SF})\frac{\mathrm{tr}\,\varepsilon^{el}}{n_{SF}}\right)\right] + G\varepsilon : \varepsilon.$$

(32)

Assuming the initial condition to be not necessarily equal to the stress-free condition, the total trace of strains results from (27). For this case, the initial condition has to be incorporated in the strain energy function by using the total trace of strains $\mathrm{tr}\,\varepsilon^{tot}$ instead of $\mathrm{tr}\,\varepsilon^{el}$ for the compression term. The energy function yields to

$$W = \frac{1}{2}\lambda\left[\frac{1}{2}(\mathrm{tr}\,\varepsilon^{el})^2 + n_{SF}^2\left[\frac{\mathrm{tr}\,\varepsilon^{tot}}{n_{SF}} - \ln\left(1 + \frac{\mathrm{tr}\,\varepsilon^{tot}}{n_{SF}}\right)\right]\right] + G\varepsilon : \varepsilon. \qquad (33)$$

The derivation of stresses follows from differentiating with respect to the strains

$$\sigma = \frac{1}{2}\lambda\left[\mathrm{tr}\,\varepsilon^{el} + n_{SF}\left(1 - \frac{n_{SF}}{\mathrm{tr}\,\varepsilon^{tot} + n_{SF}}\right)\right]1 + 2G\varepsilon. \qquad (34)$$

Figure 3 shows the relation of stresses and $\mathrm{tr}\,\varepsilon^{tot}$. It can be seen that there exists an initial porosity which not necessarily equals the stress free porosity. The difference is given by Δn. Assuming the stress depending on the volumetric strains, the Δn is incorporated in the model as a kind of volumetric strain which is added to the elastic volumetric strains by the definition of $\mathrm{tr}\,\varepsilon^{tot}$. This incorporates some initial stresses to the model. The figure also shows that the given stress function leads to a significant increase of stresses in the compression area. This leads to limited strains in this range. Here the material converges to the incompressible state due to the lack of pore space.

Another differentiation leads to the material matrix

$$\mathbb{C}_{tang} = \frac{\partial\sigma}{\partial\varepsilon} = \frac{1}{2}\lambda\left(1 + \frac{n_{SF}^2}{(\mathrm{tr}\,\varepsilon^{tot} + n_{SF})^2}\right)1\otimes 1 + 2G\mathbf{I}. \qquad (35)$$

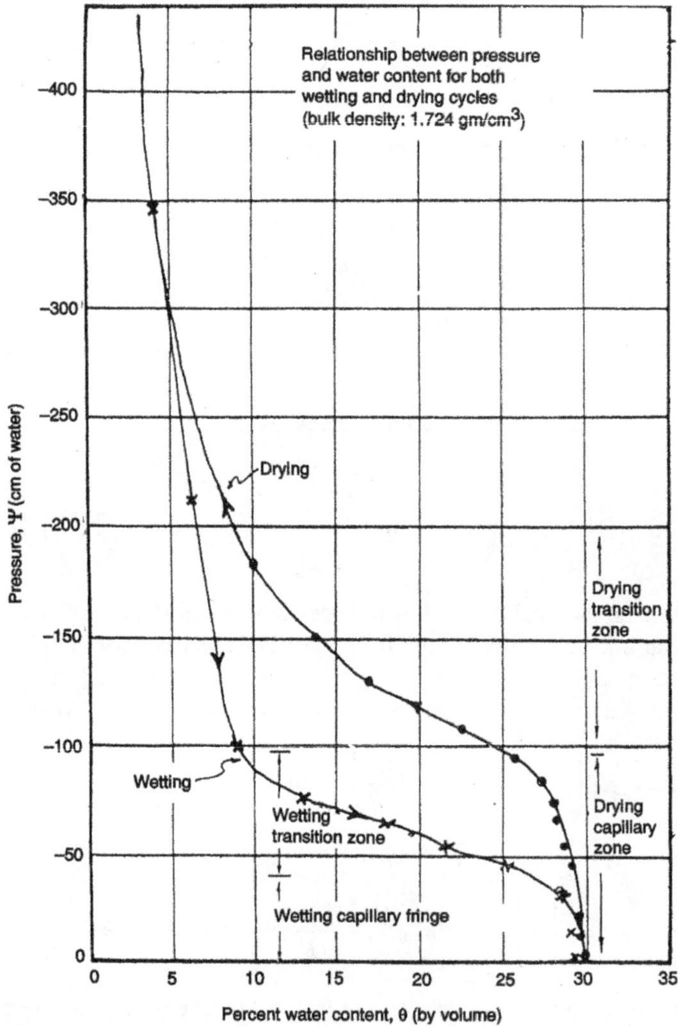

Fig. 4 Hydraulic head versus water content, with $\Psi = p/\rho^w g$ and $\theta = nS$ [35].

The validation of the presented model due to the comparison with experimental data can be found in [67].

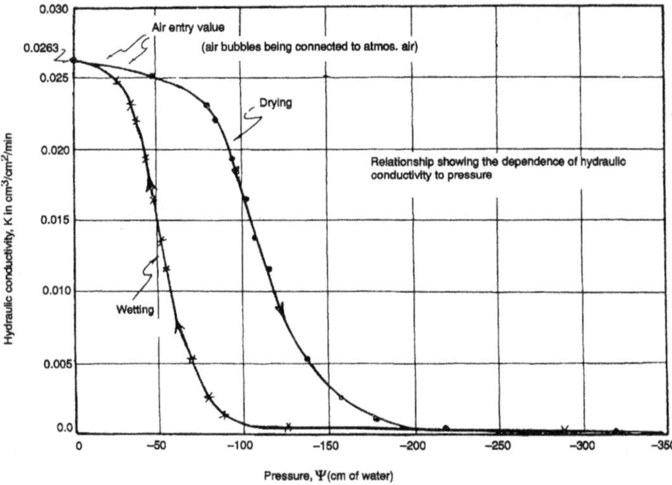

Fig. 5 Hydraulic conductivity versus pressure, with $K = k\rho^{w}g/\mu^{w}$ and $\Psi = p/\rho^{w}g$ [35].

3 Validation

3.1 Liakopoulos Drainage Test

In this section, multiphase flow in a deforming porous medium is studied. The presented test example is a drainage test based on an experiment by Liakopoulos [35]. Desaturation takes place due to gravitational effects. This example was studied previously by several authors, for example Liakopoulos [35], Narasimhan and Witherspoon [42], Zienkiewicz et al. [68] or Schrefler and Zhan [54]. Therefore, this example is well suited as benchmark, despite the lack of any analytical solutions for this type of coupled, non-linear problems.

The physical experiment of Liakopoulos was conducted in a column packed with so-called Del Monte sand. Moisture content and tension at several points along the column were measured with tensiometers (cf. Figs. 4 and 5). The capillary pressure $p_c(S)$ is a function of the water saturation S and can be given as

$$p_c = \left(\frac{1-S}{1.9722} \times 10^{11} \right)^{1/2.4279} \text{Pa} \tag{36}$$

as well as the relative permeability relationship $k_{rel}(S)$

$$k_{rel} = 1 - 2.207(1-S)^{1.0121} \text{ m}^2. \tag{37}$$

These equations fit the measured data in case of saturations larger than 0.84 and are therefore suitable for the following numerical simulation. The model set-up is

Table 1 Liakopoulos experiment. Material properties.

Parameter		Value	Unit
Young's modulus	E	1.3	MPa
Poisson's ratio	v	0.4	
Solid grain density	ρ^s	2000	$\mathrm{kg\,m^{-3}}$
Porosity	n	0.2975	
Permeability	k	4.5×10^{-13}	m^2

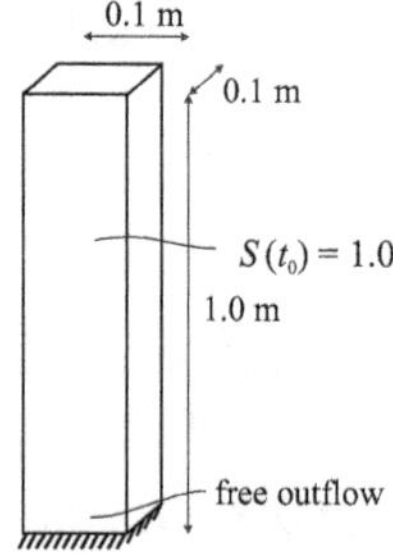

Fig. 6 Liakopoulos experiment. Set-up.

depicted in Fig. 6. The material parameters, taken from Lewis and Schrefler [34], are summarized in Table 1.

In Fig. 7 the numerical results are shown: water saturation S and vertical solid displacement u_z^s along the column height. The results are very close to the reference values taken from [34].

3.2 Step-wise Compression Test

The impact of the non-linear elastic compression behavior on the resulting strains and pressure evolution of a compression test is investigated within this section. A step-wise compression of an initially unloaded sample is simulated. The process is simulated as a coupled hydro-mechanical problem and consequently every load step leads to a classical consolidation problem with a time-dependent behavior until the final equilibrium stage is reached. For this simulation the non-linear compressibility model, already introduced in Section 2.5 is used. As the consolidation process is a classical example of a coupled hydro-mechanical problem, the resulting deformation and pressure evolution are given in various literature. Kohlmeier [28] uses this example to present a verification of the linear elastic model implemented in Rock-Flow due to the comparison with an analytical solution.

Comparing the linear elastic model with the non-linear approach, a relatively high permeability is chosen, as the focus is laid on the pure mechanical process. Afterwards, the results derived with a significantly lower permeability are presented, leading to an increase of the water pressure and a more dominant time-dependent behavior. Concerning this example, the initial porosity is chosen to be equal to the

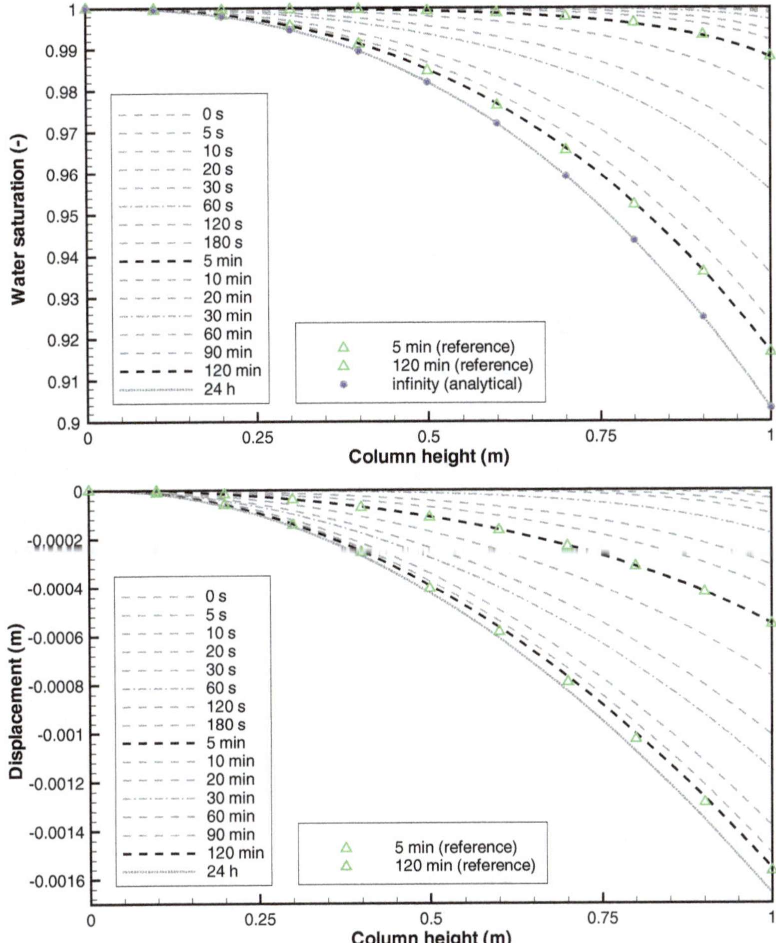

Fig. 7 Numerical results. Water saturation S (top) and vertical displacement u_z^s (bottom) (ordinate) versus height (abscissa) of deformable column ($t = \{5\,s, 10\,s, 20\,s, 30\,s, 1\,min, 2\,min, 3\,min, 5\,min, 10\,min, 20\,min, 30\,min, \ldots, 120\,min$ and $24\,h\}$).

stress-free porosity. The influence of a preloaded initial state is also investigated in [67].

3.2.1 Model Set-up

The soil column has a height of 0.10 m and a width of 0.02 m and is mechanically fixed and impermeable at the sides and at the bottom. At the top exists a permeable boundary and a time-dependent load is applied. The load results from a constant value of -125.0 kN/m multiplied with a load factor which increases in 12 steps

Table 2 Material properties for the step-wise compression test.

Parameter	Value
Young's modulus, E	250 kPa
Poisson's ratio, v	0.2
Initial porosity, n_{init}	0.05
Stress-free porosity, n_{SF}	0.05
Simulation with a higher intrinsic permeability k	$5.0 \times 10^{-10} \text{m}^2$
Simulation with a lower intrinsic permeability k	$5.0 \times 10^{-12} \text{m}^2$

from 0.0 to 0.3. The initial pore water pressure within the whole domain is assumed to be zero. The material properties are presented in Table 2.

Additionally, a strain dependent porosity as well as a strain dependent permeability are assumed. The porosity results from Eq. (22) and the current relative permeability depending on the porosity is given by the following linear relationship:

$$k_{\text{rel,n}} = 0.5n + 0.5 \tag{38}$$

3.2.2 Results: Linear versus Non-linear Model

Every load step leads to a classical consolidation problem.

In the following the temporal evolution of the volumetric strains is investigated. Within this framework, the linear and the non-linear model as well as set-ups with a higher ($k = 5.0 \times 10^{-10}$ m^2) and a lower ($k = 5.0 \times 10^{-12}$ m^2) permeability are compared.

Due to the load controlled type of boundary condition at the top, the stresses increase step-wise with constant step increments. The evolution of strains depends on the type of constitutive model. In contrast to the constant strain increments arising by the linear elastic model, the non-linear material model leads to step-wise increasing but quantitatively varying strain increments. The results of both models are given in Fig. 8. The dashed line in the figures presents the stress-free porosity, which equals the initial porosity for this example.

Whereas the simulations pictured on the left are performed with the linear elasticity model, the results of the non-linear model are given on the right side. Concerning the strains one can state that the final strains are bordered by the porosity of the material if the non-linear model is used. Especially if high compressive strains lead to a significant reduction of the pore space, the results of the non-linear model differ significantly from the results derived by the linear approach. Consequently, the simulation of problems including high compressions up to the compression point of the material have to be analyzed with the proposed model. As the compression point depends directly on the stress-free porosity of the material, the non-linear elastic compressibility model is of special interest for materials with low porosities.

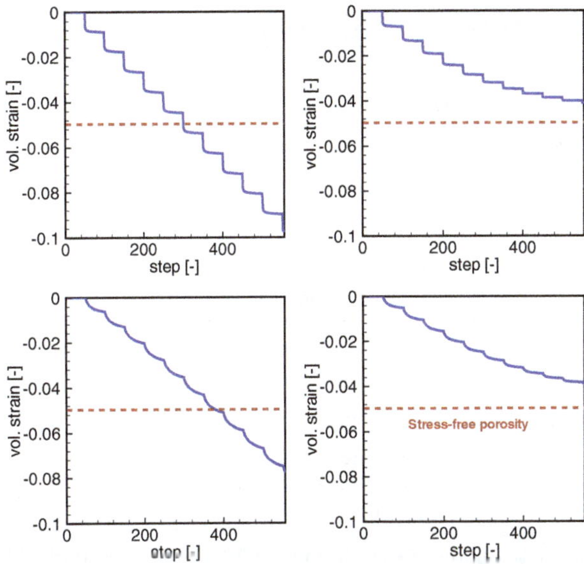

Fig. 8 Temporal evolution of the volumetric strains. Left: Linear model. Right: Non-linear model. Top: Higher permeability leads to a low time-dependent impact. Bottom: Lower permeability leads to a high time-dependency.

Concerning the pressure field, it becomes clear that lower strain increments lead to lower water pressures. Consequently, the impact of the material model increases significantly if the compression strains converge to the compression point.

4 Application

4.1 Thermo-Hydro-Mechanical Simulation of a Generic Repository

4.1.1 Introduction

The application presented in this section originated in the framework of the DECOVALEX-THMC international project. It is a multi-disciplinary interactive and co-operative research effort in modeling thermo-hydro-mechanical-chemical (THMC) processes in fractured rocks and buffer materials. In addition to coupled code development, the project investigates the role of THMC processes in the performance assessment for radioactive waste storage. The THM modeling work within preceding DECOVALEX phases covered two large-scale in-situ heater experiments: the FEBEX experiment at Grimsel in Switzerland and the drift scale test at Yucca

Table 3 Summary of Yucca Mountain type repository scenario [2].

Scenario Detail	Definition
Complete Period	10,000 years
Initial Heat Load	1,450 W/m
	Ventilation effects reduce this load during 50-year period
Tunnel	Open, no buffer
Flow in Tunnel	Gas flow
Rock	Densely fractured volcanic rock
Flow in Rock	Unsaturated
	Two-phase flow under thermal gradient

Mountain in the USA. The project DECOVALEX-THMC applies the knowledge gained from modeling the above mentioned short-term in-situ tests, with a test period between one and eight years, to the evaluation of long term processes. Two generic repository types are considered according to the FEBEX and Yucca Mountain experiments. The regulatory compliance periods in these types of repositories span over thousands to ten-thousands of years.

The work presented here concentrates on the thermo–hydro–mechanical simulation of Yucca Mountain type within sub-task D_THM. For more details concerning the other sub-task, the interested reader is referred to [2].

4.1.2 Model Set-up

Description of a Repository Scenario

The Yucca Mountain type of a repository is supposed to be located in volcanic rock and the emplacement is defined to be proceeded in open gas-filled tunnels. The scenario's details are summarized in Table 3.

Geometric Data

A schematic description of the model geometry, the boundary conditions, the specific areas of focus, and the profile for which simulation outputs should be derived is presented in Fig. 9. Due to symmetry conditions only a single drift has to be considered, representing a repository of infinite length and width. Consequently, this approach represents an extreme setting as interactions with the surrounding area are restricted to top and bottom boundaries.

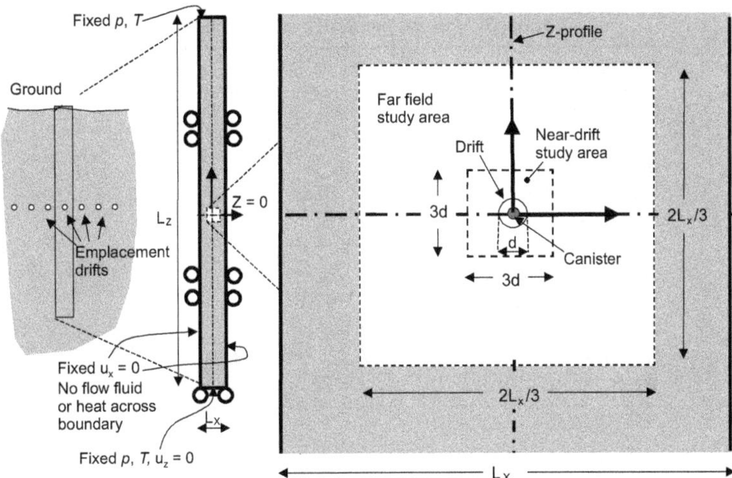

Fig. 9 Model description for Task D, THM. Model geometry, boundary conditions and line for model output [2].

Table 4 Model dimensions.

Dimension	Value
Vertical length, L_z	1,000 m
Horizontal length, L_x	35 m
Drift diameter, d	2.28 m
Diameter of waste canister	0.9 m

Model Dimensions and Material Properties

According to the definitions given in Fig. 9, the model dimensions are summarized in Table 4.

The material properties of the rock are summarized in Table 5. The bentonite buffer material is of FEBEX type, its material properties are summarized in Tab. 6. The water retention curves are a modified and a standard van Genuchten function for the bentonite and the rock, respectively.

Heat Output

The thermal power emitted by a reference pressurized water reactor (PWR) element is depicted in Fig. 10. Assuming that the waste is 30 years old at emplacement time, the current heat output is 400 W per PWR element. Considering an alignment of four PWR elements per canister of 4.54 m length and a canister spacing of 2 m results in an average thermal power per meter drift of $4 \times 400 \, \text{W}/6.54 \, \text{m} = 245 \, \text{W/m}$. The

Table 5 Material properties of the rock.

Parameter	Value
Density, ρ	$2{,}700 \text{ kg m}^{-3}$
Porosity, n	0.01
Biot's constant, α	1.0
Young's modulus, E	35 GPa
Poisson's ratio, v	0.3
Specific heat, c	$900.0 \text{ J kg}^{-1}\text{ K}^{-1}$
Thermal conductivity, λ	$3.0 \text{ W m}^{-1}\text{K}^{-1}$
Thermal expansion coeff., α_T	$1.0 \times 10^{-5} \text{ K}^{-1}$
Permeability, k	$1.0 \times 10^{-17} \text{ m}^2$
Rel. Permeability, k_{rel}	$k_{\text{rel}} = \sqrt{S}\,(1 - (1 - (S^{1/0.6})^{0.6})^2$
Water retention, $S(p_c)$	$S = (1 + (\frac{p_c}{1.47\,\text{MPa}})^{2.5})^{-0.6}$

Table 6 Material properties of the bentonite buffer.

Parameter	Value	
Dry density, ρ	$1{,}600 \text{ kg m}^{-3}$	
Porosity, n	0.41	
Biot's constant, α	1.0	
Young's modulus, E	100 MPa	
Poisson's ratio, v	0.35	
Moisture swelling coefficient, β_{sw}	0.238	
Dry specific heat, c_s	$767 \text{ J kg}^{-1}\text{K}^{-1}$,	$((1.38 \text{ K}^{-1} \cdot T + 732.5) \text{ J kg}^{-1}\text{K}^{-1})$
Thermal conductivity, λ_m	$1.3 \text{ Wm}^{-1}\text{K}^{-1}$,	$((1.28 - \frac{0.71}{1+e^{(S-0.65)/0.1}}) \text{ Wm}^{-1}\text{K}^{-1})$
Thermal expansion coeff. α_T	$1.0 \times 10^{-5} \text{ K}^{-1}$	
Permeability, k	$1.0 \times 10^{-21} \text{ m}^2$	
Rel. permeability, k_{rel}	$k_{\text{rel}} = S^3$	
Water retention, $S(p_c)$	$S = 0.01 + 0.99\,(1 + (\frac{p_c}{35\,\text{MPa}})^{1.43})^{-0.3}\,(1 - \frac{p_c}{4000\,\text{MPa}})^{1.5}$	

temporal evolution of this value is according to the decay curve depicted in Fig. 10 in consideration of the initial disposal time of 30 years.

In-situ Stress Field and Modeling Sequence

The in-situ stress field is assumed to depend linearly on the depth D. The horizontal total stress is prescribed by the function $\sigma_h = 0.055 \text{ MPa/m} \cdot D + 4.6 \text{ MPa}$ while the vertical stress is evaluated by the stress of the overlying rock mass. Thus, at the drift axis the initial value of the horizontal stress is 32.1 MPa while the vertical stress is about 13.5 MPa.

The demanded modeling sequence is concerned with (a) the pre-excavation conditions, (b) the simulation of the excavation, (c) the installation of bentonite buffer and finally (d) the transient simulation of the post-closure thermo–hydro–mechanical behavior of the repository. The sequences and the associated initial and

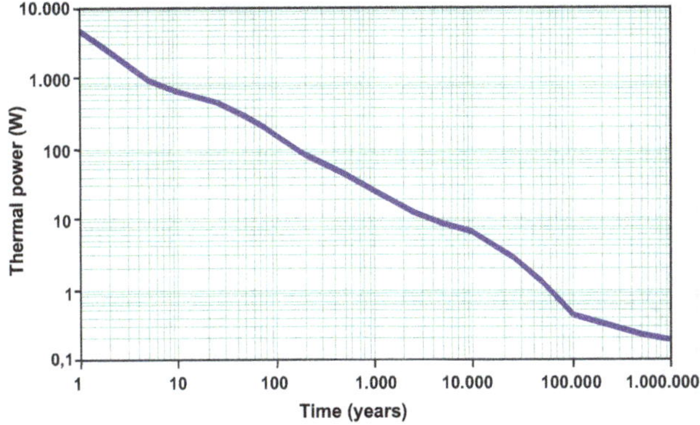

Fig. 10 Thermal power decay function of a reference fuel PWR element [2].

boundary conditions are depicted in Fig. 11. The numerical predictions for these sequences are presented in the next sections.

4.1.3 Numerical Simulation of the Pre-excavation Conditions

The pre-excavation is simulated in order to check the correctness of the initial values of the numerical model. As the stress and fluid pressure fields are correct and the strain is equal to zero, the simulation of the next phases can be initiated. The results are presented in the following three sections.

4.1.4 Numerical Simulation of the Excavation

With the pre-excavation model at hand, the excavation is simulated by disregarding the rock mass elements. The deformation and the stress increase due to excavation is depicted in Fig. 12. It also shows the decrease of fluid pressure around the tunnel causing an overall settlement of the drift and the overlying rock mass.

4.1.5 Numerical Simulation of the Installation of the Bentonite Buffer

The installation of the bentonite buffer is finally simulated by replacing the rock mass elements by bentonite elements. No initial stresses are applied. The initial saturation is 65%. This phase immediately runs over into the transient phase which is presented in the next section.

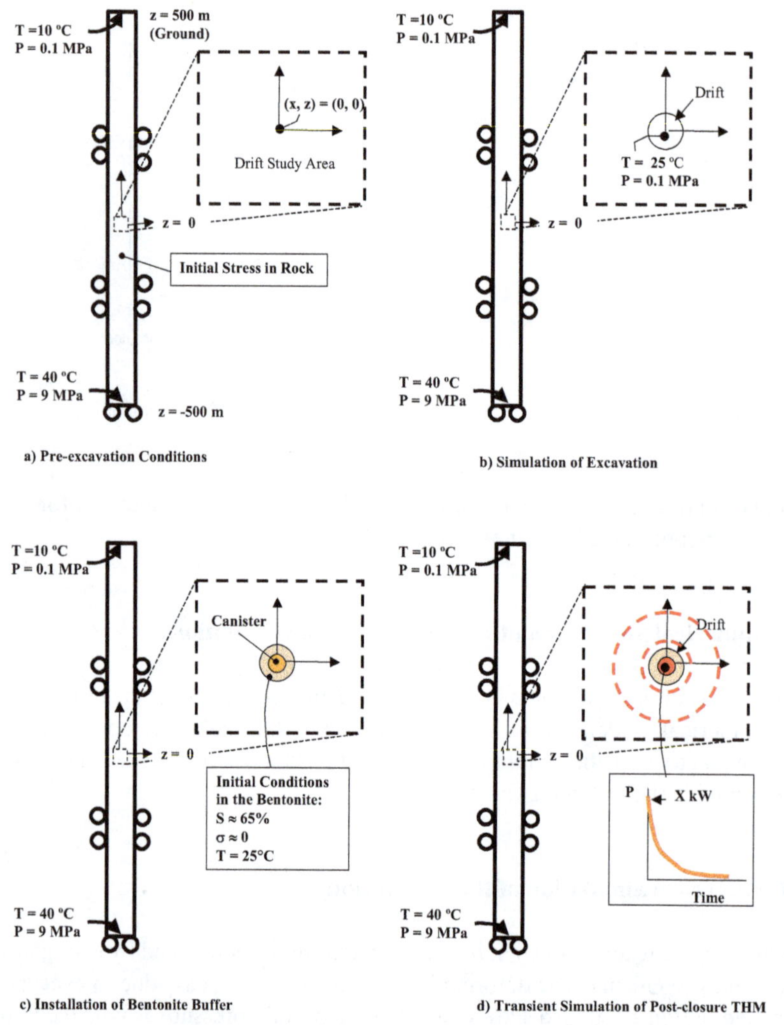

Fig. 11 Modeling sequence, initial and boundary conditions [2].

4.1.6 Numerical Simulation of the Transient Thermo–hydro–mechanically Response

The transient simulation predicts the thermo-hydro-mechanical response of the repository for a time range of 100,000 years. The RockFlow results, presented in the following, are compared to predictions carried out with two different codes, namely TOUGH2 and ROCMAS. The reference values obtained with TOUGH2 and ROCMAS are taken from Birkhölzer et al. [7]. The presentation of the resulting values is according to the model output specifications given in Fig. 9.

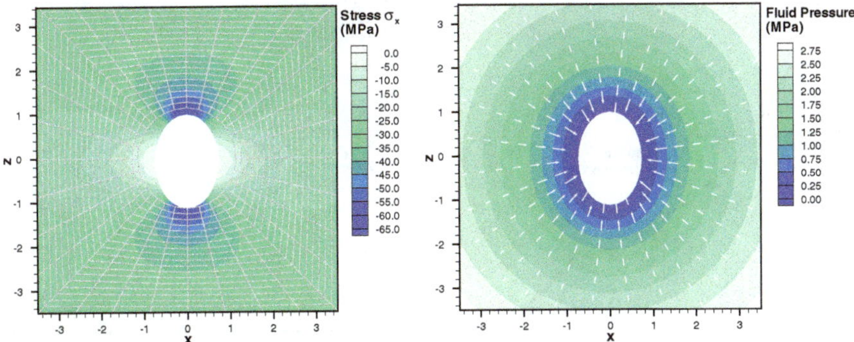

Fig. 12 Horizontal stress (left) in the deformed near field due to excavation. The undeformed mesh represents the initial configuration. The settlement of the drift is due to the transient pressure field (right) influenced by the drainage caused by the atmospheric pressure in the drift wall. The displacements are scaled by factor 250.

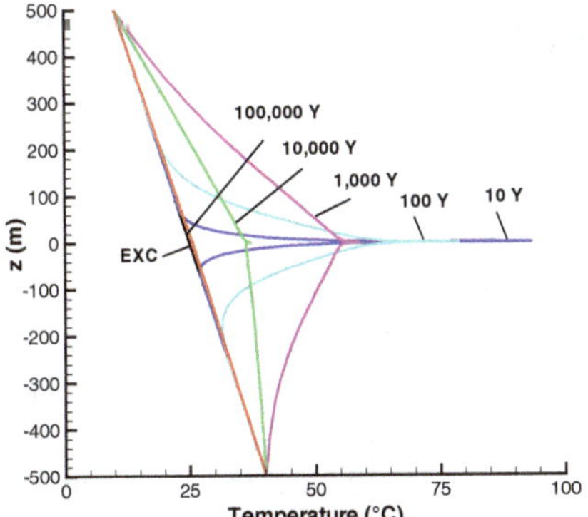

Fig. 13 RockFlow simulation results of vertical temperature profiles.

Fig. 14 Comparison of TOUGH-FLAC (TOUGH2) and ROCMAS simulation results of vertical temperature profiles [7].

Temperature Evolution

The evolution of the temperature is depicted in Figs. 13 and 14. The peak temperature of 93.2°C is reached after 11 years. As the temperature strongly depends on the heat output, the maximum temperature is very sensitive to the interpolation of the thermal power decay function depicted in Fig. 10. Its logarithmic decrease has to be reflected in the time step size in order to ensure a correct integration in time. As the temperature is significantly below the boiling point of water, the effect of evaporation and the moisture transport forced by the thermal gradient can be neglected as being done in this simulation. Nevertheless, it might be of considerable importance, if high temperatures occur at the canister. Regarding the evolution of temperature,

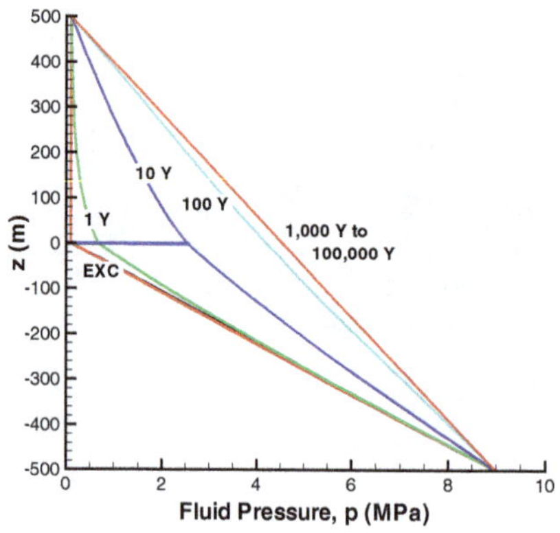

Fig. 15 RockFlow simulation results of vertical pressure profiles.

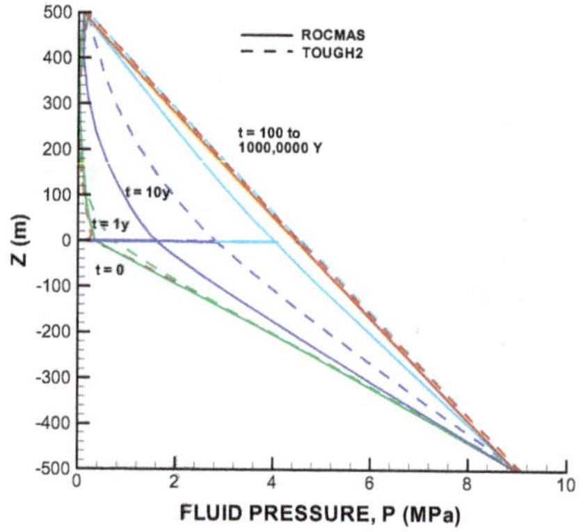

Fig. 16 Comparison of TOUGH2 and ROCMAS simulation results of vertical pressure profiles [7].

the length of the time period for the reestablishment of initial state conditions is more than 100,000 years.

Evolution of Water Saturation and Fluid Pressure

The fluid pressure profiles of the RockFlow, TOUGH2 and ROCMAS simulations are shown in Figs. 15 and 16. During the steady-state analysis of the excavation sequence, the overburden is drained due to prescribed atmospheric pressure in the drift (cf. Fig. 12). After emplacement of the PWR canisters and backfilling the tunnel with a partially saturated bentonite mixture, the closing of the drift is initiated. In the model, the closing is described by releasing the fluid pressure boundary condition at the wall of the drift. Thus, the hydrostatic gradient of fluid pressure is slowly reestablished within the whole domain of bentonite and rock mass. The resaturation of the bentonite takes about 30 years, but it has to be mentioned that no evaporation and moisture transport has been taken into account. The partially saturated bentonite buffer and the time period of resaturation are of special interest. It becomes obvious that the resaturation is limited by the low conductivity of the barrier material. The influence of swelling pressure on the porosity and permeability could be of interest. Regarding the fluid pressure field, the length of the time period for the reestablishment of the initial state is more than 100 years.

Evolution of Stress

The vertical profiles of the horizontal component of the total stress are depicted in Fig. 17. The total stress is defined in Eq. (2), whereas $\chi = S$.

In the total stress plots not only the variation of thermal stresses becomes obvious but changes in fluid pressure can also be identified as well. After excavation, the

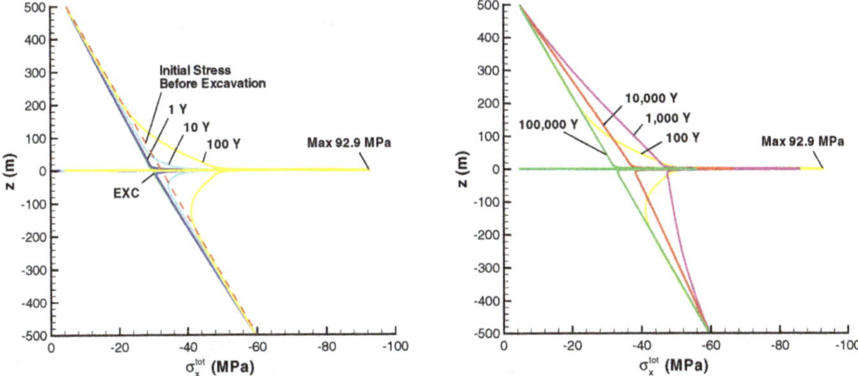

Fig. 17 RockFlow Simulation results of vertical profiles of total horizontal stress.

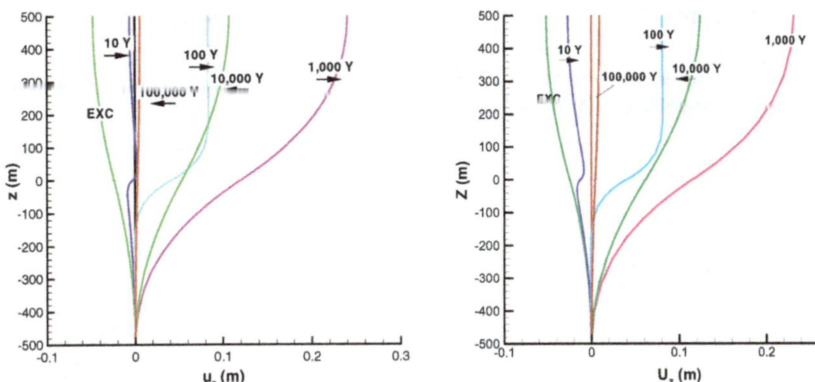

Fig. 18 RockFlow simulation results of vertical displacement profiles.

Fig. 19 ROCMAS simulation results of vertical displacement profiles [7].

total stress is lower than the initial stress, which is induced by the decrease of water pressure due to the drainage process. At the end of the simulation time of 100,000 years, when the initial water pressure field is build up again, the stresses increase linearly with depth. Neglecting the disturbance by the drift, the stresses at level $z = 0$ m would result in the initial stress values of -32.1 MPa in horizontal direction and about -13.5 MPa in vertical direction.

Evolution of Displacement

The profiles of the vertical displacement are presented in Figs. 18 and 19. The initial settlement of the entire column is caused by the drainage of water into the open drift during the excavation phase. During the excavation phase, the water pressure above the drift is assumed to be equal to atmospheric pressure. After closing the drift, the reestablishment of water pressure and the temperature increase due the prescribed heat power output cause an upwardly directed displacement of the drift and the whole column.

4.2 Long-term Saturation Process (TDR-Test)

It is planned to dispose high-level radioactive waste in underground repositories to be erected in very low permeable bedrock. Besides the barrier caused by the surrounding geology, an engineered barrier system (EBS) is planned. The investigation of the permeability of such a system is the aim of the gas migration test (GMT). This test is accomplished in the Felslabor Grimsel in Switzerland and comprises several experiments. One of the laboratory experiments is the TDR-test which covers a long-term saturation process in a bentonite-sand mixture which is measured by several pressure cells and a TDR tube (time domain reflectometry), which causes the name TDR-test. The measurements (TDR-signal, pressure, water volume and stress) are registrated since February 2002. Due to the well defined set-up including all boundary conditions and a comprehensive measuring instrumentation, the results of this experiment can be used to improve and validate the finite element code RockFlow by the simulation of the TDR-test.

4.2.1 Experimental Design

Within the TDR-test a cylindrical column of a bentonite-sand-mixture with the percentage of 20:80 is investigated (see Fig. 20). This soil column is surrounded by an impermeable and mechanically fixed metallic cylinder and has a height of 88 cm and a diameter of 20 cm. At the bottom of the bentonite-sand-mixture a layer of gravel is constructed, whereas the top is bordered by a metal plate with a small tube. The column contains different pressure cells, which are situated in three different heights. Measured from the top of the gravel they are located at 13.0, 45.0 and 75.0 cm. Another measuring probe, called TDR, is adjusted in the central axis of the cylinder. The water injection takes place at the bottom of the column and is increased step-wise over a time period of about 4.5 years.

4.2.2 Model Set-up

The presented TDR-test is characterized by a classical saturation process combined with the swelling of the material, which influences the mechanical process of the problem. The coupled hydro-mechanical simulation given in Section 2.1 additionally incorporates the swelling model presented in Section 2.3 as well as the nonlinear compressibility model given in Section 2.5.

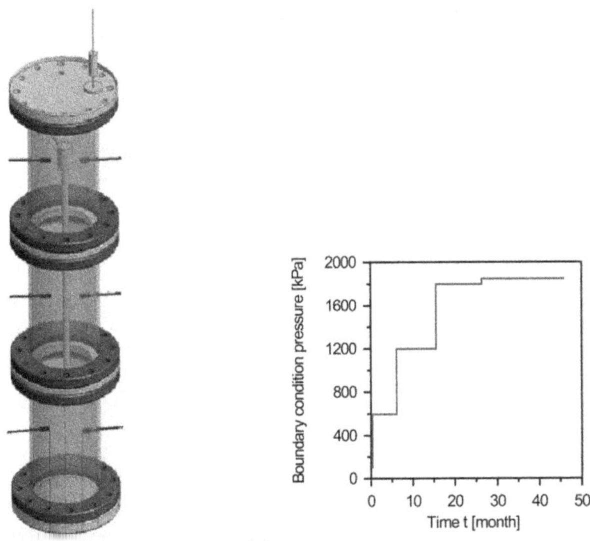

Fig. 20 Experimental set-up of the TDR-test and pressure boundary condition at the bottom of the cylinder.

- *Geometry*: Caused by the one-dimensionality of the processes, the cylinder will be simplified to a mesh with a height of 0.88 m and a width of 0.02 m. It is divided in 176 quadratic elements with side-lengths of 0.01 m.
- *Initial Conditions*: The simulation starts with a uniform saturation of 69% in the whole area. There are no initial stresses and the gravity is assumed to be negligible.
- *Boundary Conditions*: All boundaries are mechanically fixed. The hydraulic boundary conditions are given by impermeable boundaries at the left and at the right. The bottom is build by a time-dependent pressure that initiates a saturation process (see Fig. 20). Special attention has to be paid to the hydraulic boundary condition at the top. In the experiment, the top of the bentonite-sand-mixture is covered by a metal plate. This plate contains a small tube, where water can flow out. Consequently, the numerical model features a closed boundary condition for the first phase of the experiment, which covers the saturation process and the evolution of a pressure field. When the pressure at the top comes to a positive range, the boundary condition at the top is set to zero. As a matter of fact, the water can flow out.
- *Hydro-Mechanical Coupling*: In bentonite materials highly negative pressures may occur especially at a low saturation level. For this situation the effect represented by Terzaghi's approach vanishes. A modification of the effective stress law is presented in [36]. It incorporates an additional parameter, the effective stress coefficient χ which is already introduced in (2). This parameter is a function of the current and residual degree of water saturation S and S_{res} and the constant exponent κ

Table 7 Material properties for the TDR-test.

Parameter	Value
Dry density assuming 0.11 Wc, ρ	$1.869\,\mathrm{g\,cm^{-3}}$
Initial porosity from dry density, n_{init}	0.296
Initial permeability, k	$5\times10^{-19}\,\mathrm{m^2}$
Capillary pressure versus saturation	
van Genuchten α_{vG}	$0.209\,\mathrm{kPa^{-1}}$
van Genuchten n_{vG}	1.094
Residual saturation, S_{res}	0.0
Maximal saturation, S_{max}	1.0
Linear swelling model	
Volumetric swelling coefficient, β^{sw}	0.029
Initial water saturation, S_{init}	0.69
Young's modulus, E	$50\times10^6\,\mathrm{Pa}$
Poissons ratio, ν	0.33

$$\chi = \left(\frac{S - S_{\mathrm{res}}}{1 - S_{\mathrm{res}}}\right)^{\kappa}. \tag{39}$$

Various applications presented in [38] indicate that this approach give reasonable results. The simulations are done with an exponent of κ=2.0.

- *Material Properties*: Within the GMT test, many in-situ and laboratory tests have been carried out. As most of the material properties depend on the test conditions like water content, dry density, origin of the bentonite material and others, the results of these tests are directly related to the given experiment. Although the interpretation of these data have to be done carefully, some material properties of the bentonite-sand-mixture are relatively well known. Other values or dependencies which are needed for the numerical simulation like the coupling parameters have to be calibrated. The experiments leading to the material properties used here are published in various project reports (see for example [37, 50–52]) and are summarized in Table 7.

4.2.3 Results

Comparing the simulated with the measured data the resulting pressure evolution is presented in Fig. 21 and the evolution of the inflow at the bottom of the column over the time in Fig. 22.

The process at each point starts with a saturation of the material. Then, the pressure rises and the saturation front climbs the column. Finally, at a time of 22.33 months, the saturation front reaches the top and the column is fully satur-

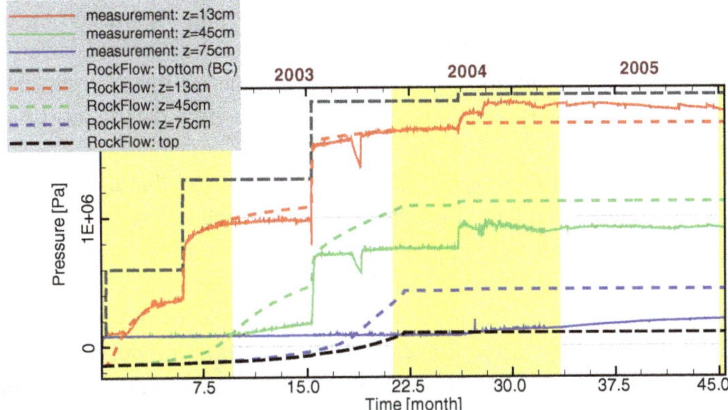

Fig. 21 Temporal evolution of the pressure within the TDR-test compared with the measurements.

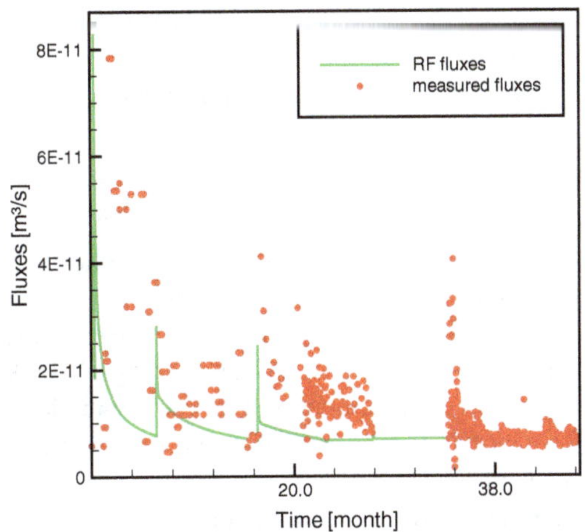

Fig. 22 Comparison of the fluxes derived from the numerical simulation and the measurements.

ated. At this time, the pressure boundary condition at the top is changed to an open boundary. The water can flow out and the equilibrium state occurs.

The resulting pressure evolution fits the measured data quite well. But there is a significant difference of the quality of the numerical results in the three different heights of the column. While the pressure in the lower measuring point fits the data very well, the results in the upper part of the column show some differences to the measured data. Here the simulated pressure overestimates the measurements. The evolution of fluxes shows that the modeled inflow rate is in the right range, whereas the temporal evolution cannot be verified.

Focusing on the time-dependency of the pictured processes, it can be seen that there are some inconsistencies in the results. In contrast to the pressure measurements, the numerical simulation as well as the measured fluxes indicate a time-dependent effect after every increase of the pressure boundary condition till the equilibrium stage is reached.

These differences indicate some additional effects in the upper part of the column. These might be induced by some kind of material damage due to high compressive stresses or an influence of the TDR tube, which might lead to some horizontal inhomogeneities. Further developments on the model set-up might be a detailed model of the column, incorporating the TDR tube and possible flow path at their surface. Additionally, some more aspects of material modeling might be implemented. However, the impact of permeability changes as well as coupling parameters or a rough model to incorporate some kind of fingering due to the pressure boundary condition indicate only moderate influence. Further developments incorporating the (saturation dependent) elasto-plastic material behavior, the effect of micro-fracturing or a non-linear swelling model might be interesting.

4.3 Modeling of Excavation Induced Processes in Claystone

4.3.1 Introduction

The objects of investigation in the following section are a century old tunnel and two new galleries built in indurated clay at the Tournemire site in the south of France (Fig. 23). The aim is an improvement in understanding the processes, leading to the observed deferred failure mechanism around the openings. This knowledge could assist in the construction and management of a repository in claystone. For the assessment of the performance and longterm safety, the development of an excavation disturbed zone around excavated openings is an important issue. Alteration of physical properties in this zone could effect the hydraulic conductivity and thus create short-cut pathways [8].

For the French safety authority, the Institute of Radioprotection and Nuclear Safety (IRSN), France, has selected the Tournemire site in order to be able to expertise the safety report and the conception options for a repository in an argillaceous media. Within the international DECOVALEX project (cf. Section 4.1.1) IRSN defined a task concerning the development of model concepts for the time-dependent behavior of the excavation disturbed zone.

4.3.2 Site Characterization

The Tournemire site is located in a Mesozoic marine basin at the southern edge of the French Massif Central. The sedimentary formations are characterized by three main Jurassic layers. The argillaceous medium of the Tournemire region consists of

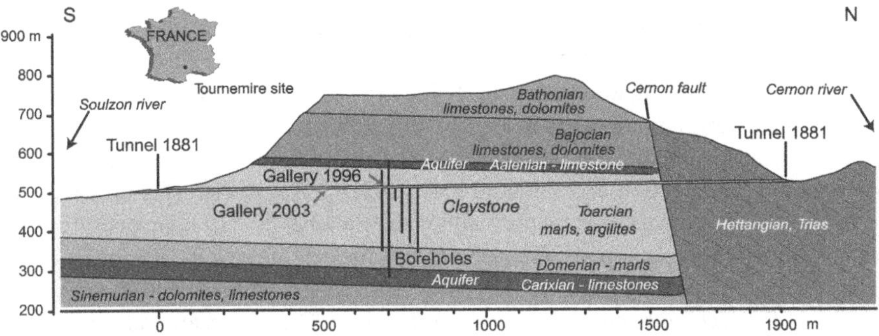

Fig. 23 Location and geological cross-section of the Tournemire site, based on [10].

a sub horizontal layer (250 m thick) located between two layers of limestone and dolomite (300 and 500 m thick), which constitute two aquifer layers (Fig. 23).

The argillaceous media is well compacted due to lithostatic pressure and diagenesis. It can be characterized as indurated clay (claystone), composed of thinly bedded minerals, corresponding to an anisotropic texture. The porosity is between 3 and 14% with an extremely small pore size of about 2.5 nm. The grain density amounts to between 2.7 and 2.8×10^3 kg/m^3 [10, 48].

The present water circulation takes place along the lower and upper limestone aquifer layers and along the Cernon fault. The water content is very low (1–5%) but the saturation seems to be around 100%. The argillaceous medium exhibits very low permeabilities.

Daupley [12] studied the relation between capillary pressure p_c and liquid saturation. The results can be well approximated with the following van Genuchten function

$$S = \left(\left(\frac{p_c}{48 \text{ MPa}} \right)^{1/1-0.41} + 1 \right)^{-1}. \tag{40}$$

The argillaceous formations of the Tournemire site are well indurated, its mechanical properties are between those of elasto-plastic clay and crystalline rock. Thus, the galleries, which have been excavated in 1996 and 2003, are mechanically stable. Laboratory uniaxial and triaxial tests have brought out the strong transversely isotropic elastic properties. The compressive strength depends on loading direction and varies between $R_{\text{comp}} = 20$ and 57 MPa. The tensile strength parallel to the bedding plane amounts to $R_{\text{ten}} = 3.6$ MPa [10, 48].

The failure mechanism has been analyzed with the Mohr–Coulomb model. A friction angle of $\phi = 20°$ and a cohesion C between 6.6 and 10.8 MPa, depending on the loading direction, have been determined. The anisotropy of the failure strength is not very important [43].

A distinctive anisotropic swelling/shrinkage behavior has been observed [62]. The measured parameters are summarized in Table 8.

Excavation Disturbed Zone

During an excavation new surfaces are being created, which leads to a significant change in the surrounding stress field. The radial stresses (normal to the wall) vanish as long as no support is built. The tangential stress components increase, which can lead to a compaction and a failure mechanism, respectively. This process can be influenced by hydro-mechanical coupling phenomena [8]. The compaction could lead to an increase of the pore water pressure, which influences the mechanical reaction by the effective stresses, as defined in (2). Further effects can arise due to the contact of the rock with atmospheric or ventilated air. The region around an excavation, which is affected by this kind of processes, is called excavation disturbed zone (EDZ), which may include a damaged zone.

The Tournemire site provides the opportunity to study the EDZs around three openings at different time scales. The EDZ around all openings have been studied by geological mapping, measurements of ultrasonic velocity, permeability, and water saturation. Around the 1881 excavated old tunnel a distinctive fractured zone can be observed. The zone of tangentially oriented fractures around the opening ("onion shape") extends 0.7 m on average, but extensions up to 3 m have been observed [47]. In this zone the permeability is increased up to five orders of magnitude [39]. Around the new galleries no damaged zone like this has been found. Shortly after the excavation of the 1996 and 2003 excavated galleries, small fissures occur at the sides and working faces of the gallery. A strong correlation between the aperture of these fissures with the humidity ($h_{rel} = 40$–100%) and temperature ($\vartheta = 6$–$16°C$) of the ambient air has been observed [47]. In the wet summer time, the fractures are nearly closed, in the dry winter time, they are open. The cores extracted around the galleries possess microcracks, which are mainly oriented parallel to the bedding plane. The fractured zone measures about 40 cm.

4.3.3 Model Set-up

The modeling set-up is founded on observations and measurements at galleries at the argillaceous Tournemire site. A delayed failure mechanism around the excavations and seasonally influenced fracture apertures have been observed in the EDZ. The goal is to model this delayed mechanism, coming along with fracturing.

The excavation of galleries at the Tournemire site leads to contact of the initially saturated claystone with the ambient air. The induced desaturation process yields tensile strains and as a consequence desaturation cracks. In a linear-elastic mechanical approach, the arising tensile stresses are not restricted and will exceed the tensile strength of the rock. Furthermore, the degradation of the material due to the occurring cracks is not considered. In order to simulate this tensile failure mechanism of the rock, a damage model is applied (cf. Section 2.4).

Two main processes constitute the basis of the applied model: deformation (mechanics) and fluid flow (hydraulics) (Fig. 24). The mechanical problem is stated as an elastic-damage model, the hydraulic problem as a Darcy type flow, consider-

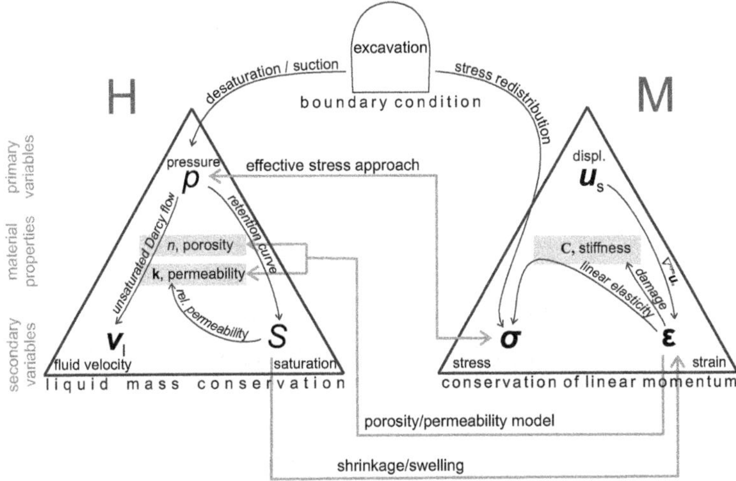

Fig. 24 Modeling approach.

ing unsaturated conditions by the Richards' approximation. Several coupling phenomena are taken into account: Terzaghi's effective stress concept (2), mass conservation of the liquid in deformable porous media (5), drying induced shrinkage (cf. Section 2.3), and a deformation and damage dependent permeability. Additionally, transverse isotropy is considered in the material properties of elasticity, shrinkage/swelling, and fluid flow (cf. Section 2.2).

The permeability is an important parameter describing the fluid flow through porous media. The most important influences come from the diagenesis of the rock, by chemical, thermal, and mechanical impacts. In the framework of safety assessment of a repository, the permeability and its development is of special interest, whereby the anisotropy has to be taken into account [61]. The applied permeability model, as presented in detail in [38], has the capacity to investigate an orthotropic change of permeability, induced by damage, elastic and swelling strains. It is based on a tensorial description of the pore space by the void fabric tensor [25, 40] and basic flow models, as the cubic law [56, 57]. The approach leads to similar relationships between porosity/strains and permeability as can be found in literature [26, 66]. A quantitative comparison proves the accuracy of the formulation and implementation [38]. With this approach, the change of permeability in a damaged (fractured) rock mass is much higher than the permeability change in the intact rock mass, even if the same strains are applied. This is a typical behavior of rock.

Two-dimensional hydraulic-mechanically coupled simulations under plane strain and plane flow conditions are used. The lengths of the galleries permit this reduction of dimension. The domain of the numerical model is 80.0 m times 80.0 m and just represents the argillite. In Fig. 25 the hydraulic and mechanical boundary and initial conditions are shown on one half of the symmetric domain. Furthermore, the meshes for the simulation of the 1881 excavated tunnel and the 1996 and 2003 excavated galleries are depicted. The time-dependent boundary condition of pore water pres-

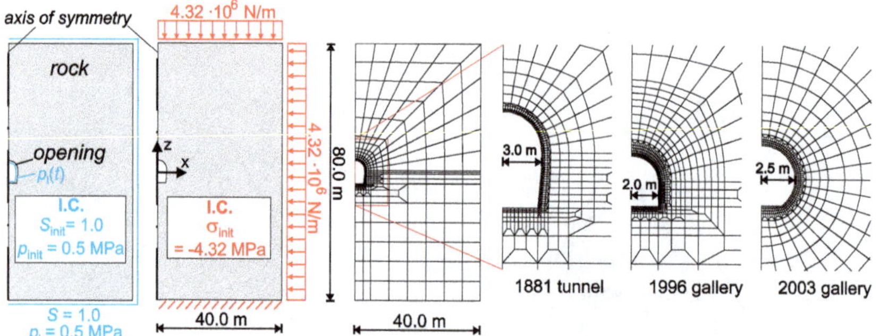

Fig. 25 Model set-up for the simulation of excavation induced processes; hydraulic and mechanical boundary and the initial conditions (left) as well as meshes (right).

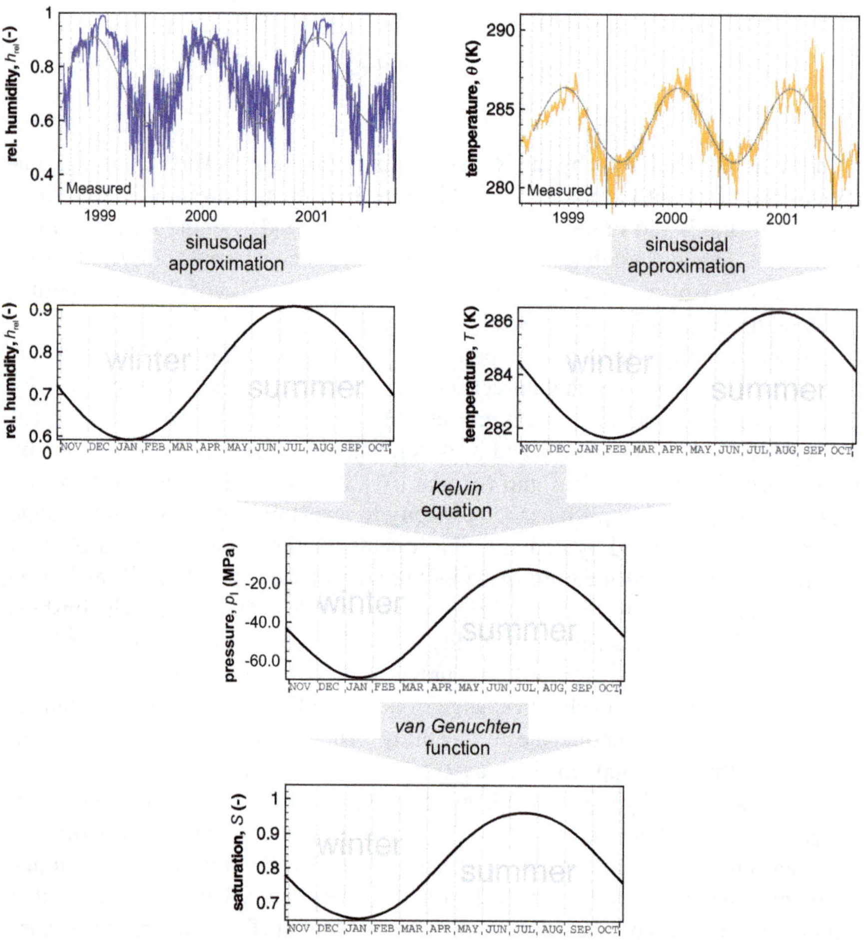

Fig. 26 Determination of the saturation at the wall of the opening by measured temperature and relative humidity profiles. Measurements by IRSN [46].

Table 8 Summary of material and simulation parameters.

Parameter	Measured	Used for simulation	Unit
Hydraulics			
Initial permeability, k	$10^{-22}\ldots10^{-17}$	0.2×10^{-20}	m^2
Initial porosity, n_{init}	$0.03\ldots0.14$	0.09	–
Initial pore water pressure, p_{init}	$0.14\ldots0.6$	$0.2\ldots0.5$	MPa
Mechanics			
Young's modulus			
In bedding plane, $E_x = E_y$	$20.15\ldots28.23$	24.19	GPa
Vertical to bedding plane, E_z	$5.18\ldots13.36$	9.27	GPa
Poisson's ratio			
$\quad \nu_{xz}$	$0.14\ldots0.20$	0.17	–
$\quad \nu_{zy}$	$0.17\ldots0.23$	0.20	–
Shear modulus, G_{xz}	$3.46\ldots4.42$	3.94	GPa
Mohr–Coulomb criterion			
$\quad$ Cohesion, C	$6.6\ldots10.8$	6.6	MPa
$\quad$ Angle of internal friction, ϕ	20.0	20.0	$^\circ$
Initial stress, σ_{init}	$-1.1\ldots-6.0$	$-4.32\ldots\ 5.0$	MPa
Swelling/shrinkage			
Power-law model:*			
$\quad$ Initial swelling strain, $\varepsilon^{sw}_{\mathrm{init}}$		0.0	–
$\quad$ Reference saturation, S^{sw}_{init}		1.0	–
$\quad$ Swelling coefficient, $\beta^{sw}_{x} = \beta^{sw}_{y}$		0.0022	–
$\quad$ Swelling coefficient, β^{sw}_{z}		0.0068	–
$\quad$ Swelling exponent, $\gamma^{sw}_{x} = \gamma^{sw}_{y}$		4.55	–
$\quad$ Swelling exponent, γ^{sw}_{z}		6.51	–

*Power-law swelling/shrinkage model (cf. Section 2.3)

sure at the opening is calculated with the help of humidities and temperatures. The measured relative humidity h_{rel} and temperature T [46] can be approximated by sinusoidal functions, as shown in Fig. 26. Based on these functions, the value for the capillary pressure p_c at the boundary can be determined with the Kelvin equation [17] as a function of relative humidity h_{rel}:

$$p_c = \frac{\rho_l R T}{M_v} \ln h_{\mathrm{rel}} \tag{41}$$

with the density of the liquid ρ_l, here water (≈ 1000.0 kg m^{-3}), the perfect gas constant R (≈ 8.314 J(mol k)$^{-1}$), and the molecular weight of vapor M_v (≈ 0.018 kg mol^{-1}). The temperature is assumed to be constant ($T = 301$ K).

The material properties for the calculation of the reference solution are listed in Table 8.

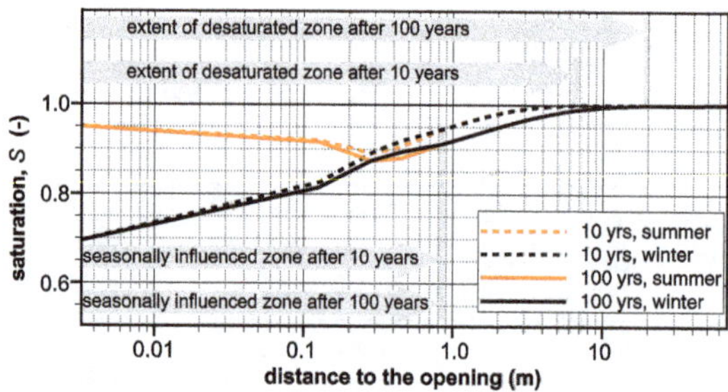

Fig. 27 Simulated horizontal saturation profiles at different times and seasons.

The excavation is simulated by an instantaneous change in the mechanical boundary conditions at the opening, afterwards the seasonal hydraulic boundary condition is applied at the wall of the opening.

4.3.4 Results

Desaturation Process

The contact of the claystone with ambient air creates a desaturated area of nearly circular shape around each excavation. The modeling work brings out a distinction of two zones (Fig. 27): The first zone near the opening (about 1 m) is characterized by a seasonally affected saturation. In the second zone the saturation is decreased, but changes take place more slowly. The size of both zones strongly depends on the material properties. The most significant parameters are the intrinsic permeability and the relative permeability.

A comparison with measured saturation profiles [38] has indicated that the developed model has the capacity to simulate the desaturation process only partly. The wide spread of measured saturation profiles at the three openings can only be covered by simulations with a wide range of permeability ($k = 2 \times 10^{-21} \ldots 2 \times 10^{-18}$ m^2). In this context, a consideration of minor significant effects, such as the hydro-mechanical coupling or the tunnel lining, is not meaningful.

The wide spread of measured saturations could be caused by an inhomogeneous distribution of permeability. Even if the rock is nearly homogeneous in the undisturbed configuration, the desaturation fissures could lead to preferential flow paths. Along these fissures the desaturation process would take place much faster than in the undisturbed matrix material. Thus, the in-situ measured permeability is not sufficient to characterize the desaturation process in the fractured zones.

Mine-by-Test

In order to investigate the response of the rock to the excavation of underground openings, a mine-by-test was carried out by the IRSN in 2003 [48]. For this experi-

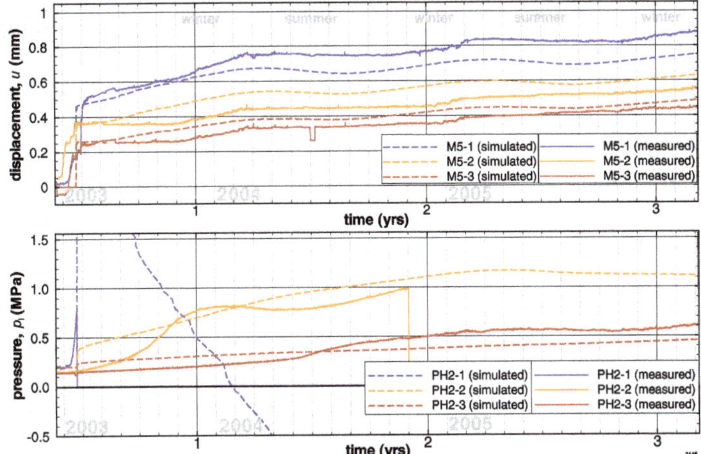

Fig. 28 Comparison between measured (solid lines) and simulated (dashed lines) displacements (top) and pore water pressures (bottom) at different distances to the excavation (M5-1: 1 m, M5-2: 2 m, M5-3: 3.5 m, PH2-1: 1 m, PH2-2: 4.5 m, PH2-3: 10 m). A hydraulic-mechanically coupled model is applied, using a transversely isotropic linear-elastic model, considering damage, transversely isotropic shrinkage, and seasonally influenced desaturation.

ment, a new gallery with a length of 40 m and a width of 4.4 m has been excavated perpendicular to the pre-existing tunnel. The deformations and pore water pressures have been measured in the host rock at different distances to the gallery (1 ... 10 m) during and after the excavation. The analyses of the measurements and in-situ observations show that the response of the Tournemire argillite to excavation occurs in two phases: First, the main hydraulic-mechanical response is governed by the rearrangement of stresses in the rock mass, inducing linked variations in deformation and pore water pressure. During this phase, neither failure nor damage is observed around the openings. The second phase begins some time after excavation with the desaturation/resaturation process at the uncovered walls of the openings and an initiation of microcracks.

This general response of the argillite can be simulated well by the applied model: Directly after the excavation, the stress field around the excavation changes, which initiates deformations towards the opening and a pressure increase close to the opening; after approximately three months, the desaturation process, simulated by a seasonally changing hydraulic boundary condition, dominates the pressure field. Shrinkage and damage is predicted in the near field of the opening.

The quantitative comparison of the measured data from the mine-by-test experiment with the simulations, as depicted in Fig. 28, indicates that the developed coupled model can be used to simulate the general behavior of the argillaceous rock due to mechanical excavation. The measured displacements directly after the excavation can be predicted by linear elasticity, using material properties as determined in laboratory. The measured increasing pore water pressure can be simulated with a hydro-mechanically coupled model considering transversely isotropic linear elasticity.

Seasonal shrinkage could be an explanation for the measured seasonal variation of the displacements. With the transversely isotropic shrinkage parameters, as determined in laboratory, the seasonal influence is overestimated, but in combination with continuum damage mechanics, a good agreement can be reached.

In the long term behavior (3 years), the main trends of the measurements can be predicted. The aberrations could be explained by several reasons. Inhomogeneity may be one reason. Even if the rock was initially homogeneous, the desaturation could induce inhomogeneities due to fracturing. Another reason for the aberrations may be given by additional processes, which are not considered in the applied model. This could be for instance non-linear elasticity, viscoplasticity, material hardening due to desaturation, fatigue due to recurring desaturation/resaturation cycles, and discrete fracture propagation.

The good agreement of the linear mechanical model with the displacement measurements directly after excavation indicates that non-linear material behavior becomes accountable in the long term behavior only.

The simulation results are very sensitive to the used permeability.

Analysis of the Excavation Disturbed Zone

The state of stresses around the excavated opening is analyzed to investigate potential failure mechanisms. Therefore, a Mohr–Coulomb criterion [27] is used. Furthermore, a continuum damage model (Section 2.4) is applied. Several excavation induced processes have been mentioned. They act on different scales of time and space. The simulation predicts an EDZ, which is characterized by tension induced damage and an increased radial permeability. Its extent increases slowly with time and equals approximately the observed one (Fig. 29). The simulated strong anisotropic shape of the damaged zone is not coherent to the Mohr–Coulomb failure criterion and is also not observed. The interaction between anisotropic shrinkage, permeability and damage could lead to this overestimation. The simulated and observed size of the desaturated zone is also similar, solely the simulated extent after 100 years seems to be overestimated.

All observed radially oriented fractures can be explained by the shrinkage process and can be well predicted with the used model.

Around the century old tunnel, observations suggest the assumption of a second failure mechanism, which is not predicted by the model. An explanation of the observed tangentially oriented fractures could be given by the presumption of a loss of integrity in the long term as a consequence of the desaturation induced shrinkage. This process is not proved so far.

All in all, the shrinkage seems to be the governing process for the development of the EDZ and consequently coupled modeling is indispensable.

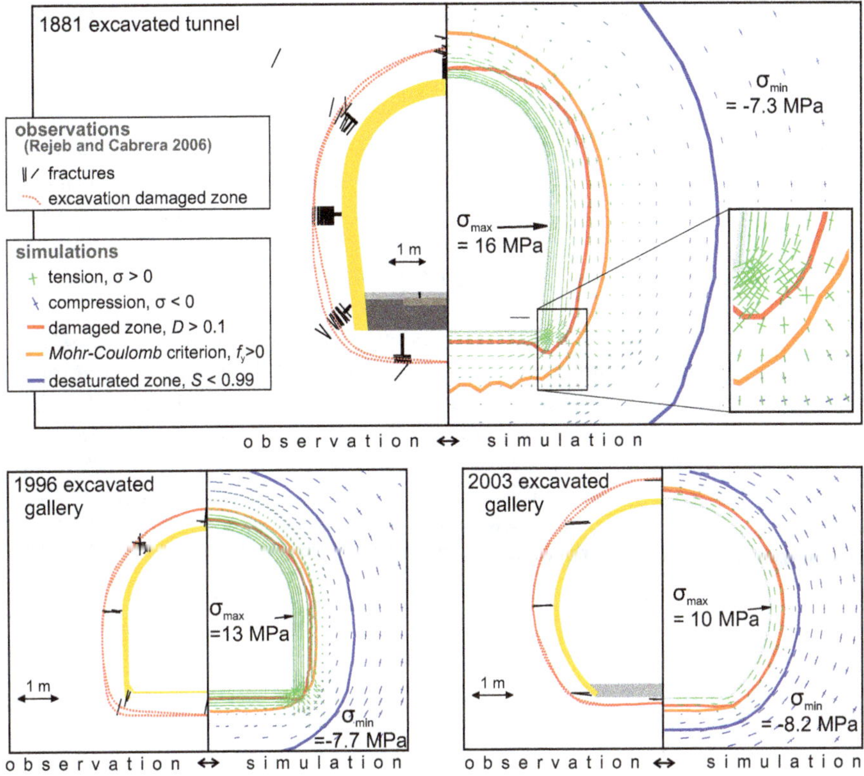

Fig. 29 Comparison of the observed fractured zones [47] with simulation results. The simulated time corresponds to the time period between excavation and core drilling at each opening (tunnel: 124 years; 1996 gallery: 9 years; 2003 gallery: 1.5 years). Since the measurements are approximately isotropic, the right hand side has been flipped horizontal.

5 Conclusions and Outlook

The investigation of coupled thermo-hydro-mechanical processes is part of current research in various fields of engineering applications. The work at hand focuses on geotechnical problems related to radioactive waste disposal, where the behavior of materials with very low permeability like clayey materials is of special interest. In order to improve the safety assessment, ongoing enhancements of numerical models are needed. Within this context, model concepts have to be developed, implemented and validated. Caused by the interaction of a great variety of effects, a classification of the impact of various processes is important to achieve a robust and clear model, which might be solved numerically.

Based on the developed coupled thermo-hydro-mechanical model this contribution focuses on the development as well as on the application of some model extensions. Shrinkage, mechanical and hydraulic anisotropy, damage and deformation dependent permeability have been investigated to allow the study of excavation induced processes in argillite.

The purely mechanical non-linear compressibility model is developed to investigate the compression behavior of materials with very low porosities. Within this field of applications, the model pictures the behavior of a wide range of materials. It is valid for the geometric linear case and verified for the compression state between very low negative stresses up to stresses close to the compression point. Furthermore, it incorporates an initial state which might be different from the stress-free state.

The finite element model RockFlow is being verified continuously against analytical solutions. Here, a validation of the hydro-mechanical coupling in unsaturated conditions against an experiment by Liakopoulos is presented in detail and, pointing out the effect of the non-linear model extension, a step-wise compression test is numerically simulated.

Three comprehensive case studies prove the applicability of the new implementations. The first one treats the thermo-hydro-mechanical coupled simulation of a generic repository of heat producing radioactive waste. The numerical predictions achieved in the RockFlow simulations are in good agreement with the results predicted by the other codes. The partially saturated bentonite buffer and the time period of resaturation are here of special interest. It becomes obvious that the resaturation is limited by the low conductivity of the barrier material. The influence of swelling pressure on the porosity and permeability could be of interest. As the temperature is significantly below the boiling point of water, both the effect of evaporation and the moisture transport forced by thermal gradients have been neglected.

In the second case study a long-term saturation process in a cylindrical column has been simulated and the results have been compared with measurements. Here it can be seen that the quality of the results varies significantly over the height of the modeled column. In some parts, the simulation results fit the measurements very well, whereas other parts show some differences. This fact leads to the assumption that additional effects which are not yet incorporated in the numerical simulation proceed in that column.

The third study is concerned with excavation induced processes in claystone. With the enhancements of the computer code RockFlow new possibilities of analysis and prediction have been provided. In particular, the observed delayed development of an excavation damaged zone in the argillaceous Tournemire site can be predicted correctly for a time period of at least ten years. Even for longer periods, a good estimation can be given, though not all observed failure mechanisms are understood so far. Further investigations on the long term material behavior, especially under unsaturated conditions, are needed in order to improve the understanding and in consequence the further numerical modeling. Of special interest is the coupling between shrinkage, damage and permeability, considering transverse isotropy.

In the context of the performance assessment of a repository in argillite a number of lessons have been learned. The most important one is, that it is not sufficient to consider the mechanical properties on its own, since the key process for the long term behavior is a hydraulic one: the dehydration.

Finally, the coupled thermo-hydro-mechanical simulation tool RockFlow as well as its theoretical background and some model extensions are presented and validated within this publication. Various applications indicate good results, but they also point out, that there is a great variety of processes and effects depending on the specific situation implied in geotechnics. The simulation of such a kind of problem requires a profound understanding and a precise classification of the impact of all the processes involved. As a matter of fact further research will be of great interest in various fields of application.

Nowadays the authors are concerned with land subsidence, wind energy or geothermal energy. Within these fields of application similar questions arise and some of them can be analyzed with RockFlow.

Acknowledgements

This research was supported by the German Research Foundation (DFG) within the GRK 615. The participation in the DECOVALEX project and the associated research was funded by the Federal Institute for Geosciences and Natural Resources (BGR). In this regard the authors thank Dr. Hua Shao for the continuous cooperation and helpful assistance. Furthermore, the authors acknowledge the Institute of Radioprotection and Nuclear Safety, France, for kindly providing the measurements concerning the Tournemire site.

References

1. C. Barlag. Adaptive Methoden zur Modellierung von Stofftransport im Kluftgestein. PhD Thesis, Institut für Strömungsmechanik und Elektronisches Rechnen im Bauwesen, Universität Hannover, Report No. 52, 1997.
2. D. Barr, J. Birkhölzer, J. Rutqvist, and E. Sonnenthal. Draft Description for DECOVALEX-THMC. Task D: Long-Term Permeability/Porosity Changes in EDZ and Near Field, due to THM and THC Processes in Volcanic and Crystalline-Bentonite Systems. REV02 December 2004, Office of Repository Development, U.S. Department of Energy; Earth Sciences Division, LBNL, USA, Berkeley, 2004.
3. Z.P. Bazant (Ed.). *Mathematical Modelling of Creep and Shrinkage of Concrete.* John Wiley & Sons, 1989.
4. J. Bear. *Dynamic of Fluids in Porous Media.* Elsevier, New York, 1972.
5. BfS. Endlagerung radioaktiver Abfälle als nationale Aufgabe. Technical Report, Bundesamt für Strahlenschutz, Salzgitter, Germany, 2005.
6. BGR. Endlagerung radioaktiver Abfälle in Deutschland. Untersuchung und Bewertung von Regionen mit potenziell geeigneten Wirtsgesteinsformationen. Technical Report, Federal Institute for Geosciences and Natural Resources (BGR), April 2007.
7. J. Birkholzer, J. Rutqvist, E. Sonnenthal, and D. Barr. DECOVALEX-THMC. Task D: Long-Term Permeability/Porosity Changes in EDZ and Near Field due to THM and THC Processes in Volcanic and Crystalline-Bentonite Systems. Status Report October 2005, Earth Sciences Division, LBNL, USA; Office of Repository Development, DOE, USA, Berkeley, 2005.

8. P. Blümling, F. Bernier, P. Lebon, and C.D. Martin. The excavation damaged zone in clay formations time-dependent behaviour and influence on performance assessment. *Physics and Chemistry of the Earth*, 32:588–599, 2007.

9. R.M. Bowen. Incompressible porous media models by use of theory of mixtures. *International Journal of Engineering Science*, 18:1129–1148, 1980.

10. J. Cabrera. Caractérisation des discontinuités en milieu argileux (station expérimentale de tournemire). Scientific and Technical Report, Institute of Radioprotection and Nuclear Safety (IRSN), France, 2002.

11. A.S. Chiarelli, J.F. Shao, and N. Hoteit. Modeling of elastoplastic damage behavior of a claystone. *International Journal of Plasticity*, 19:23–45, 2003.

12. X. Daupley. Etude du potentiel de l'eau interstitielle d'une roche argileuse et de relations entre ses propriétés hydriques et mécaniques – Application aux argilites du Toarcien dans la région de Tournemire (Aveyron). PhD Thesis, Paris School of Mines, 1997.

13. R.O. Davis and A.P.S. Selvadurai. *Elasticity and Geomechanics*. Cambridge University Press, 1996.

14. R. de Boer. Development of porous media theories – A brief historical review. *Transport in Porous Media*, 9:155–164, 1992.

15. W. Ehlers. Foundations of multiphasic and porous materials. In W. Ehlers and J. Bluhm (Eds.), *Porous Media*. Springer, Berlin, 2002.

16. G. Eipper. Theorie und Numerik finiter elastischer Deformationen in fluidgesättigten porösen Festkörpern. PhD Thesis, Institut für Mechanik der Universität Stuttgart, 1998.

17. L.R. Fisher and J.N. Israelachvili. Direct experimental verification of the Kelvin equation for capillary condensation. *Nature*, 277:548–549, 1979.

18. S. Flügge (Ed.). *Die nicht-linearen Feldtheorien der Mechanik*, volume III/3 of *Handbuch der Physik*. Springer, Berlin, 1965.

19. A. Habbar. Direkte und inverse Modellierung reaktiver Transportprozesse in klüftig-porösen Medien. PhD Thesis, Institut für Strömungsmechanik und Elektronisches Rechnen im Bauwesen, Universität Hannover, Report No. 65, 2001.

20. R. Helmig. Theorie und Numerik der Mehrphasenströmungen in geklüftet-porösen Medien. PhD Thesis, Institut für Strömungsmechanik und Elektronisches Rechnen im Bauwesen, Universität Hannover, Report No. 34, 1993.

21. J.A. Hudson, O. Stephansson, J. Andersson, C.-F. Tsang, and L. Jing. Coupled T-H-M issues relating to radioactive waste repository design and performance. *International Journal of Rock Mechanics and Mining Science*, 38:143–161, 2001.

22. IAEA. Geological disposal of radioactive waste: Safety requirements. Technical Report WS-R-4, International Atomic Energy Agency (IAEA), Vienna, 2006.

23. J.W. Ju. On energy-based coupled elastoplastic damage theories: constitutive modeling and computational aspects. *International Journal of Solids and Structures*, 25(7):803–833, 1989.

24. R. Kaiser. Gitteradaption für die Finite-Elemente-Modellierung gekoppelter Prozesse in geklüftet porösen Medien. PhD Thesis, Institut für Strömungsmechanik und Elektronisches Rechnen im Bauwesen, Universität Hannover, Report No. 63, 2001.

25. K.-I. Kantani. Stereological determination of structural anisotropy. *International Journal of Engineering Science*, 22(5):531–546, 1984.

26. P.C. Kelsall, J.B. Case, and C.R. Chabannes. Evaluation of excavation-induced changes in rock permeability. *International Journal of Rock Mechanics and Mining Science*, 21(3):123–135, 1984.

27. A.S. Khan and S. Huang. *Continuum Theory of Plasticity*. John Wiley & Sons, New York, 1995.

28. M. Kohlmeier. Coupling of thermal, hydraulic and mechanical processes for geotechnical simulations of partially saturated porous media. PhD Thesis, Institut für Strömungsmechanik und Elektronisches Rechnen im Bauwesen, Leibniz Universität Hannover, Report No. 72, 2006.

29. O. Kolditz. Stoff- und Wärmetransport im Kluftgestein. Habilitation, Institut für Strömungsmechanik und Elektronisches Rechnen im Bauwesen, Universität Hannover, Report No. 47, 1996.

30. K.-P. Kröhn. Simulation von Transportvorgängen im klüftigen Gestein mit der Methode der Finiten Elemente. PhD Thesis, Institut für Strömungsmechanik und Elektronisches Rechnen im Bauwesen, Universität Hannover, Report No. 29, 1991.

31. T. Lege. Modellierung des Kluftgesteins als geologische Barriere für Deponien. PhD Thesis, Institut für Strömungsmechanik und Elektronisches Rechnen im Bauwesen, Universität Hannover, Report No. 45, 1995.

32. J. Lemaitre. *A Course on Damage Mechanics*. Springer, Berlin, 1992.

33. J. Lemaitre and R. Desmorat. *Engineering Damage Mechanics. Ductile, Creep and Brittle Failures*. Springer, Berlin, 2005.

34. R.W. Lewis and B.A. Schrefler. *The Finite Element Method in the Static and Dynamic Deformation and Consolidation of Porous Media*. John Wiley & Sons, Chichester, 1998.

35. A. Liakopoulos. Retention and distribution of moisture in soils after infiltration has ceased. *Bulletin of the International Association for scientific hydrology*, 10:58–69, 1965.

36. N. Lu and W.J. Likos. *Unsaturated Soil Mechanics*. John Wiley & Sons, New Jersey, 2004.

37. P. Marschall, M. Fukaya, J. Croise, S. Yamamoto, and G. Mayer. Project Report 00-20 GMT/IR 00-01: Laboratory data compilation report. Technical Report, National Cooperative for the Disposal of Radioactive Waste (NAGRA), 2001.

38. J. Maßmann. Modeling of excavation induced coupled hydraulic-mechanical processes in claystone. PhD Thesis, Institut für Strömungsmechanik und Elektronisches Rechnen im Bauwesen, Leibniz Universität Hannover, Report No. 77, 2009.

39. J.M. Matray, S. Savoye, and J. Cabrera. Desaturation and structure relationships around drifts excavated in the well-compacted Tournemire's argillite (Aveyron, France). *Engineering Geology*, 90:1–16, 2007.

40. B. Muhunthan and J.L. Chameau. Void fabric tensor and ultimate state surface of soils. *Journal of Geotechnical and Geoenvironmental Engineering*, 123(2):173–181, 1997.

41. NAGRA. Project opalinus clay. Technical report, National Cooperative for the Disposal of Radioactive Waste, 2002.

42. T.N. Narasimhan and P.A. Witherspoon. Numerical model for saturated-unsaturated flow in deformable porous media. *Water Resources Research*, 14:1017–1034, 1978.

43. H. Niandou, J.F. Shao, J.P. Henry, and D. Fourmaintraux. Laboratory investigation of the mechanical behaviour of Tournemire shale. *International Journal of Rock Mechanics and Mining Science*, 34(1):3–16, 1997.

44. M. Ortiz. A constitutive theory for the inelastic behaviour of concrete. *Mechanics of Material*, 4:67–93, 1985.

45. R. Ratke, O. Kolditz, and W. Zielke. ROCKFLOW-DM2 – 3-D Dichteströmungsmodell. Technical Report, Institut für Strömungsmechanik und Elektronisches Rechnen im Bauwesen, Universität Hannover, 1996.

46. A. Rejeb and J. Cabrera. Description for Task C, Excavation Disturbed Zone (EDZ) in the argillaceous Tournemire site (France). Technical Report, Institute of Radioprotection and Nuclear Safety (IRSN), France, 2004.

47. A. Rejeb and J. Cabrera. Time-dependent evolution of the excavation damaged zone in the argillaceous Tournemire site (France). In Weiya Xu (Ed.), *Proceedings of the 2nd International Conference on Coupled T-H-M-C Processes in Geosystems and Engineering (GeoProc2006)*, Nanjing, P.R. China, HoHai University, 2006.

48. A. Rejeb, K.B. Slimane, J. Cabrera, J.M. Matray, and S. Savoye. Hydro-mechanical response of the Tournemire argillite to the excavation of underground openings: Unsaturated zones and mine-by-test experiment. In J.F. Shao and N. Burlion (Eds.), *Thermo-Hydromechanical and Chemical Coupling in Geomaterials and Applications: Proceedings of the 3rd International Symposium GeoProc'2008*, pp. 649–656. John Wiley & Sons, June 2008.

49. A. Rejeb and O. Stephansson. DECOVALEX IV THMC. Excavation Damaged Zone (EDZ) in the Argillaceous Tournemire Site (France). Report of Task C1. May 2005, Institute of Radioprotection and Nuclear Safety, France; GeoForschungsZentrum-Potsdam, Germany, 2006.

50. E. Romero and E. Castellanos. Project Report 04-07 GMT/IR 03-03: Complementary tests on compacted sand/bentonite/lead nitrate buffer material for the GMT emplacement project.

Technical Report, National Cooperative for the Disposal of Radioactive Waste (NAGRA), 2004.

51. E. Romero, E. Castellanos, and E.E. Alonso. Project Report 02-24 GMT/IR 02-01: Laboratory tests on compacted sand/bentonite/lead nitrate buffer material for the GMT emplacement project. Technical Report, National Cooperative for the Disposal of Radioactive Waste (NAGRA), 2003.

52. E. Romero, I. Garcia, and E.E. Alonso. Project Report 03-03 GMT/IR 02-02: Laboratory gas tests on compacted sand-bentonite buffer material used in the GMT in-situ emplacement. Technical report, National Cooperative for the Disposal of Radioactive Waste (NAGRA), 2003.

53. A. Rouabhi, M. Tijani, and A. Rejeb. Triaxial behaviour of transversely isotropic materials application to sedimentary rocks. *International Journal for Numerical and Analytical Methods in Geomechanics*, 2006.

54. B.A. Schrefler and X. Zhan. A fully coupled model for water flow and air flow in deformable porous media. *Water Resources Research*, 190(29):155–167, 1993.

55. H. Shao. Simulation von Strömungs- und Transportvorgängen in geklüftet porösen Medien mit gekoppelten Finite-Elemente- und Rand-Element-Methoden. PhD Thesis, Institut für Strömungsmechanik und Elektronisches Rechnen im Bauwesen, Universität Hannover, Report No. 37, 1994.

56. D.T. Snow. Parallel plate model of fractured permeable media. PhD Thesis, University of California, Berkeley, 1965.

57. D.T. Snow. Anisotropic permeability of fractured media. *Water Resources Research*, 5(6):1273–1289, 1969.

58. C. Thorenz. Model adaptive simulation of multiphase and density driven flow in fractured and porous media. PhD Thesis, Institut für Strömungsmechanik und Elektronisches Rechnen im Bauwesen, Universität Hannover, Report No. 62, 2001.

59. D. Tikhomirov. Theorie und Finite-Element-Methode für die Schädigungsbeschreibung in Beton und Stahlbeton. PhD Thesis, Institut für Baumechanik und Numerische Mechanik, Universität Hannover, 2000.

60. C. Truesdell and R.A. Toupin. The classical field theories. In S. Flügge (Ed.), *Handbuch der Physik*, volume III/1, pp. 226–902. Springer, Berlin, 1960.

61. C.-F. Tsang, F. Bernier, and C. Davies. Geohydromechanical processes in the excavation damaged zone in crystalline rock, rock salt, and indurated and plastic clays - in the context of radioactive waste disposal. *International Journal of Rock Mechanics and Mining Science*, 42:109–125, 2005.

62. F. Valés, D. Nguyen Minh, H. Gharbi, and A. Rejeb. Experimental study of the influence of the degree of saturation on physical and mechanical properties in Tournemire shale (France). *Appl. Clay Sci.*, 26:197–207, 2004.

63. K. von Terzaghi. *Theoretical Soil Mechanics*, 7th edition. John Wiley & Sons, New York, 1954.

64. W. Wittke. *Felsmechanik. Grundlagen für wirtschaftliches Bauen im Fels*. Springer, Berlin, 1984.

65. J. Wollrath. Ein Strömungs- und Transportmodell für klüftiges Gestein und Untersuchungen zu homogenen Ersatzsystemen. PhD Thesis, Institut für Strömungsmechanik und Elektronisches Rechnen im Bauwesen, Universität Hannover, Report No. 28, 1990.

66. J. Zhang, W.B. Standifird, J.-C. Roegiers, and Y. Zhang. Stress-dependent fluid flow and permeability in fractured media: from lab experiments to engineering applications. *Rock Mechanics and Rock Engineering*, 40(1):3–21, 2007.

67. G. Ziefle. Modeling aspects of coupled hydraulic-mechanical processes in clay material. PhD Thesis, Institut für Strömungsmechanik und Elektronisches Rechnen im Bauwesen, Leibniz Universität Hannover, Report No. 74, 2008.

68. O.C. Zienkiewicz, Y.M. Xie, B.A. Schrefler, A. Ledesma, and N. Bicanic. Static and dynamic behaviour of soils: A rational approach to quantitative solutions, II. Semi-saturated problems. *Proceedings of the Royal Society of London Series A-Mathematical Physical and Engineering Sciences*, 429:11–21, 1990.

Multibody Contact Algorithms for Fracturing Solids

Peter Wriggers and Sven Reese

Abstract Processes in which localizations lead to fracture are common for brittle materials. To simulate these problems methods have to be designed that allow for crack detection and propagation. Within this study a finite element program was developed that allows for fast contact detection of multiple deformable bodies and is able to automatically introduce new surfaces for cracks and the parts being cut out. Within the simulation inertial effects have to be considered that occur during the time dependent solution process. Due to the complexity of the simulations, a new open software tool was developed.

1 Computational Fracture Mechanics

The incipiencies of the classical stiffness hypothesis were stated in the end of the 19th, respectively the beginning of the 20th century. A distinction between ductile and brittle material behavior is made, depending on the inelastic deformation before the failure process starts. In a material model related to concrete, brittle material behavior can be observed, contrary to ductile metallic materials like steal.

In linear fracture mechanics a solid containing cracks is modeled as linear elastic body in the entire fracture domain. Within this linear approach, the fracture process is limited to the crack tip field. Thereby, the analytical as well as the experimental determination of the so called stress intensity factors (K-factors) and the energy dissipation rate $\mathscr{G}$ are essential, being the dominating parameters in the process zone. This established method for elastic processes is also applicable for three dimensional problems. For elastic-plastic fracture processes, e.g. the determination of the

P. Wriggers
Institute for Continuum Mechanics, Leibniz Universität, Appelstraße 11, 30167 Hannover, Germany; e-mail: wriggers@ikm.uni-hannover.de

S.R. Reese
E-ON AG, Hannover, Germany; e-mail: sven.reese@eon.de

E. Stephan & P. Wriggers (Eds.): Modelling, Simulation and Software Concepts, LNACM 57, pp. 75–116.
springerlink.com © Springer-Verlag Berlin Heidelberg 2011

$\mathcal{J}$-Integral introduced by Rice in 1968 is necessary. It is based on the deformation theory of plasticity.

Classically the previously described dominating fracture parameters are obtained either by analytical models or by experimental investigations. This procedure is only feasible for few particular cases. For more complex components including multi-dimensional loading cases or nonlinear material models, today's engineers apply numerical simulation tools.

During the last years, mainly pushed by the rapid development of computer technology, the increasing demand to shorter development times for complex components and last but not least the establishment of the Finite Element Method, complete three dimensional simulations have become state of the art. Many methods for an accurate simulation of fracturing processes were developed in recent years. For example the eXtended Finite Element Method (XFEM) and the Strong Discontinuity Approach (SDA), which will be described later, should be mentioned in this context. The fundamental idea of both methods is the extension of the standard finite element approximation by an additional enhancement of the displacement field. Nevertheless, many other methods like gradient enhanced models, cohesive models at interface, r-adaptive schemes or meshless and particle methods are established.

1.1 Properties of Concrete

In the covered work reinforced concrete, being a very important construction material in modern civil engineering, will be used as the underlying material. In general, concrete consists of a ceramic matrix, filler materials[1] and reinforcement. The material cohesive coherence is ensured by hydrated cement paste (HCP) as the ceramic matrix material. Aggregates and filler materials like gravel or basalt provide a mechanical compression strength. Finally, the inclusion of steel or nowadays glass fibre materials provide the structure's tensile strength.

In order to model the failure process of concrete is is necessary to understand the procedures inside the cracking material. There can be different reasons for the initiation and development of a crack. Every solid contains initial imperfections in its micro-structure like pores. These are for example defectives or dislocations in the material atomic structure, impurities based on the production process or defects of natural kind. Due to mechanical, chemical or thermal loading states, these micro-faults may accumulate and result into a transcrystalline or intercrystalline crack that will extend under certain conditions to the macro-level (Fig. 1). Based onto these pores and micro-faults, the crack progresses continuatively resulting into a accumulating decrease of the strength.

---

[1] In general all filler material particles are surrounded by a cohesive zone including very low stiffness properties. Due to this fact damage mainly starts in this cohesive zone between the ceramic matrix and filler materials.

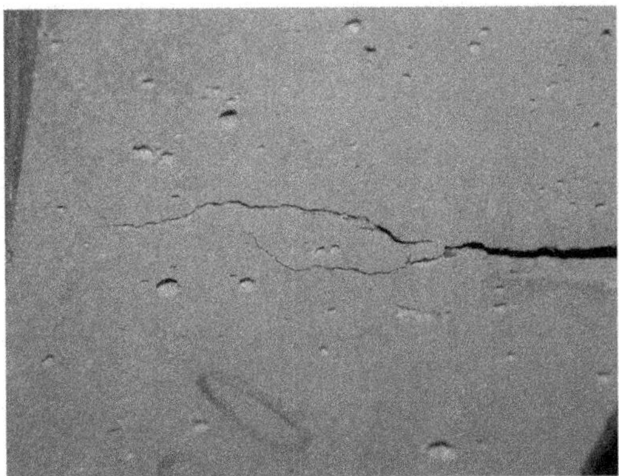

Fig. 1 Experimental concrete cracking (Institute of Building Materials Research, Aachen University).

The resulting macro-crack is visible with the naked eye. With the method described in the following it is possible to take micro-mechanical phenomena into account and simulating crack propagation on macro-level.

1.2 Continuum Mechanical Fracture Effects

In quasi-brittle materials like concrete the fracture process takes place through a transition process that involves formation and coalescence of micro-cracks as mentioned before. Thus, according to Oliver and Huespe [49], it can be distinguished between three distinct failure zones (Fig. 2).

The initiation of the dissipation phenomena starts in the *diffuse failure* zone. It results into an increase and concentration of strains where material discontinuities appear. The displacements, the strains and subsequently the stresses are spatially smooth and remain continuous. By means of a *weak discontinuity* regime the diffuse failure zone narrows. The concentration of the strains accumulates up to a localization into a discontinuous strain field. The displacement field remains still continuous. When the weak discontinuity develops into a band whose width is decreasing up to a zero thickness discontinuity band at collapse. This zone is denoted the *strong discontinuity* zone. Thus the displacement field is discontinuous and experiences a jump. The strains become unbounded.

To capture these phenomena at each distinct state of the failure process is one of the main aims of today's numerical failure analysis.

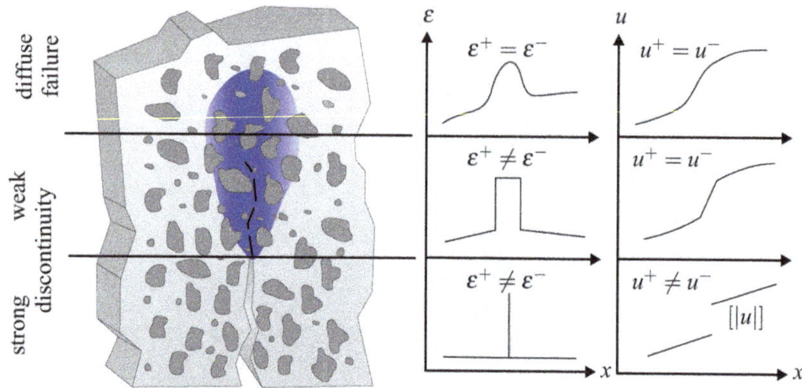

Fig. 2 Failure zones.

1.3 Numerical Fracture Approaches

The fracture processes described above can be modeled by different numerical methods. Here continuous numerical fracture approaches will be discussed in more detail. Discontinuous methods like remeshing strategies will be considered later in the context of a combined continuous-discontinuous model. A general classification is depicted in Fig. 3. Thus the experimental preserved crack path can be modeled within a non-geometrical numerical approach in different ways.

First numerical techniques in the area of *continuous numerical fracture approaches* are incipient discontinuity methods, where all elements represent potential discontinuities. Here a potential cohesive element is placed at the interface between each two adjacent finite element in the region of the crack tip reflecting a traction separation [22, 57]. This method was firstly introduced in [78] where cohesive elements are placed everywhere inside the finite element domain. It states an adequate choice for the implementation in commercial codes while keeping the programming tasks inside the finite element kernel as low as possible. The entire nonlinear process zone is represented by a single discontinuity. Hence, the opening of the crack represents the cumulative effect of micro-cracking integrated across the width of the process zone. An extension of the cohesive zone model to higher-order theories was presented in [10]. Classical damage approaches like plastic damage models for concrete [14, 32] or a strain-stress based continuum damage model [67, 68] are also state of the art.

Another method in the context of numerical fracture mechanics is the Variational Arbitrary Lagrangian Eulerian (VALE) approach, also known as variational r-adaptivity, see [42]. This method seeks to minimize the energy function by equilibrating energetic forces acting on the nodes with respect to the finite element mesh over the reference configuration of the body. It is applicable in the context of continuum fracture mechanics as shown in [75].

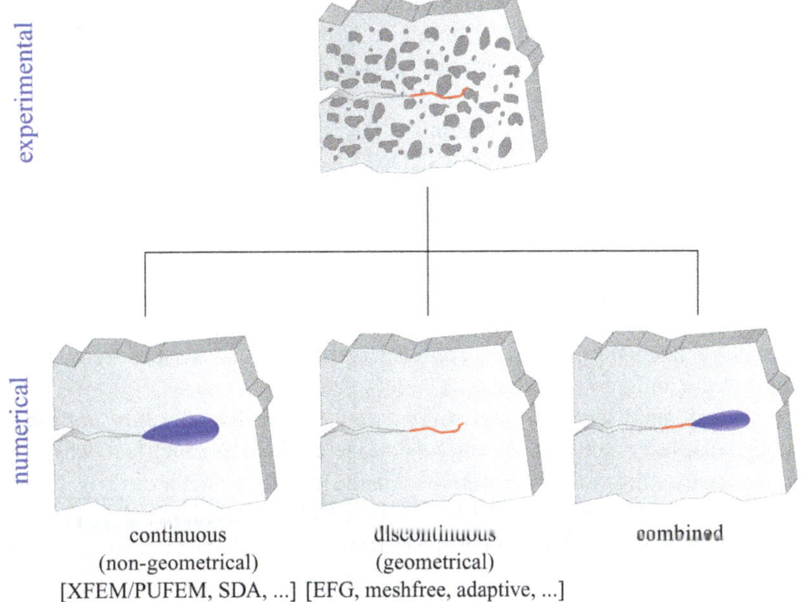

Fig. 3 Numerical fracture approaches.

One widely used approach is the eXtended Finite Element Method. This continuous crack presentation is based on the Partition of Unity Method, introduced by [4], [5] and [35]. Extensions to three-dimensional crack propagation [15] and a development of a crack representing interface element [72] were developed. Within the context of the eXtended Finite Element Method discontinuities are introduced by a discontinuous Partition of Unity, representing a particular case of the Partition of Unity Method. The eXtended Finite Element Method is discussed in various publications like [8, 30, 31, 39, 40, 74, 80]. The general idea is to enrich the approximation space spanned by a Partition of Unity by products of the standard basis functions, e.g. the incorporation of a priori knowledge of the solution behavior. All needed additional degrees of freedom are inserted globally or otherwise in a certain region of interest. As a result of this global enrichment, the total degrees of freedom, which are associated with the nodes, increases. The displacement interpolation is conforming with no incompatibilies between the elements. The strains on both sides of the stress free crack are fully decoupled. Hence discontinuities in the displacement field are captured exactly. Within this approach continuative studies for arbitrary branched and intersecting cracks [12], and explicit time stepping schemes [36] were performed. An extension to adaptive refinement techniques in the region of interest was presented in [3]. Extensions to three-dimensional crack propagation can be found in [3, 74, 80].

Among other established approaches are Enriched Element Methods like the Strong Discontinuity Approach, where the discontinuities are embedded in each finite element. An overview and comparative studies related to the eXtended Finite Element Method and the Strong Discontinuity Approach can be found in [21,22,38]. Elements with embedded discontinuities are part of publications like [23, 24] in the scope of smeared cracks.

Another model for embedded discontinuities is the Statically and Kinematically Optimal Nonsymmetric formulation (SKON). A general version for an arbitrary type of parent elements is the Strong Discontinuity Approach developed by Simo et al. [71] while it was introduced to make a standard continuum plasticity model compatible with the discontinuous displacement field [71]. It was completed in various publications, see for instance [28,41,44,48,49]. The method of Enhanced Assumed Strains plays an important role for modeling discontinuities in today's computational fracture mechanics. Thus this approach is used in the described framework of an explicit time integration scheme.

The second main category are *discontinuous numerical fracture approaches* where the displacement discontinuity is modeled in a geometrical way. Consequently, the existing crack is incorporated in the model geometry directly. Here for example meshfree methods like the Element Free Galerkin Method (EFG) [7,25,61], Meshless Finite Element Methods (MFEM) using an extended Delaunay tessellation to build a mesh or methods generating interfaces like in contact (see [27, 76]) can be mentioned. A particular case of the Finite Element Method incorporating general polyhedral shape also belongs to this category of methods [20].

Additionally, particle methods for the representation of cracks as introduced in [58] or [79] build another approach for handling numerical fragmentation processes.

Finally, complete adaptive Finite Element Methods for the numerical modeling of fracture processes in the context of cohesive elements [66] and adaptive remeshing techniques for forging processes based on a posteriori ZZ error estimator[2] [81] in the context of large deformations [11] were developed in recent years. Geometrical three-dimensional fragmentation procedures were presented in [9] in the context of real time cutting for tetrahedral meshes or in [59] for complete three-dimensional fragmentation procedures.

Methods comprising the advantages of both approaches will be the scope of the following sections.

2 Strong Discontinuity Analysis

In this section the mathematical and mechanical background including basic concepts and ideas of the Strong Discontinuity Approach will be given. This overview is based on [44–46] fundamentally. The basic equations of continuum mechanics are summarized below. The equilibrium equation[3] disregarding dynamic effects can

---

[2] The denomination of this simple but nevertheless very efficient error estimator is based on the first letter of both inventors surname – Zienkiewicz and Zhu.

[3] In the following context the subscript $(\cdot)_e$ denotes the enriched body. The superscript $(\cdot)^h$ identifies the discontinuity band, while the subscript $(\cdot)_h$ describes the approximated solution anymore.

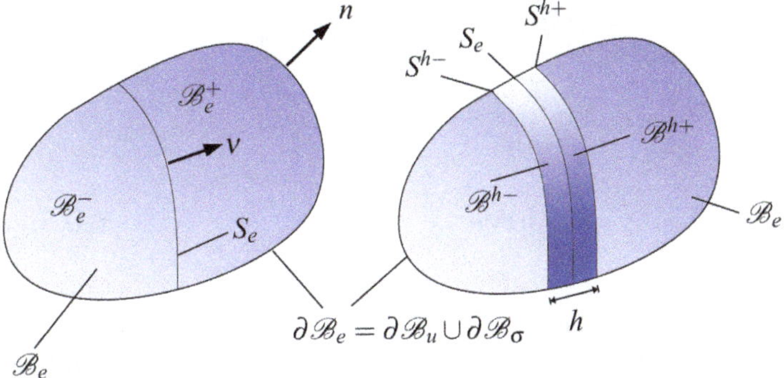

Fig. 4 Model problem: (a) weak discontinuities, (b) strong discontinuities.

be stated as follows:

$$\nabla\sigma + f = 0 \qquad \text{in} \qquad \mathscr{B}_e \backslash S_e \tag{1}$$

while limiting the validity of the equation to the body $\mathscr{B}_e$ excluding the discontinuity surface S_e. This surface subdivides the enriched body into two distinct regions. Part $\mathscr{B}_e^-$ describes the domain at one side of S_e and $\mathscr{B}_e^+$ the region at the other side. The classical Dirichlet and Neumann boundary conditions can be formulated for the boundary of the enriched body $\partial\mathscr{B}_e$

$$u = \bar{u} \qquad \text{on} \qquad \partial\mathscr{B}_u \tag{2}$$

$$\sigma \cdot n = \bar{t} \qquad \text{on} \qquad \partial\mathscr{B}_\sigma. \tag{3}$$

were σ is the Cauchy stress. For weak discontinuities, as depicted in Fig. 4a, the outer traction vector continuity condition has to be fulfilled

$$\sigma^+ \cdot v = \sigma^- \cdot v \qquad \text{in} \qquad S_e \tag{4}$$

yields. For strong discontinuities inside the body $\mathscr{B}_e$, the following equation has to be fulfilled:

$$\sigma_{S_e} \cdot v = \sigma^+ \cdot v = \sigma^- \cdot v \qquad \text{in} \qquad S_e \tag{5}$$

where $\sigma_{S_e} \cdot v$ denotes the stress at the discontinuity surface. In context of a strong discontinuity, a shearband of size $h(t)$ develops. It is bounded by the subdomains $\mathscr{B}^{h-}$ resp. $\mathscr{B}^{h+}$ limited by the discontinuity surface boundaries S^{h-} behind and S^{h+} ahead the discontinuity.

The weak formulation of equilibrium including a discontinuity surface has the form

$$\int_{\mathscr{B}_e} \sigma : \nabla \eta \, dv = \int_{\mathscr{B}_e \setminus S_e} \sigma : \nabla \eta \, dv$$

$$= - \int_{\mathscr{B}_e \setminus S_e} \nabla \sigma \cdot \eta \, dv + \int_{\partial \mathscr{B}_e^+ \cup \partial \mathscr{B}_e^-} v \cdot \sigma \, da$$

$$= - \int_{\mathscr{B}_e \setminus S_e} \nabla \sigma \cdot \eta \, dv + \int_{\partial \mathscr{B}_\sigma} n \cdot \sigma \cdot \eta \, da - \int_{S_e} v \cdot (\sigma^+ - \sigma^-) \cdot \eta \, da. \qquad (6)$$

This form satisfies equations (1), (3) and (4) in a weak sense including the discontinuities. Thus only the traction vector continuity condition (equation 5) has to be enforced at the discontinuity surface S_e locally.

2.1 Kinematics

The discontinuities are handled by introducing a special enhancement of the displacement field. As proposed in [45] the displacement field can be decomposed into a regular Galerkin part u and an irregular enhanced part $\hat{u}$. This additive decomposition yields for small deformations

$$\begin{aligned}
\tilde{u}(x,t) &= u(x,t) + [H_{S_e}(x) - \Phi(x)]\,\alpha_e(t) \\
&= u(x,t) + M_{S_e}(x)\alpha_e(t) \\
&= \underbrace{u(x,t)}_{\text{regular}} + \underbrace{\hat{u}(x,t,\alpha_e)}_{\text{enhanced}}
\end{aligned} \qquad (7)$$

and models a strong discontinuity in S_e. The function Φ is arbitrary as stated in [46] and can be chosen as

$$\Phi = \begin{cases} 0 & \forall x \in \mathscr{B}_e^- \setminus \mathscr{B}^{h-} \\ 1 & \forall x \in \mathscr{B}^h \\ \text{arbitrary} & \forall x \in \mathscr{B}_e^+ \setminus \mathscr{B}^{h+} \end{cases} \qquad (8)$$

Furthermore, H_{S_e} is the Heaviside function restricted to the domain $\mathscr{B}_e$. M_{S_e} can be interpreted as the incompatible mode corresponding to element e. The displacement jump is denoted as α_e corresponding to the element e. It is defined to be the difference of the displacements on both sides of the discontinuity surface

$$[u^+(x,t) - u^-(x,t)]_{x \in S} = [|u|](x,t)|_{x \in S} = \alpha_e(t). \qquad (9)$$

Figure 5 depicts linear ansatz functions for each node including a discontinuity constructed as described inside the element.

The enhanced formulation of the strain field follows directly from the displacements. The bracketed parts of the displacement and jump description will be omitted for the sake of simplicity. Again the strains are decomposed in an additive manner

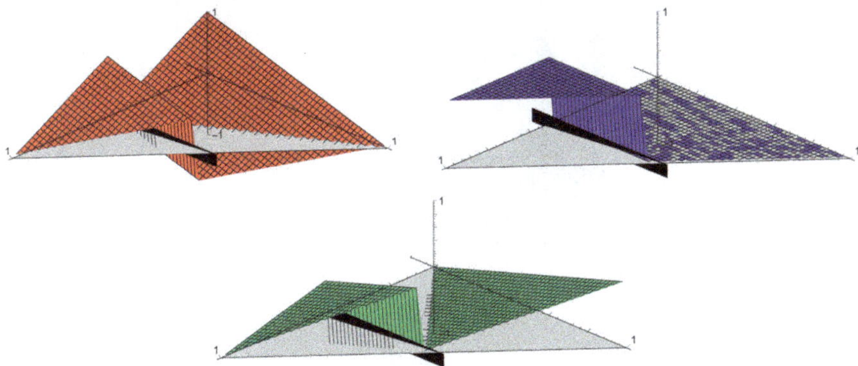

Fig. 5 Enhanced ansatz functions for two-dimensional tetrahedron element including displacement jump function.

into a regular and an enhanced part as follows:

$$\tilde{\varepsilon} = (\nabla \tilde{u})^S = \underbrace{(\nabla u)^S + (M_{S_e} \nabla \alpha_e)^S}_{\varepsilon \text{ (regular)}} + \underbrace{(\nabla M_{S_e} \alpha_e)^S}_{\hat{\varepsilon} \text{ (enhanced)}}. \tag{10}$$

Here the second term is zero, because the gradient of the displacement jump is zero.

2.2 Finite Element Approximation

The finite element approximation for the kinematics described above has to be adapted to the enhancement of the ansatz functions. Therefore, all enhanced elements are initially related to the subset $\mathscr{J} := \{e \in 1, 2, \ldots, n_e \mid \mathscr{B}_e \subset \mathscr{B}\}$. The standard finite element approximation is now enhanced by the enrichment of the displacement field

$$\tilde{u}_h = N \cdot a + \sum_{e \in \mathscr{J}} M_{S_e}^h \cdot \alpha_e. \tag{11}$$

Here N denotes the matrix of shape functions and a contains the nodal displacements. If an element is not enriched, the nodal displacement jump $\alpha_e = 0 \, \forall \, e \notin \mathscr{J}$ can be easily set to zero for all elements not participating to the crack . Because of the fact that Φ is arbitrary in the region of $\mathscr{B}_h$ it can be chosen as a certain part of the nodal ansatz functions. Hence it is determined to be the standard linear shape function corresponding to the node to which the normal vector of the discontinuity surface points. This node is called the solidary node N_{k_e} as depicted in Fig. 6 acting as the node of nodal enrichment

$$\Phi^h(x) = N_{k_e}(x). \tag{12}$$

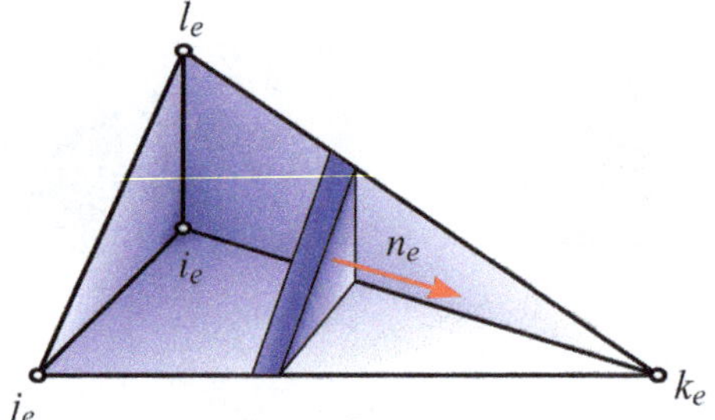

Fig. 6 Solidary node.

The enriched kinematic description of the strain field (equation 10) it yields in the context of the finite element approach

$$\tilde{\varepsilon}_h = \left(\nabla u_h\right)^S + \left(\sum_{e=1}^{n_{el}} \nabla M_{S_e}^h \alpha_e\right)^S .$$

The gradient of the incompatible mode corresponding to the enriched element e is given by

$$\nabla M_{S_e}^h = \nabla H_{S_e} - \nabla \Phi_{S_e}^h = \delta_{S_e} n_e - \nabla N_{k_e} = G_e$$

with the explicit form of matrix G_e

$$G_e = \begin{bmatrix} \delta_{S_e} n_x - \dfrac{\partial N_{k_e}}{\partial x} & 0 & 0 \\[2ex] 0 & \delta_{S_e} n_y - \dfrac{\partial N_{k_e}}{\partial y} & 0 \\[2ex] 0 & 0 & \delta_{S_e} n_z - \dfrac{\partial N_{k_e}}{\partial z} \\[2ex] \delta_{S_e} n_y - \dfrac{\partial N_{k_e}}{\partial y} & \delta_{S_e} n_x - \dfrac{\partial N_{k_e}}{\partial x} & 0 \\[2ex] 0 & \delta_{S_e} n_z - \dfrac{\partial N_{k_e}}{\partial z} & \delta_{S_e} n_y - \dfrac{\partial N_{k_e}}{\partial y} \\[2ex] \delta_{S_e} n_z - \dfrac{\partial N_{k_e}}{\partial z} & 0 & \delta_{S_e} n_x - \dfrac{\partial N_{k_e}}{\partial x} \end{bmatrix} . \tag{15}$$

The approximation of the strains in the finite element formulation leads to

$$\tilde{\varepsilon}_h = B \cdot a + \sum_{e=1}^{n_{el}} G_e \cdot \alpha_e$$

where B is the standard differential operator matrix.

2.3 Discretized Set of Enhanced Equations

The discretized weak form is then given by

$$\int_{\mathscr{B}_e} B^T : \tilde{\sigma}_h \, dv = \int_{\mathscr{B}_e \backslash S_e} B^T : \tilde{\sigma}_h \, dv = f_{\text{ext}}. \tag{17}$$

Now the weak formulation has to be completed by the local enforcement of the traction continuity condition

$$\sigma_{S_e} \cdot n = \sigma^+ \cdot N_{k_e} = \sigma^- \cdot N_{k_e} \quad \text{in} \quad S_e. \tag{18}$$

Instead the following set of equations can be applied to locally enforce the traction vector continuity condition, see [45]

$$\int_{\mathscr{B}_e} G_e^{*T} \cdot \tilde{\sigma}_h \, dv = 0 \qquad e = 1 \ldots n_{el}, \tag{19}$$

$$G_e^* = \begin{cases} \left(\delta_{S_e} - \dfrac{l_e}{|\mathscr{B}_e|} \right) N_{k_e} \; \forall x \in \mathscr{B}_e, \\[2mm] 0 \qquad\qquad\qquad \text{otherwise.} \end{cases} \tag{20}$$

Herein l_e measures S_e and $|\mathscr{B}_e|$ denotes $\mathscr{B}_e \backslash S_e$. It can be shown that

$$\frac{1}{l_e} \int_{S_e} \tilde{\sigma}_h \cdot N_{k_e} \, da = \frac{1}{|\mathscr{B}_e|} \int_{\mathscr{B}_e \backslash S_e} \tilde{\sigma}^h \cdot N_{k_e} \, dv \tag{21}$$

enforces the traction vector continuity condition in an average sense. Including these conditions the corresponding weak form has to be additionally fulfilled in every enriched finite element . It can be written as

$$\int_{\mathscr{B}_e} G_e^{*T} \cdot \tilde{\sigma}^h \, dv = 0 \qquad \forall e \in \mathscr{J}. \tag{22}$$

This leads to a discretized set of equations, including dynamic effects,

$$\rho \int_{\mathscr{B}} N^T N \ddot{u}_h \, dv + \int_{\mathscr{B}} B^T : \tilde{\sigma}_h \, dv = P, \tag{23}$$

$$\int_{\mathscr{B}_e} G_e^{*T} \cdot \tilde{\sigma}_h \, dv = 0 \qquad \forall e \in \mathscr{J}. \tag{24}$$

The resulting jump function can be condensed at element level as stated in [21]. Hence, this method provides an appropriate platform for explicit time integration schemes. The variables related to the displacement jump can be evaluated directly

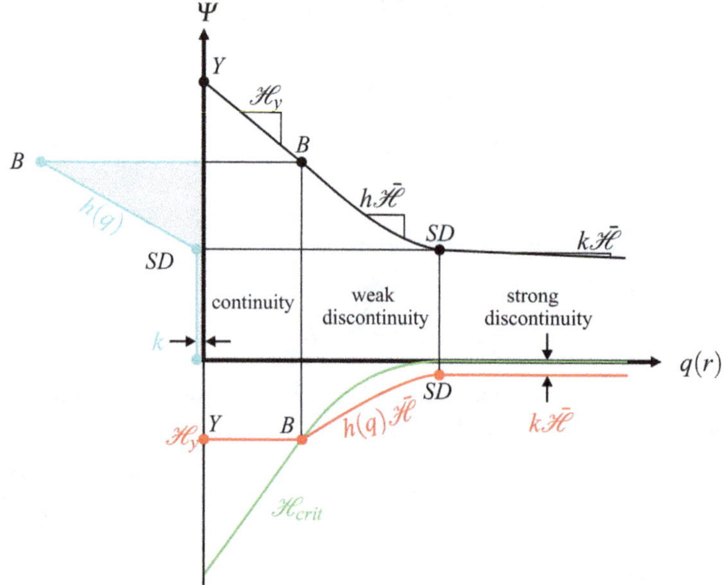

Fig. 7 Variable bandwidth model.

at element level

$$\alpha_e = -(G_e^{*T} \cdot \mathbb{C} \cdot G_e)^{-1} \cdot G_e^{*T} \cdot \sigma^h. \tag{25}$$

By considerating the displacement jump α_e, the enhanced assumed stress follows for an enriched finite element by addition of the Galerkin stress σ^h and discontinuity stress $\hat{\sigma}^h$, respectively

$$\tilde{\sigma}^h(x,t) = \sigma^h(x,t) + \hat{\sigma}^h(x,t,\alpha_e,\sigma^h). \tag{26}$$

Finally, three points regarding the implementation have to be considered.

2.3.1 Variable Bandwidth Model

As mentioned above, the localization bandwidth varies over the simulation time and depends mainly on an internal variable q_t. The variable bandwidth model was introduced in [48]. Figure 7 shows the main coherences of the variable bandwidth[4] $h(q)$, the softening parameter $\mathcal{H}$, the position of the failure initiation (point Y), the material bifurcation (point B) and the existence of a strong discontinuity (point SD). The critical softening parameter can be applied as the unique variable for the detection of material instability, see [53]. The onset of a strong discontinuity at the

---

[4] In the computational model the localization bandwidth variable is updated with one time step delay.

material instability point (q_B) can be determined using the internal variable q_t by

$$q_{sd} = (1 + \gamma)q_b \quad \gamma \in [0, 1] \tag{27}$$

where γ denotes the width of the weak discontinuity regime and varies from $\gamma = 0$ for a zero-valued discontiniuity bandwidth to $\gamma = 1$ for full bandwidth dimension. This discontinuity bandwidth expansion describing parameter is set to $\gamma = 0.8$ for every numerical simulation in the following, see [53]. The localization bandwidth varies over the simulation time linearly. According to [53] it can be determined from

$$h(q) = \begin{cases} h_0 & q_t < q_b \\ h_0 + \dfrac{k - h_0}{q_{sd} - q_b}(q_t - q_b) & q_b \leq q_t < q_{sd} \\ k & q_t \geq q_{sd}. \end{cases} \tag{28}$$

Additional to the use in the context of the Strong Discontinuity Approach, the variable bandwidth can be applied to identify the element state change. If the discontinuity bandwidth of a distinct element reaches the limit value k, the discontinuity inside the element is of strong kind. This variable can be used to decide when the elemental representation has to be switched from the continuous (non-geometrical) to the discontinuous (geometrical) level. This important modeling feature will be re-addressed later.

2.3.2 Regularization

The Dirac's delta function (δ_{S_e}) are present in the formulation. Since they cannot be introduced directly a regularization, by defining a delta-sequence instead, is necessary. Therefore a discontinuity band $\mathcal{B}_e^h$ of mutable width $h(q_t)$ is considered

$$\delta_{S_e} = \lim_{h(q_t) \to 0} \delta_{S_e}^h(x), \quad \delta_{S_e}^h = \frac{1}{h(q_t)}\mu_{S_e}, \quad \mu_{S_e} = \begin{cases} 1 \forall x \in \mathcal{B}_e^h \\ 0 \forall x \notin \mathcal{B}_e^h. \end{cases} \tag{29}$$

With this the Dirac's delta function is expressed by means of the bandwidth of the discontinuity band $h(q_t)$ that varies over the simulation time and depends directly on the internal stress inside the enriched finite element. If the width of the discontinuity differs from zero $(h(q_t) \neq 0)$ the discontinuity is of weak kind. Otherwise the parameter tends to a limit zero-equal[5] value $(h(q_t) \to k)$. The state where the bandwidth is approximately zero denotes the initiation of a strong discontinuity representation as depicted in Fig. 8 [53]. Additionally the softening parameter $\mathcal{H}$ of the material model has to be the regularized

$$\mathcal{H} = \begin{cases} \infty & \forall x \notin \mathcal{B}_e^h \quad \text{(elastic behavior)} \\ h(q_t)\bar{\mathcal{H}} & \forall x \in \mathcal{B}_e^h \quad \text{(inelastic behavior).} \end{cases} \tag{30}$$

---

[5] Due to computational requirements the minimum discontinuity bandwidth is not equal to zero, but equals a very small positive double value.

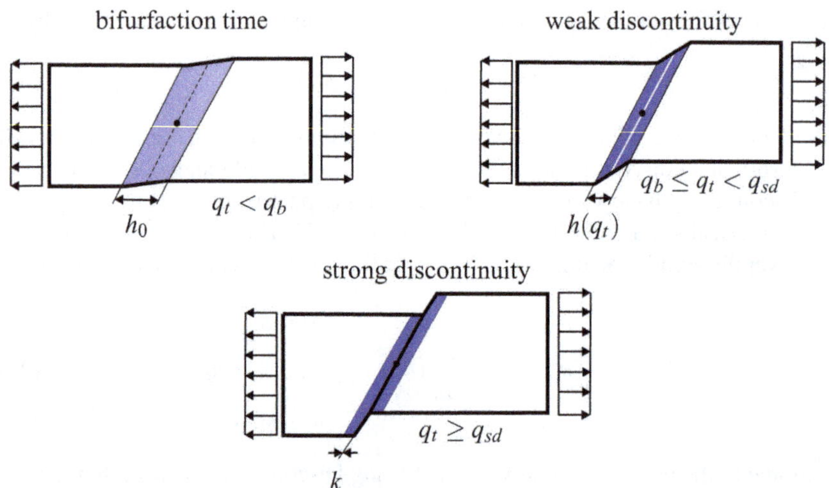

Fig. 8 Discontinuity zones.

If the discontinuity tends to zero, the softening parameter tends to zero as well. That is equivalent to a perfect plasticity material or to perfect damage ($\mathscr{H} \approx 0$).

2.3.3 Integration Rule

Due to the nature of the regularized discontinuity band $\mathscr{B}_e^h$ (see Fig. 6), a specific numerical integration is needed. The strains and the stresses are piecewise constant in both domains $\mathscr{B}_e^h$ and $\mathscr{B}_e \setminus \mathscr{B}_e^h$ in a tetrahedral element with linear ansatz functions. Thus no specific locations for the integration points at the corresponding domain need to be specified. Subsequently, the following conditions for the weighting point by means of Gauss integration can be used

$$
\begin{aligned}
w_{Q_1} &= |\mathscr{B}_e| - h(q_t) \cdot l_e, & \mathscr{H}_1 &= \infty, & Q_1 &\in \mathscr{B} \setminus \mathscr{B}_e^h, \\
w_{Q_2} &= h(q_t) \cdot l_e, & \mathscr{H}_2 &= h(q_t)\bar{\mathscr{H}}, & Q_2 &\in \mathscr{B}_e^h.
\end{aligned}
\tag{31}
$$

The softening parameter reflects an elastic behavior outside the discontinuity domain and inside the localization band it incorporates softening behavior. If the element is in the regime of a diffuse failure and no discontinuity band exists, standard one point Gauss integration is used. Once a material bifurcation state is reached the second integration point is introduced and applied to the model.

3 Material Model

Due to the fact that deformations in concrete are small, a material model for small deformations is used to represent the physical behavior of the material. The constitutive equation used in the elastic regime is the Hooke material. Its material stiffness tensor is given by, see e.g. [77],

$$\mathbb{C}_{el} = \lambda \mathbf{1} \otimes \mathbf{1} + 2\mu \mathbb{I}, \tag{32}$$

$$\mathbb{C}_{ijkl} = \lambda \delta_{ij}\delta_{kl} + \mu(\delta_{il}\delta_{jk} + \delta_{ik}\delta_{jl}). \tag{33}$$

with the Lamé parameters λ and μ.

3.1 Isotropic Damage Model

For simplicity an isotropic damage model with linear strain softening is used to include inelastic damage effects. Here a scalar damage parameter d is introduced that varies from $d = 0$ for the undamaged material to $d = 1$ for the theoretically fully damaged state. An extension of this isotropic damage model to the method of Enhanced Assumed Strains was first introduced in the context of the Strong Discontinuity Approach in [45, 46].

The stress-strain relation for the isotropic damage model yields

$$\sigma = (1-d)\mathbb{C} : \varepsilon, \qquad d \in [0,1[. \tag{34}$$

The evolution of the scalar damage parameter can be integrated in closed form at the distinct time t, see [44],

$$^{t}d = G(^{t}r) \qquad ^{0}r = \max\{^{0}r, \tau^{\varepsilon}\}. \tag{35}$$

Herein r_t denotes the size of the elastic domain which is defined by an adequate norm of the elastic energy rate of the strains τ^{ε} and an initial value r_0. This value is mainly defined by the maximum uniaxial stress σ_u

$$\tau^{\varepsilon} = \sqrt{\varepsilon : \mathbb{C}_{el} : \varepsilon} \qquad \text{and} \qquad ^{0}r = \sigma_u/\sqrt{E}. \tag{36}$$

For a model including linear strain softening law the scalar damage function ^{t}d can be directly determined including the regularized hardening-softening parameter $\mathscr{H} = h(^{t}q)\bar{\mathscr{H}}$. For further purposes the discrete softening parameter $\bar{\mathscr{H}}$ has to be regularized as described before. The damage function can be explicitely written and calculated through the internal stress like variable $^{t}q(^{t}r)$ and the regularized softening parameter, respectively

$$G(^{t}r) = 1 - \frac{^{t}q}{^{t}r} = \frac{1}{1+\mathscr{H}}\left(1 - \frac{^{0}r}{^{t}r}\right) \qquad \forall \quad ^{0}r < {}^{t}r < r_{\max} = -\frac{1}{\mathscr{H}}{}^{0}r. \tag{37}$$

The non-regularized softening parameter $\bar{\mathcal{H}}$ is defined by the specific fracture energy, considering the following relation including linear strain softening [52]:

$$\bar{\mathcal{H}} = -\frac{\sigma_u^2}{2EG_f} \quad \text{or} \quad \bar{\mathcal{H}} = -\frac{\sigma_u^2}{EG_f}\exp\left(-\frac{r_0}{G_f}(r_t - r_0)\right) \tag{38}$$

for exponential strain softening, where G_f denotes the discrete fracture energy.

For the determination of the displacement jump inside the enriched finite element, a consistent tangential operator of the material stiffness matrix is needed. Standard linearization procedures yield the tangential operator

$$\mathbb{C} = (1 - {}^t d)\mathbb{C}_{\mathrm{el}} - \frac{\partial^t d}{\partial C}\sigma \otimes \sigma. \tag{39}$$

According to [52] the partial derivative of the scalar damage parameter ${}^t d$ with respect to the strain tensor yields

$$\frac{\partial^t d}{\partial \varepsilon} = \frac{{}^t q - \mathcal{H}({}^t q)\,{}^t r}{{}^t r^3}. \tag{40}$$

3.2 Identification of the Fracture Plane Normal

For the application of the Strong Discontinuity Approach as well as for the introduction of discrete fracture surfaces, the fracture plane normal has to be determinated. Fully localized elements have to be divided with respect to localization direction resulting from the acoustic tensors analysis. The distinct time of the material instability, the bifurcation point, can be calculated by analysis of the acoustic tensor Q^{loc} as follows:

$$\det(Q^{\mathrm{loc}}) = 0, \qquad Q^{\mathrm{loc}}(x,n,t) = n \cdot \mathbb{C} \cdot n. \tag{41}$$

A material point looses its positive definiteness and so the strong ellipticity if the determinant of the acoustic tensor is equal to zero

$$t \cdot \varepsilon \cdot t = 0 \qquad \forall \quad t | t \cdot n = 0. \tag{42}$$

Unfortunately the solution of the system of equations is computationally very costly. Particularly in consideration of en explicit time integration scheme where the system has to be solved at every time step for every three-dimensional finite element, thus a numerical iterative solution is too expensive. A closed form solution for the determination of the material instability was developed in [50]. This will be applied within the explicit solution scheme, see also [64].

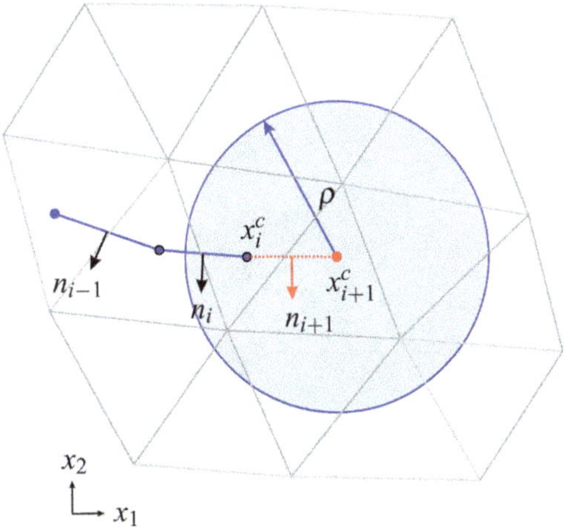

Fig. 9 Extended local smoothing procedure.

3.2.1 Extended Local Smoothing

Since cracks in a fracturing solid will open and close, contact of the crack surfaces
has to be considered. For an efficient and stable contact search algorithm as well as
for the potential Discrete Element interaction, sufficient smoothness of the surface
is necessary. Here an extended local smoothing procedure is introduced for the direction of the normal fracture plane vector. Based on the two-dimensional smoothing
method described in [27] a set of elements $\mathscr{B}_{nw}$ will be used for a weighted determination of the normals. This procedure can be defined with the help of the discrete
softening parameter

$$\mathscr{B}_{nw} : \left\{ x \in \mathscr{B} \quad \frac{\mathscr{H}^{crit}(x)}{\mathscr{H}_e^{cur}} \leq \upsilon, \qquad \forall c(x) \leq c_{max} \right\} \tag{43}$$

$$\text{with} \quad c(x) = |x_i^c - x_{i+1}^c|. \tag{44}$$

Herein the current state of the critical softening parameter is observed continuously.
The distance to the nearby elements $c(x)$ is calculated by the midpoints of the elements (x_i^c resp. x_{i+1}^c) and being limitted by c_{max}.

If the variable υ is chosen to be $\upsilon < 1$, elements which are near to the material
instability point, can be taken into account additionally. $\mathscr{H}_e^{cur}$ has to be determined in every time step within the explicit time integration scheme for every finite
element . The extended local smoothing procedure is depicted in Fig. 9 for the two-
dimensional case. Finally, the subsequent equation is used for the determination of
the new modified and smoothed fracture plane normal. The extended local smoothing is done via a Gaussian bell function weighting over the distance to other nearby
fractured elements

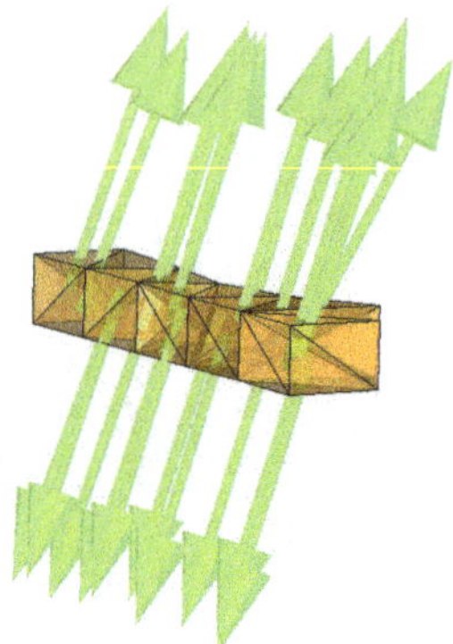

Fig. 10 Smoothed localization vectors in initial crack tip.

$$\bar{n}_{i+1}^{\text{crit}} = \frac{\int_{\mathscr{B}_{\text{nw}}} n^{\text{crit}}\,\varpi\,\mathrm{d}v}{\int_{\mathscr{B}_{\text{nw}}} \varpi\,\mathrm{d}v}, \qquad \varpi = e^{-\frac{r(x)^2}{2\rho^2}}. \tag{45}$$

Here ρ denotes the standard deviation of the weighting function. Localization directions n^{crit} and $-n^{\text{crit}}$ are treated equivalently. Even for linear tetrahedron elements, that are in known to be direction dependent, consistent directions can be obtained by this method (Fig. 10). Remarks containing time step sizes of the explicit computations can be found in [64].

4 Numerical Example

This numerical example will show the abilities of the Strong Discontinuity Approach in scope of an explicit time integration scheme. The example models a simplified compact tension test leading to a mode I failure. Figure 11 depicts the model geometry and concrete material parameters according to Areias and Belytschko [3]. Additionally, the specific density of concrete material is used. The specimen is pulled by 0.00125 mm within a simulation time of 3 ms. The maximum admissible time step size depends on the state of calculation. According to this fact an adaptive time stepping scheme is used, see [64], When the loading process is finished, all finite element depicted in Fig. 12 are fractured. Figure 12(b) shows all bifurcated elements that include a material instability while making all unfractured elements transparent. All marked elements have a displacement discontinuity. The determination of the crack path and the bifurcated elements is computed without any further assumptions expect the extended local smoothing procedure.

Computational results for the displacement and stress field are shown in Fig. 13. In Fig. 13(a) the displacement discontinuity in the region of the failured elements (cf. Fig. 12(a)) is depicted and can be identified very clearly. A closer inspection of

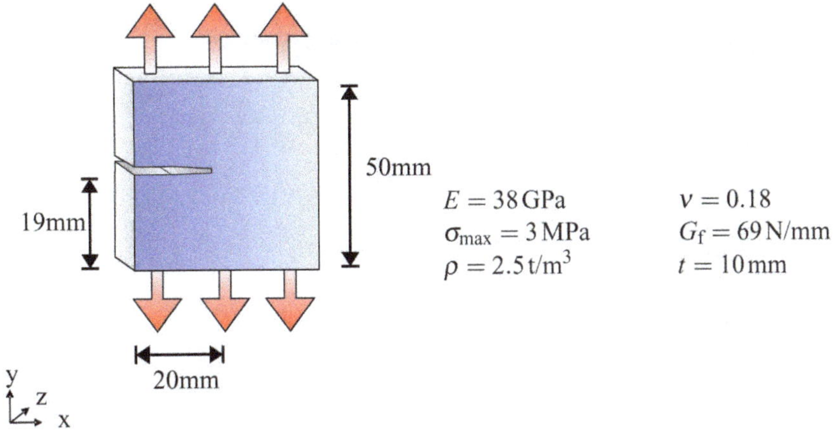

Fig. 11 Computational continuous SDA model, model geometry and material parameters.

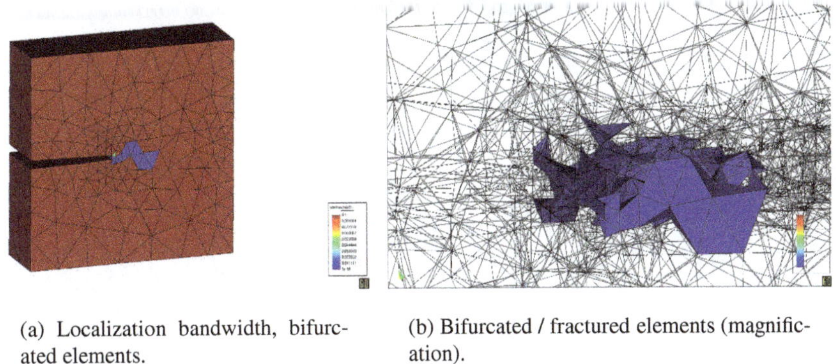

(a) Localization bandwidth, bifurcated elements.

(b) Bifurcated / fractured elements (magnification).

Fig. 12 Computational continuous SDA model, localization bandwidth.

Fig. 13 shows the lack of a distinct shearband development. This fact is explained by inertia effects within the extremely short loading period. If loading is applied within a longer time interval, then a shearband develops as in a quasi-static analysis. Figure 13(c) shows the relocation of the stresses in loading direction, where the highest values are located at the crack tip of the developing discontinuity.

A distinct fracture surface cannot be computed within this appraoch, as shown in this example, which motivates the introduction of a continuous-discontinuous model.

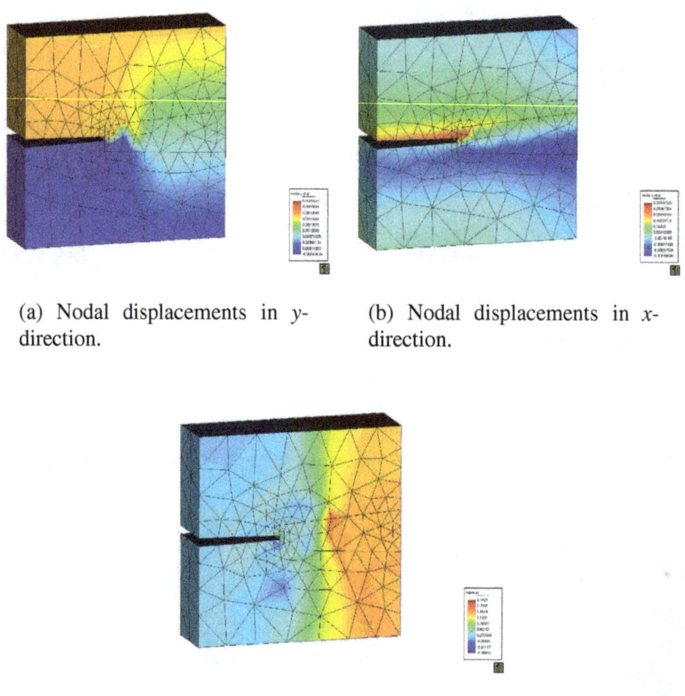

(a) Nodal displacements in y-direction.

(b) Nodal displacements in x-direction.

(c) Stresses in yy-direction.

Fig. 13 Computational continuous SDA model, displacements and stresses.

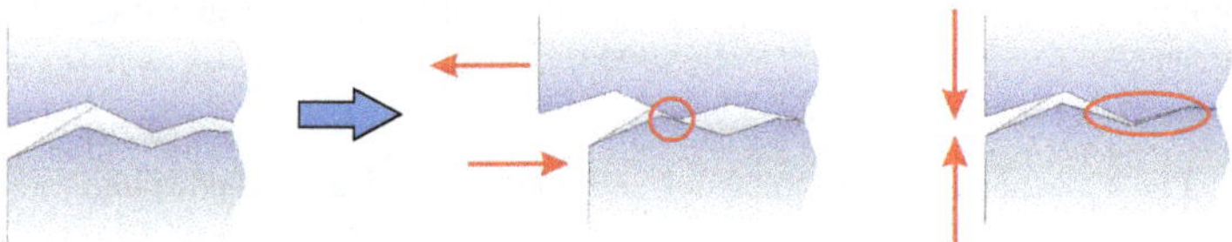

Fig. 14 Post fracture contact.

5 Continuous-Discontinuous Model

A continuous, non-geometrical modeling by the SDA was introduced in the previous section in the context of an explicit integration scheme. For reliable modeling of loaded structures, the knowledge of the post critical behavior is of interest frequently. This incorporates the simulation of secondary loading processes of the previous damaged and fractured body as shown schematically for example in Fig. 14. As depicted, the generated fracture surfaces will reach a contact state. The continuous model is sufficient for normal contact forces, whereas the determination

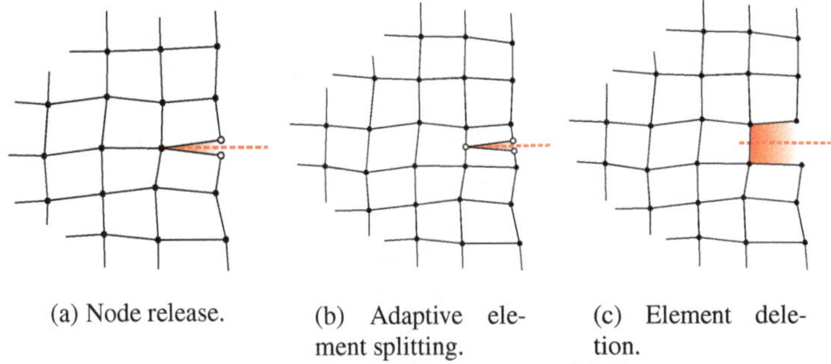

(a) Node release.   (b) Adaptive element splitting.   (c) Element deletion.

Fig. 15 Crack propagation techniques (two-dimensional).

of tangential contact forces fails mostly and needs additional considerations, see e.g. [18, 26].

5.1 Overview of Fragmentation Techniques

Before introducing different fragmentation techniques including various approaches such as connectivity release methods or adaptive schemes, it is necessary to take a closer view to the finite element topology. For example a linear tetrahedral volume element consists in general of four nodes, six edges and four surfaces. Thus one three-dimensional volume element can be decomposed into surfaces (two-dimensional elements), edges (one-dimensional elements) and nodes (zero-dimensional elements) accordingly. Assuming a set of elements, one node (or any other subdimensional component) is shared by a number of other volume elements (or by its superior dimensions). Consequently, it is possible to generate neighborhood relationships such as node-neighbors, edge-neighbors or surface-neighbors for each finite element. Within the context of fragmentation techniques theses connectivities have to be divided and/or restructured. In Fig. 15 some crack propagation methods like node release, adaptive element splitting and element deletion techniques are depicted.

Furthermore, crack path continuity is one important task. It is relatively simple to handle in two dimensions but extends to a very challenging task in three dimensions. The emphasis lies on schemes that result in a complete adaptive refinement including corresponding subdimensions. Here we restrict ourselves to the usage of three-dimensional tetrahedral elements for the implementation of a geometrical fracture approach. As stated in [18], different fragmentation procedures will be briefly presented in the following. Also new procedures will be introduced and further extensions to the developed framework will be stated.

5.1.1 Simple Release Methods

As a first approach different simple release methods can be applied for the geometric modeling of developing cracks. These element reconnection schemes do not need a full adaptive mesh refinement. Nevertheless, some subdimensional elements such as nodes, edges or surfaces have to be doubled and reassigned in order to obtain a geometric fracture surface representation. Consequently, the computational model has toinclude such restructuring, see [18]. Remodeling requirements are relatively low in comparison to full adaptive refinement schemes, but remain a challenging tasks also. Here various commercial and research codes reach their limits. This lack of usability is often based on the structure of data management such as storage of history and node variables and connectivity aspects. Thereby, the benefit of a complete object oriented framework for the computational finite element modeling, see [64].

- *Node Release* (zero-dimensional) The basic idea behind node release strategy is quite simple and is based on a zero-dimensional primitive. Within one undirected release the element failure criterion is projected to the nodes which are subsequently released including all other elements referencing this node. This is one applicable approach if the material model provides a scalar undirected failure criterion only and no distinct fracture surface description can be obtained. Propagating cracks result into a huge amount of releasing energy causing eventually dynamic unloading waves. Furthermore, this strategy can result in a number of unconnected elements. Here the incorporation of very stable and efficient multi body contact algorithms are inevitable especially for the simulation of large three-dimensional models.
- *Edge Subdivision* (one-dimensional) One other possible fragmentation procedureis based on dividing of the element edges. Thus a new node has to be introduced additionally and all elements sharing this edge have to be refined accordingly. The way the edges are split frequently leads to non-physical crack directions. Furthermore, the splitting direction is highly mesh dependent and has to be checked regarding the plausibility of possible new introduced geometrical fracture surfaces.
- *Surface Release* (two-dimensional) Consequently, the next technique is a two-dimensional release where the surfaces are decoupled. For linear tetrahedral elements this procedure is equivalent to the node release problem. For quadratic shape functions using hexahedron elements the crack construction procedure is explained in, e.g., [59]. The introduction of higher order shape functions enables the release of an inner node without affecting the nodes at the edges.
- *Element Deletion* (three-dimensional) As the last simple approach the elements reaching a limiting damage value can be removed from the model. In this technique a material model providing only a scalar damage parameter suffices. In various commercial codes crack propagation and fracture is handled in this way. As one can imagine simple removal of elements out of the finite element model is equally to the loss of material. For some dissipative processes, this behavior can be reasonable motivated by material phenomena. But in general this method

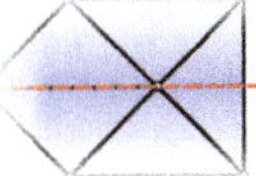 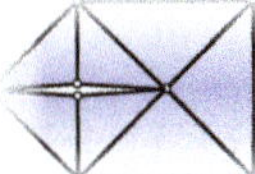 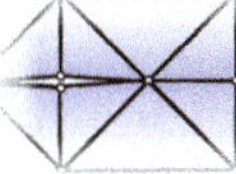 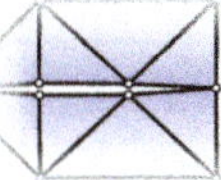

Fig. 16 Two-dimensional crack propagation by means of adaptive element refinement and node doubling.

is non-physical. The element deletion approach can be applied in the context of cohesive elements as well. Thus the introduced cohesive elements [59] can be deleted after the surface tractions reach a limiting value. This model reproduces the continuum mechanical effects in a much better way but leads to multiple fracture surfaces.

5.2 Adaptive Mesh Refinement

Adaptive mesh refinement in the context of finite element fracture mechanics can be based on most common adaptive mesh refinement techniques, see [83]. I can be extended easily to discrete fracture mechanics. For fracture processes at least one node is inserted twice and the corresponding connected elements like edges and surfaces are handled accordingly. Thus both newly generated surfaces represent the fracture surface, see Fig. 16. As one can imagine the issues of data management and the maintainance of mesh integrity at every time of the simulation remains a big challenge inside every adaptive finite element program.

A first adaptive refinement technique is a complete remeshing in the domain of interest (crack) or for the whole body. Especially for large three-dimensional models and explicit time integration schemes this method is very costly and thus inapplicable.

Node positions are adjusted in *r-adaptive refinement* schemes. Especially in combination with the Variational ALE approach [42, 75] this adaptive finite element scheme provides a good choice for modeling discrete fracture processes in commercial finite element software due to relatively small computational implementation tasks. As depicted in Fig. 17 nodes are moved corresponding to the predicted crack direction. *h-adaptive refinement* schemes or element subdivision methods are used frequently and will be part of this work as well. Here all failed elements can be simply divided into smaller ones while new generated surfaces can be identified as fracture surfaces. Subsequently, some connectivity relations have to split up or rearrange in order to introduce a real crack. One important condition for an adaptive fragmentation technique in the context of an explicit time integration scheme is to keep the minimum element length as big as possible in order not to decrease time step sizes. Here a sufficient adaptive scheme is generated in order to obtain good fracture surface quality on the one hand and to keep the computational costs as low as possible on the other hand.

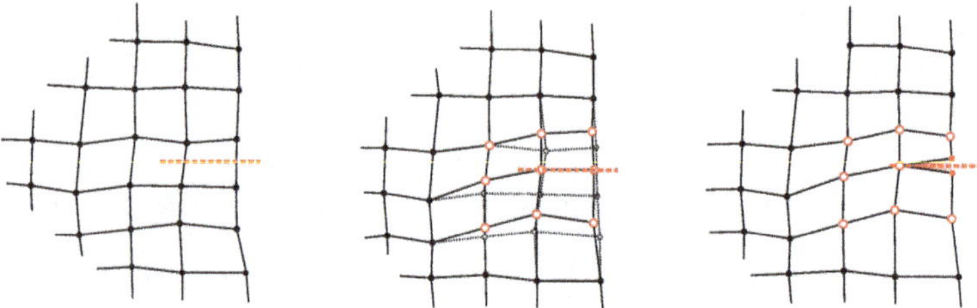

Fig. 17 Crack propagation technique (two-dimensional): Variational ALE approach including node release.

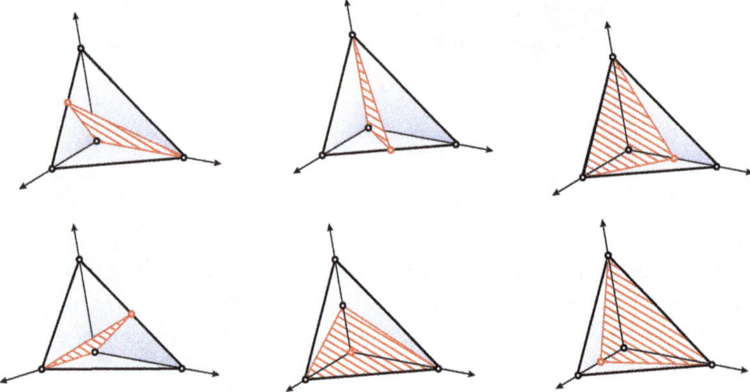

Fig. 18 Admissible crack positions with edge refinement.

Adaptive Green Element Splitting

In general two different approaches exist for adaptive subdividing of a tetrahedral element. The first technique is denoted by *red refinement*. Here the finite element is subdivided into eight new tetrahedral elements with the same volume and consequently with the same edge measure. Another approach is the *green refinement* where one to three new nodes are introduced on one or more edges of the element. Both refinement techniques have in common, that new nodes are introduced on the edges of the element. Accordingly other nearby volume elements (and subdimensional elements) have to be refined as well to maintain continuity. It is clear that the exclusive usage of the red refinement strategy results in a complete refined finite element mesh. In order to keep the refinement process as lean as possible and to limit the number of involved elements the green refinement with one irregular node will be applied. This procedure is comparable to the longest edge refinement, introduced in [65] for the two-dimensional case. Due to the framework of fracture mechanics these new nodes have to be doubled and related elements have to be reprocessed as described previously.

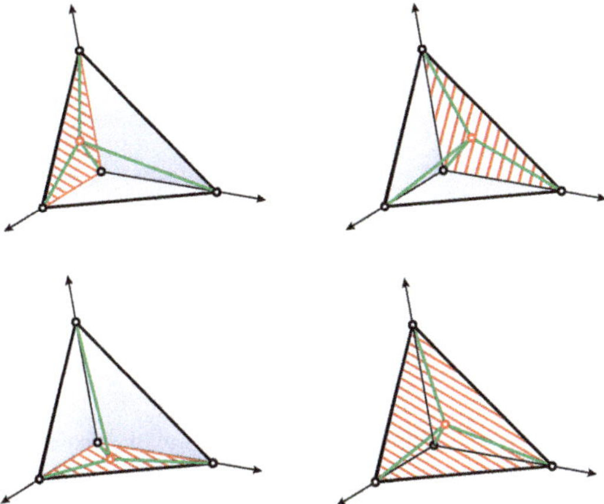

Fig. 19 Admissible crack positions with surface release.

Figure 18 shows the used refinement approach and the resulting various alignments in the context of discrete fracture mechanics. The presented refinement strategy is denoted as *adaptive green element splitting* in the following.

Advanced Surface Refinement

Figure 19 shows the applied extended approach of an *advanced surface refinement*. Herein, one outer surface of the tetrahedral element is identified as crack surface. Additionally, another node, lying on the midpoint of the surface of interest, is inserted and resulting new elements are adaptively generated. It is easy to see that this extended approach provides the opportunity for a discrete damage initiation inside the volume body motivated by Fig. 16. In this case only the element that includes a crack is processed by remeshing tasks and the crack is limited to one element inside a very small region. In the following, these two schemes for the introduction of discrete cracks will be used. The combination of both approaches provides a good approximation of the developing fracture plane without incorporating lots of refinement tasks in nearby elements while having ten different possibilities for the description of the crack direction.

6 Crack Path Continuity

In order to obtain a suitable fracture surface and to eliminate multifractured bodies, crack path continuity has to be enforced continuously which leads to a smooth fracture plane by incorporation of constraints. These constraints should be minimal to ensure a possibility of crack branching and unification. Thus global smoothing

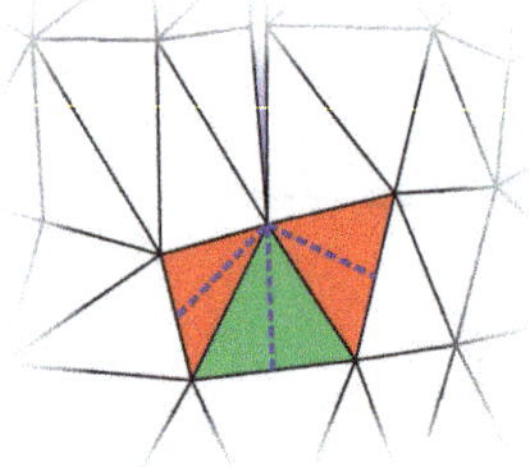

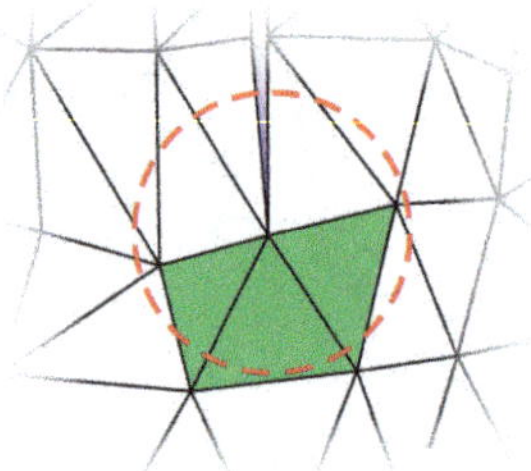

(a) Admissible splitting direction.   (b) Admissible splitting domain.

Fig. 20 Crack path continuity checks.

operators, as introduced in [56], are not applicable and local criteria have to be developed.

6.1 Continuity Checks

The material model within an element is checked at every time step in order to detect splitting or bifurcation. An introduction of a discrete crack and consequently the adaptive refinement is only admitted when the element is already bifurcated and was not split before.[6] Furthermore, adaptive splitting schemes prioritize the connection to newly introduced nodes. Hence an adaptive element splitting resulting in an already introduced feature is preferred, even if the fracture normal does not agree completely.

In order to ensure a unique crack path, three main continuity checks are necessary. The first one has to check the distance of from the element center to the fracture surface. Therefore, the minimal distance to all elements of the considered surface set is computed as depicted schematically for the two-dimensional case in Fig. 20(b). Here the check is included wether the element belongs to an already introduced fracture surface set or represents the initiation of a possible new fracture surface. The absolute distance to the crack as well as its projected value to the averaged fracture surface is determined. In computational investigations a maximum distance of 0.5 to 0.8 times the maximum element edge length gave a practicable upper limit that an element belonged to the fracture surface set. For a greater distance the element

---

[6] Here it is also imaginable to weaken this restriction by allowing additional fracturing for already fractured elements after a certain time period. Due to the numerical explicit time integration scheme this possibility is not taken into account a priori.

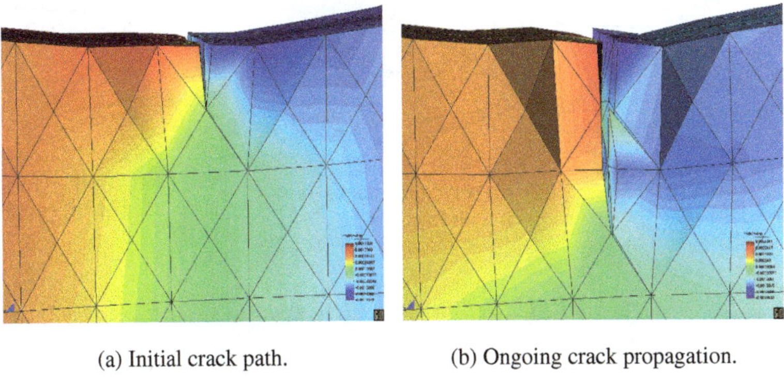

<table>
<tr><td>(a) Initial crack path.</td><td>(b) Ongoing crack propagation.</td></tr>
</table>

Fig. 21 Crack Development.

refinement request is rejected. All elements that are more than 3 times the maximum element edge length away will be associated to a new fracture surface.

After this first check, the conformity of the element fracture surface with the averaged fracture surface normal is checked (Fig. 20(a)). This direction check incorporates various parameters such as the usage of the real element fracture surface based on the chosen refinement technique or the computed fracture normal. Consecutively the continuity of the crack is controlled.

6.2 Numerical Example

Figure 21 confirms the ability of the presented approach to introduce fracture surfaces. A developing crack propagation is depicted for two states of the calculation. Figure 21(a) depicts the initial crack path while it can be seen that the crack splits the upper front element by a green adaptive splitting. Next an element is split according to the predicted fracture surface by means of a surface detaching scheme. As the calculation continues the crack proceeds through the volume body as shown in Fig. 21(b). The crack path proceeds directly downward in conjunction with the applied loads. The presented approach guarantees a unique and smooth crack path. It can be seenas a mixture of a global smoothing approach, an incorporation of a shielding zone and a direction independent local approach.

6.3 Concept of Weak Nodes

When dealing with adaptive remeshing while handling crack propagation, poorly connected nodes can appear in the finite element model. To overcome problems

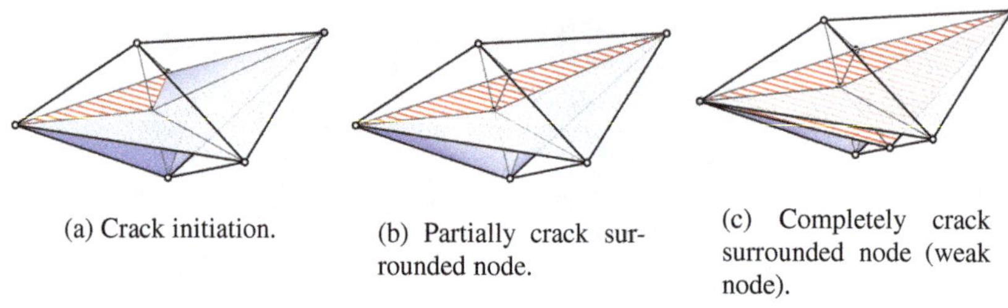

(a) Crack initiation.   (b) Partially crack sur-   (c) Completely crack
                               rounded node.   surrounded node (weak
                                                           node).

Fig. 22 Concept of weak nodes.

the concept of weak nodes is introduced. Every time an element is adaptively split, all affected nodes are initially marked as "crack tip nodes" which is equivalent to the identification of potential fracture candidates. This nodal integer type marker is increased every time an element sharing this node is split by an appropriate scheme. Combining this additional nodal information with the knowledge of the existence of a generated fracture plane in the region surrounding the node, allows a sufficient prediction of the existence of a weak node. Therefore, the surface affiliation of the node is determined by means of inner-, outer- or crack surface distinction.

The basic concept and the development of a poorly connected node is depicted in Figs. 22(a)–22(b). Herein the crack (dashed red) starts in the first element and propagates to the next one. After the node is completely surrounded by the crack surface (Fig. 22(c)), this node is of weak type based on the connection state of the element entity graphs. Now the node is doubled and the element affiliations have to be reassigned. Consequently, due to element refinements some remeshing is needed in the adjoining elements.

7 Discrete Fracture

In order to introduce a real crack elements have to be split or detached by an appropriate technique as described above. One main task is to obtain a material parameter that is significant for the initiation of cracks and can be seen as a decisive value for the initiation of a geometrical element representation.

Reviewing the ideas of the Strong Discontinuity Approach, the discontinuity bandwidth varies over the simulation time and identifies the developing discontinuity inside the finite element. Consequently, the current discontinuity bandwidth $h(q_t)$ can be seen as the crucial parameter to determine the failure process of a finite element in an adequate way and thus is suitable for the unique determination of real cracks. Hence, the element crack representation is transferred from a material to a geometrical approach if the discontinuity bandwidth reaches the limit of the strong discontinuity value as depicted in Fig. 23 and denoted in (28).

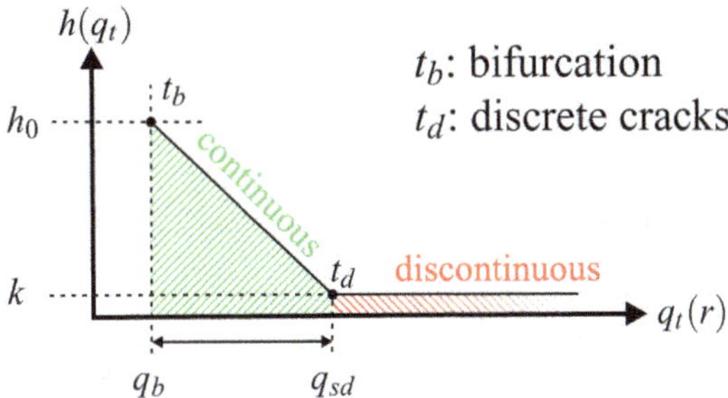

Fig. 23 Discrete fracture by means of the discontinuity bandwidth.

Together with the introduction of a discontinuous crack a unique direction of the real crack has to be specified. This is also included in the Strong Discontinuity Approach that incorporates the calculation of the fracture plane normal vector. In the discretization we restrict ourselves to cracks through the element center or along the element surface to keep the computational costs as low as possible.[7] The refinement technique that conforms best with the calculated fracture plane will be described in Section 7.2.

7.1 3D Crack Propagation

Unfortunately, the algorithmic realization of a stable, reliable and efficient dynamic crack propagation algorithm is much more complicated in three dimensions than in two. However theinitiation of a discrete crack and the fracture plane orientation can be transferred straightforward from two to three dimensions.

In two dimensions the crack path is described by lines. Thus a simple node splitting technique and the incorporation of some casual restrictions are sufficient for a qualitatively good geometrical crack representation. Therefore, a simple node releasing strategy answers the purpose inside a two-dimensional simulation like it is state of the art in various commercial and research finite element programs. In contrast the extension of a crack propagation to a complete three-dimensional simulation satisfying the model requirements is far away from a simple extension since cracks are defined are then defined by surfaces as shown in Fig. 24.

These surfaces can be described by C^0 continuous planes which are limited to a particular domain, sometimes only spanning one or two elements. One main goal is the construction of a unique and smooth fracture surface that is at least C^0 continu-

---

[7] Within the Strong Discontinuity Approach the fracture plane normal is assumed to go through the element center. By identifying an outer surface as fracture plane, a small model error may arise. This difference is minor and will be neglected. Nonetheless, the continuum mechanical part of the discontinuity is modeled correctly.

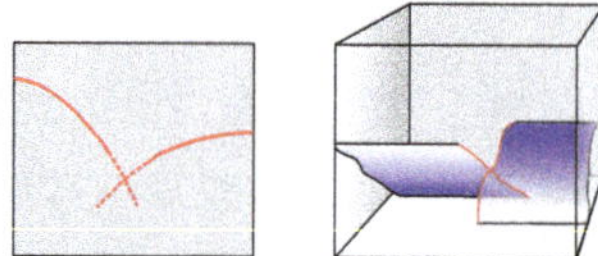

Fig. 24 Crack propagation two vs. three dimensions.

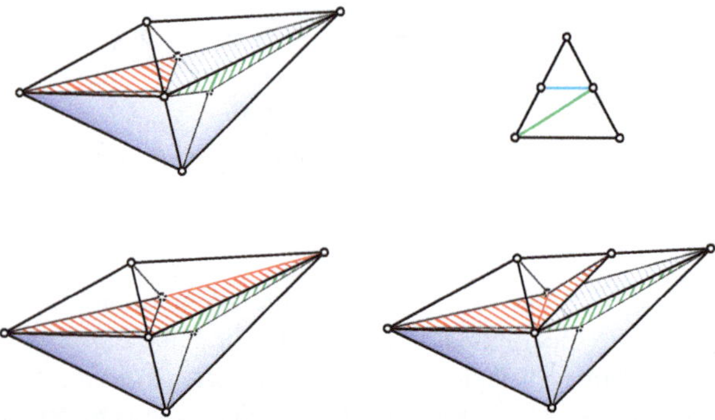

Fig. 25 3D Crack propagation schema: first splitting step.

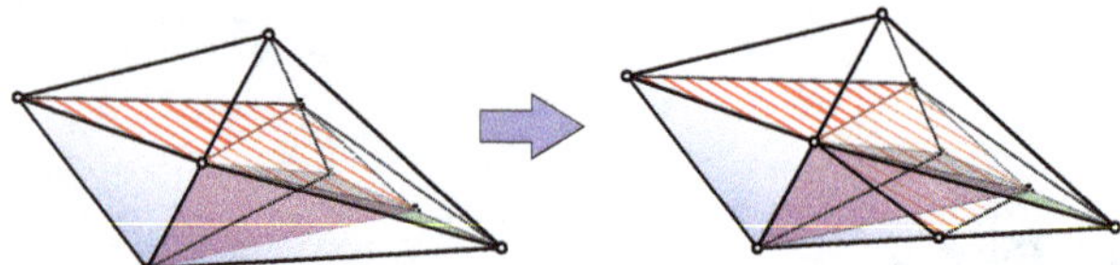

Fig. 26 3D Crack propagation schema: second splitting step.

ous. In this context it makes sense to take a closer view to the continuity properties of the introduced adaptive fragmentation schemes. The surface will be continuous for the surface refinement due to the standard finite element framework.

In contrast to this fact crack path continuity for an adaptive element splitting is not self-explanatory. Therefore, Figs. 25 and 26 show schematically C^0 crack continuity for the green adaptive element splitting scheme. The first figure clarifies the necessary refinement tasks in the nearby element (hatched green and blue surface) due to the developing fracture surface, depicted in hatched red. As the crack proceeds to the next element, Fig. 26 shows the feature that a crack can proceed through the elements changing directions.

Here the criteria for the determination of the crack direction is based on the material fracture plane normal only and thus a completely local criteria. This approach also enables a crack to start inside a solid body. Altogether these two pictures deliver an insight in the necessary refinement tasks to be performed in neighbouring elements for introducing a new fracture plane.

Another feature of the presented approach is directly related to the explicit time integration scheme and the corresponding small time step size. The elements identified as discrete fracture candidates are sorted into a list and cached at every time

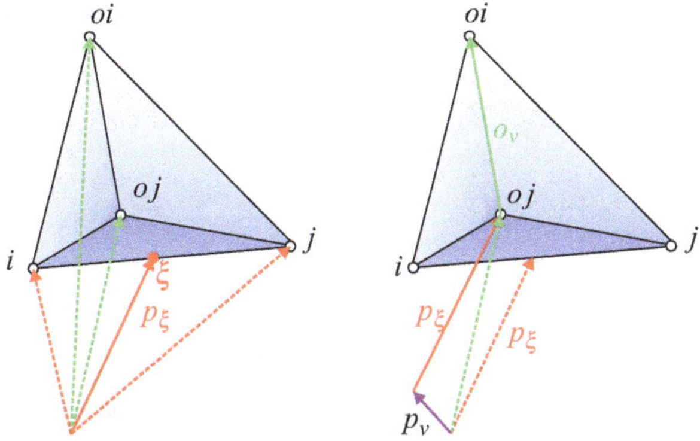

Fig. 27 Adaptive green element splitting fracture normal quality determination.

step. After having been processed, all elements are checked regarding to their discrete splitting feasibility. Consequently, additional refinement is performed in neighboring elements. As a consequence the crack propagation speed is quasi-unlimited.[8]

7.2 A Combined Strategy

Based on the features of the presented discrete crack approach, a unique determination of the adaptive element splitting technique can be designed. Therefore, the orientation of all possible fracture planes are calculated initially. Subsequently, the conformability of the possible fracture planes (resp. their normal vectors) with the previous calculated fracture normal direction is checked. The refinement with the best correlation is chosen as the fracture technique of choice. The computation of the possible fracture normal n_{ags} for the adaptive green splitting technique is stated in (46) where p_ξ denotes the position of the introduced new node at the edge defined by nodes i and j and the corresponding position vectors p_i and p_j.

$$n_{\mathrm{ags}} = p_v \times [p_i - p_j], \qquad p_v = p_{oj} - p_\xi. \tag{46}$$

Nodes on the other element sides are denoted by oi resp. oj. By cyclic permutation all nodes and all six possible fracture planes can be processed accordingly. In case of regular refined elements this point is located in the middle of the corresponding edge. Subsequently, a reformulation of the possible fracture normal vectors yields

$$n_{\mathrm{ags}} = \left[p_{oj} - \frac{1}{2}(p_{oi} + p_{oj}) \right] \times [p_i - p_j]. \tag{47}$$

---

[8] In fact the crack propagation speed is limited to the domain of fractured elements and thus only limited by the dynamic relocation of the strains and hence stresses inside the material.

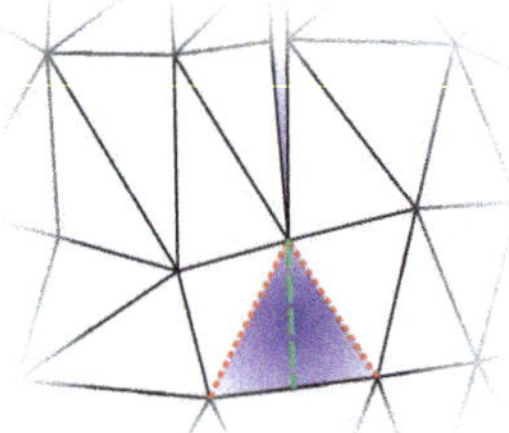

Fig. 28 Combined adaptive refinement technique: admissible crack paths (two dimensions).

So the quality q_{split} of the gathered fracture plane normal n_{ags}, considering adaptive green element splitting, can be identified as follows:

$$q_{\text{split}} = n_{\text{gs}} \cdot \bar{n}_{i+1}^{\text{crit}}. \tag{48}$$

In the framework of an advanced surface refinement strategy the determination of the fracture plane candidates is simpler than for the first refinement approach because potential fracture planes conform with the surfaces defined by the outer normal vector n_{gd}. Thus the conforming quality q_{detach} is denoted by

$$q_{\text{detach}} = n_{\text{gd}} \cdot \bar{n}_{i+1}^{\text{crit}} \tag{49}$$

for the second refinement procedure. The method providing the best approximation features (50) is chosen for the discrete splitting tasks

$$q_{\text{res}} = \min(q_{\text{split}}, q_{\text{detach}}). \tag{50}$$

Figure 28 shows the basic concept of the combined strategy where in general two splitting schemes for a failed element are possible. The element of interest can be split by adaptive green element splitting (dashed green line) or by an advanced surface refinement technique (dotted red line). Including both discrete splitting techniques at local element level, a crack direction is not fixed a priori, thus the crack propagation direction can change as the crack continues.

By combining both adaptive element splitting techniques, a qualitatively good fracture surface is obtained and ensures good correlation of the calculated and the discrete fracture plane. Figure 29 shows both applied refinement techniques. The smoothness of the generated fracture surfaces can be clearly observed.

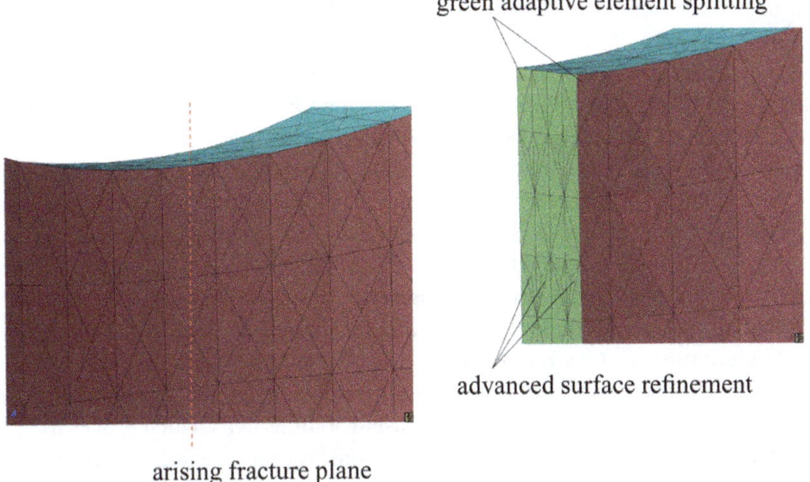

Fig. 29 Combined adaptive refinement technique: initial element configuration and cut through at the fracture surface showing refined elements.

7.3 *Extension to Explicit Time Integration Schemes*

The first restriction regards the positions of introduced new nodes in the framework of an adaptive green splitting technique. To obtain an efficient numerical model the developing fracture planes have to be limited. The limitation for adaptive procedures is based on the Courant stability criteria of the time integration scheme which relates the time step size to the element size. Thus for efficiency the constraint $l_e \geq l_{\text{limit}}$ has to be fulfilled in every state and for every element of the adaptively generated mesh. Therefore, crack positions are limited to midpoints of the edges or to the edges themselves as depicted in Fig. 30 for the two-dimensional case.

Accessory variation of the additional node position on the element edge yields a more general approach. Nevertheless, positions close to parent element nodes result in very small elements. Thus it is necessary to keep a minimum distance ε to the element nodes for an efficient numerical simulation.

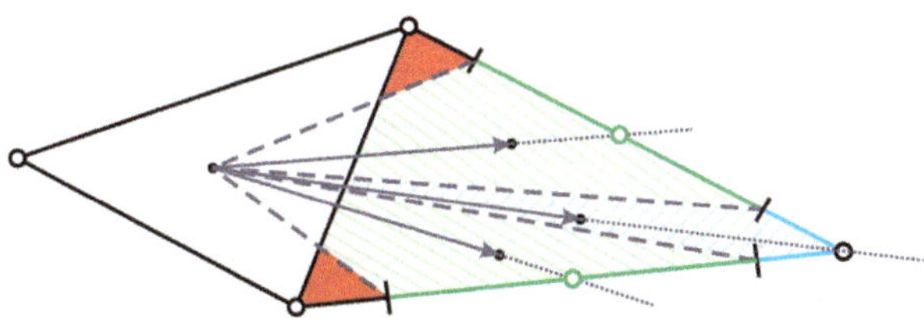

Fig. 30 Crack propagation: position limitation.

In addition damping influences the stress redistribution after adaptive element splitting. Thus the choice of a global damping factor for nodal velocity scaling, plays a considerable role for the crack propagation. On the one hand the damping factor should be high for a stable system behavior. On the other hand high damping factors influences the stresses being responsible for the damage initiation process. By numerical investigations it was clarified that the choice of high damping factors only effects the response in very short time periods. Thus for reasonable selected damping factors the influence on the complete global system behavior is negligible low. In order to minimize the influence on the global system, a local artificial stabilization of dynamic effects by ramping material constants or by introducing a nodal mass scaling can be introduced. Here nodal masses are increased rapidly at the discrete fracture time inside a region of interest. Afterwards scaled masses can be reduced by an exponential function. In the numerical example the initial mass scaling factor measures 10 and the damping time lasts 150 time steps which proves to be an adequate time duration for numerical calculations; for more details, see [64].

7.4 Incorporation of SDA Kinematics

The approach presented so far neglects the existence and the influence of developing shear bands inside the finite element as it was introduced before in the framework of a discontinuity modeling. Since the discontinuity representation inside the Strong Discontinuity Approach is smeared, the previously calculated discontinuity jump inside the finite element has to be considered. Consequently, jumps in the displacement field are summed up for both integration domains of the finite element and have to be projected to the fracture plane. The resulting projected displacement vector α_e^{pc} is added to the new fracture nodes. In the framework of the green adaptive element splitting it is added half-and-half to the new nodal coordinates. Taking these considerations into account the new positions of nodes are

$$^t X_i^{\text{new}} = {}^{t-1} X_i^p + {}^{t-1} u_i + \beta_u {}^{t-1} \alpha_e^{pc}, \qquad \beta_u \in [0,1]. \qquad (51)$$

In order not to overestimate the additional displacements, the discontinuity jump is scaled by the scalar value β_u.

Additionally, the nodal velocities are updated after a discrete element splitting by the displacement jump. This approach is adequate when dealing with relatively small problems and resulting large finite element to account for the highly dynamic processes. Thus all nodal velocities are changed as follows:

$$^t \dot{u}_i^{\text{new}} = {}^{t-1} \dot{u}_i + \frac{\beta_v}{t_{\text{loc}} \cdot \Delta t} {}^{t-1} \alpha_e^{pc}, \qquad \beta_v \in [0,1] \qquad (52)$$

where t_{loc} denotes the number of time steps related to the evolution of the localization bandwidth. Δt defines the time step size. Additionally, velocities are scaled by a scalar factor β_v motivated by the previous considerations.

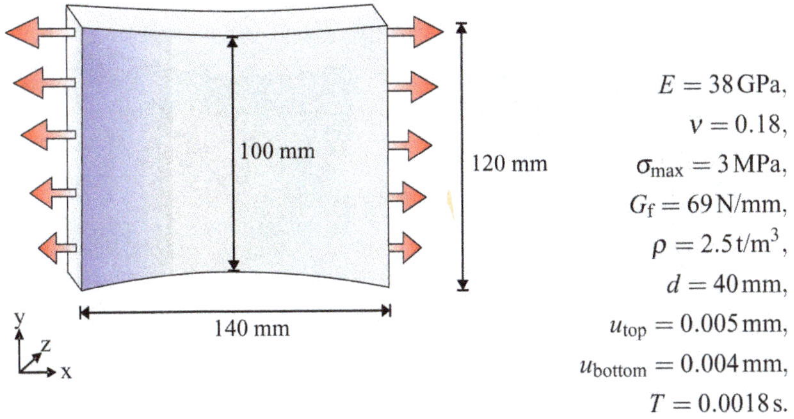

Fig. 31 Numerical model fragmentation example.

As a second possibility, the nodal positions of the fractured finite element can be relocated by an energy functional minimization. This approach is motivated by the variational ALE technique [75].

Once the nodal values are computed as described, the damage state of the finite element has to be processed. Hence, the damage value is increased to a unique global maximum value and the element representation is transferred to a standard one without an incorporation of the Strong Discontinuity Approach.

8 Numerical Example

As a first example a slightly tapered concrete bar is considered. The model geometry and loading are shown in Fig. 31. The structure is loaded in both directions with a trapezoidal distribution in order to make the crack start at the top. The problem is discretized using a relative structured mesh, serving for a crack propagation algorithm development and model evaluation. Due to the complex crack propagation algorithm and the occurring high dynamic effects this restriction to the model problem is essential. Material parameters are chosen for concrete, see also Section 4. The time step size ${}^{t}t_{s}$ varies over the simulation time T by means of the adaptive time stepping strategy: $2\%\Delta\,{}^{t}t_{s} \leq \Delta t_{\mathrm{crit}} \leq 10\%\Delta\,{}^{t}t_{s}$.

In Figs. 32(a) to 33(d) numerical results for evolution of the damage state variable and the equivalent stress distribution are presented for discrete times $t = 0.00144$ s, $t = 0.00146$ s, $t = 0.00154$ s, $t = 0.00162$ s and $t = 0.00173$ s. The depicted time steps denote the end of a crack propagation phase resulting in a redistribution of stresses caused by the propagation of energy waves inside the structure.

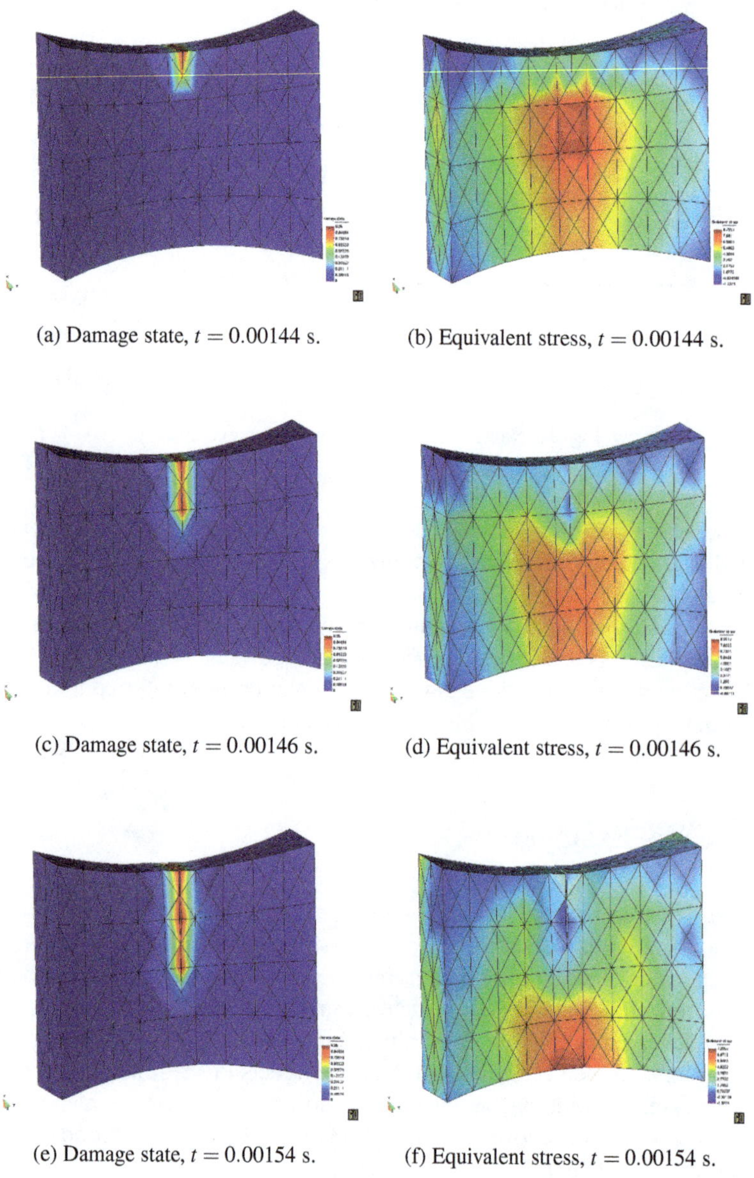

(a) Damage state, $t = 0.00144$ s.   (b) Equivalent stress, $t = 0.00144$ s.

(c) Damage state, $t = 0.00146$ s.   (d) Equivalent stress, $t = 0.00146$ s.

(e) Damage state, $t = 0.00154$ s.   (f) Equivalent stress, $t = 0.00154$ s.

Fig. 32 Numerical adaptive crack propagation model, initiation and intermediate state.

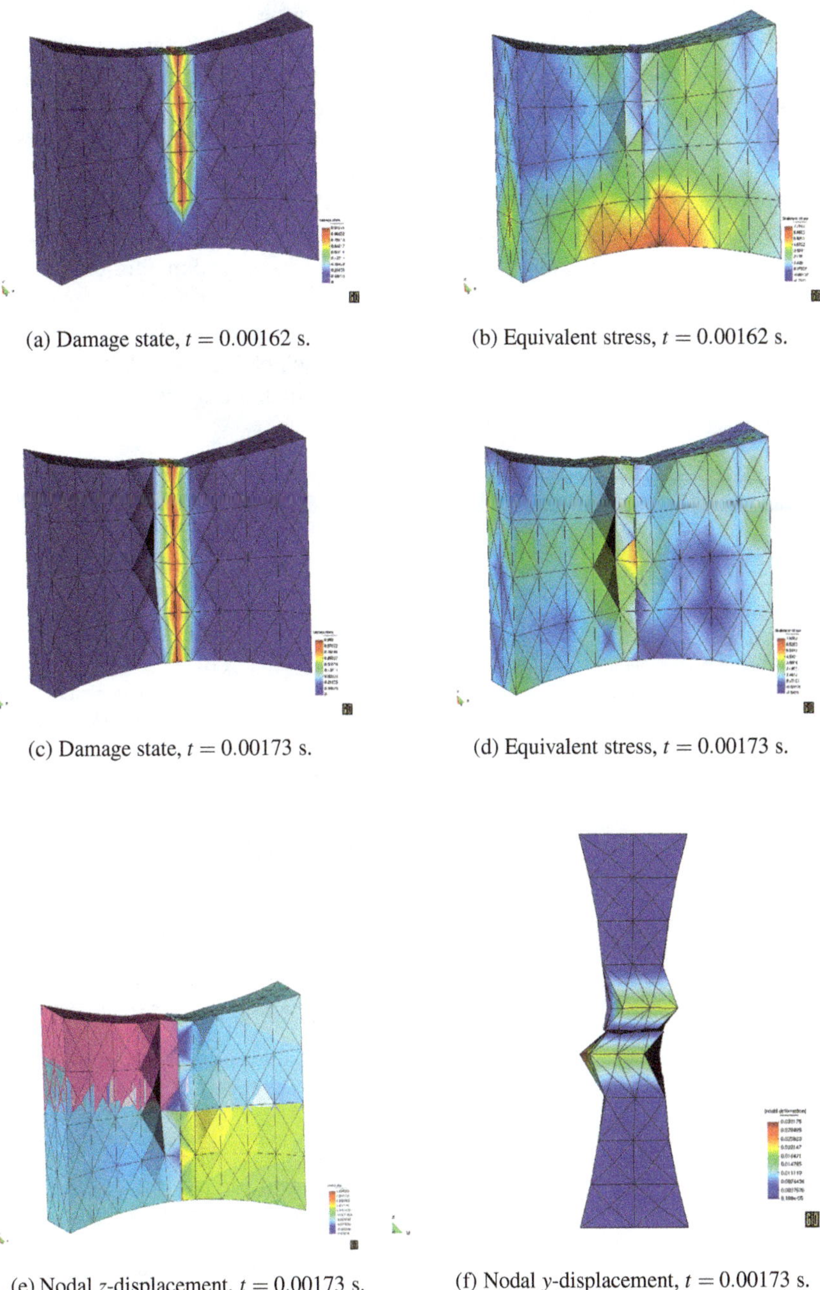

(a) Damage state, $t = 0.00162$ s.

(b) Equivalent stress, $t = 0.00162$ s.

(c) Damage state, $t = 0.00173$ s.

(d) Equivalent stress, $t = 0.00173$ s.

(e) Nodal z-displacement, $t = 0.00173$ s.

(f) Nodal y-displacement, $t = 0.00173$ s.

Fig. 33 Numerical adaptive crack propagation model, intermediate and final state.

Additionally Fig. 33(e) shows the final displacement in loading direction. The incorporation of the fracture model properties were already stated in Fig. 21 and the corresponding section. Due to the very brittle material nature of concrete only small deformations are visible in the numerical results. Furthermore, due to the fracture process very high elastic energy is released in the fragmentation process. The global damping factor is set to be as low as possible and subsequently the system behavior is highly dynamic which results in large nodal velocities and accelerations.

In the context of the Strong Discontinuity Approach numerical simulation, the absence of a distinct shear band was mentioned. In contrast an inspection of Fig. 33(f) depicts the development of a shear band within this continuous-discontinuous approach.

Finally, this model states the abilities of the developed numerical fragmentation algorithm, incorporating small restrictions to the crack path and crack development properties. Thus no global restrictions to the crack path are made a priori and the crack propagation is based solely on local criteria.

References

1. L. Angermann and P. Knabner. *Numerik partieller Differentialgleichungen: Eine anwendungsorientierte Einführung*. Springer, 2000.
2. T. Apel. An adaptive algorithm for tetrahedral meshes. Preprint, 2001.
3. P.M.A. Areias and T. Belytschko. Analysis of three-dimensional crack initiation and propagation using the extended finite element method. *International Journal for Numerical Methods in Engineering*, 63(5):760–788, 2005.
4. I. Babuska and J. Melenk. The partition of unity method. *International Journal for Numerical Methods in Engineering*, 40(4):727–758, 1997.
5. I. Babuska and Z. Zhang. The partition of unity method for the elastically supported beam. *Computer Methods in Applied Mechanics and Engineering*, 152:1–18, 1998.
6. T. Belytschko, W.K. Liu, and B. Moran *Nonlinear Finite Elements for Continua and Structures*. Wiley, Chichester, 2003.
7. T. Belytschko, Y.Y. Lu, and L. Gu. Element-free Galerkin methods. *International Journal for Numerical Methods in Engineering*, 37:229–256, 1994.
8. T. Belytschko, C. Parimi, N. Mos, N. Sukumar et al. Structured extended finite element methods for solids defined by implicit surfaces. *International Journal for Numerical Methods in Engineering*, 56:609–635, 2003.
9. D. Bielser, P. Glardon, M. Teschner, and M. Gross. A state machine for real-time cutting of tetrahedral meshes. *Graphical Models*, 66:398–417, 2004.
10. R. de Borst, M.A. Gutirrez, G.N. Wells, J.J.C. Remmers et al. Cohesive-zone models, higher-order continuum theories and reliability methods for computational failure analysis. *International Journal for Numerical Methods in Engineering*, 60:289–315, 2004.
11. R. Boussetta, T. Coupez, and L. Fourment. Adaptive remeshing based on a posteriori error estimation for forging simulation. *Computer Methods in Applied Mechanics and Engineering*, 195:6626–6645, 2006.
12. C. Daux, N. Mos, J. Dolbow, N. Sukumar et al. Arbitrary branched and intersecting cracks with the extended finite element method. *International Journal for Numerical Methods in Engineering*, 48(12):1741–1760, 2000.
13. C. Döbert. Meso- Makromechanische Modellierung von Faserverbundwerkstoffen mit Schädigung. PhD Thesis, University of Hannover, Institute of Mechanics and Computational Mechanics (IBNM), 2000.

14. R. Faria, J. Oliver, and M. Cervera. A strain-based plastic viscous-damage model for massive concrete structure. *International Journal of Solids and Structures*, 1998.
15. T.C. Gasser and G.A. Holzapfel. Modeling 3D crack propagation in unreinforced concrete using PUFEM. *Computer Methods in Applied Mechanics and Engineering*, 194(25–26):2859–2896, 2004.
16. D. Gross and T. Seelig. *Bruchmechanik mit einer Einführung in die Mikromechanik*. Springer Verlag, Berlin, 2001.
17. E. Haeirer, S. Norsett, and G. Wanner. *Solving Ordinary Differential Equations I – Nonstiff Problems*. Springer, 1987.
18. C. Hahn. Models, algorithms and software concepts for contact and fragmentation in computational solid mechanics. PhD Thesis, University of Hannover, Institute of Mechanics and Computational Mechanics (IBNM), 2005.
19. M. Hain and P. Wriggers. Numerical homogenization of hardened cement paste. *Computational Mechanics*, 42:197–212, 2008.
20. S. Idelsohn and E. Onate. To mesh or not to mesh. that is the question ... *Computer Methods in Applied Mechanics and Engineering*, 195:4681–4696, 2006.
21. M. Jirasek. Comparative study on finite elements with embedded discontinuities. *Computer Methods in Applied Mechanics and Engineering*, 188(1–3):307–330, 2000.
22. M. Jirasek and T. Belytschko. Computational resolution of strong discontinuities. In *Proceedings of 6th World Congress on Computational Mechanics*, Beijing, China, 2002.
23. M. Jirasek and T. Zimmermann. Embedded crack model. Part I: Basic formulation. *International Journal for Numerical Methods in Engineering*, 50(6):1269–1290, 2001.
24. M. Jirasek and T. Zimmermann. Embedded crack model. Part II: combination with smeared cracks. *International Journal for Numerical Methods in Engineering*, 50(6):1291–1305, 2001.
25. Y. Krongauz and T. Belytschko. EFG approximation with discontinuous derivatives. *International Journal for Numerical Methods in Engineering*, 41:1215–1233, 1998.
26. L. Krstulovic-Opara. C1 – Continuous formulation for finite deformation contact. PhD Thesis, University of Hannover, Institute of Mechanics and Computational Mechanics (IBNM), 2001.
27. C. Leppin. Ein diskontinuierliches Finite-Element-Modell für Lokalisierungsversagen in metallischen und granularen Materialien. PhD Thesis, University of Hannover, Institute of Mechanics and Computational Mechanics (IBNM), 2000.
28. D. Linero, J. Oliver, A. Huespe, and M. Pulido. Cracking modeling in reinforced concrete via strong discontinuity approach. In *Computational Modelling of Concrete Structures*, pp. 173–182, 2006.
29. S. Löhnert. Computational homogenization of microheterogeneous materials at finite strains including damage. PhD Thesis, University of Hannover, Institute of Mechanics and Computational Mechanics (IBNM), 2004.
30. S. Löhnert and T. Belytschko. A multiscale projection method for macro-/microcrack simulations. *International Journal for Numerical Methods in Engineering*, 71:1466–1482, 2007.
31. S. Löhnert and T. Belytschko. Crack shielding and amplification due to multiple microcracks interacting with a macrocrack. *International Journal of Fracture*, 145:1–8, 2007.
32. J. Lubliner, J. Oliver, S. Oller, and E. Onate. A plastic-damage model for concrete. *International Journal of Solids an Structures*, 25:299–326, 1998.
33. O.L. Manzoli and W.S. Venturini. An implicit BEM formulation to model strong discontinuities in solids. *Computational Mechanics*, online publication, http://www.springerlink.com/content/84083408 34212371/, 2006.
34. J. Mediavilla. Continuous and discontinuous modelling of ductile fracture. PhD Thesis, Technische Universiteit Eindhoven, 2005.
35. J. Melenk and I. Babuska. The partition of unity finite element method: Basic theory and applications. *Computer Methods in Applied Mechanics and Engineering*, 139:289–314, 1996.
36. T. Menouillard, J. Rthor, A. Combescure, and H. Bung. Efficient explicit time stepping for the extended finite element method (X-FEM). *International Journal for Numerical Methods in Engineering*, 68:911–939, 2006.
37. J. Mergheim. Computational modeling of strong and weak discontinuities. PhD Thesis, Technische Universitt Kaiserslautern, Maschinenbau und Verfahrenstechnik, 2005.

38. G. Meschke and P. Dumstorff. Discontinuous representation of brittle failure. In *Modelling of Cohesive-Frictional Materials 2004*, pp. 339–368, 2004.
39. G. Meschke, P. Dumstorff, W. Fleming, and S. Jox. Numerical analysis of crack propagation in concrete structures using X-FEM. In *Computational Modelling of Concrete Structures*, 2006.
40. N. Moës, M. Cloireca, P. Cartrauda, and J.F. Remacle. A computational approach to handle complex microstructure geometries. *Computer Methods in Applied Mechanics and Engineering*, 192:3163–3177, 2003.
41. J. Mosler and G. Meschke. Embedded crack vs. smeared crack models: A comparison of elementwise discontinuous crack path approaches with emphasis on mesh bias. *Computer Methods in Applied Mechanics and Engineering*, 193:3351–3375, 2004.
42. J. Mosler and M. Ortiz. On the numerical implementation of variational arbitrary Lagrangian–Eulerian (VALE) formulations. *International Journal for Numerical Methods in Engineering*, 67:1272–1289, 2006.
43. J. Mueller and J. Korvink. A genereal purpose adaptivity driver for FE software. *Softw. Pract. Exper.*, 33:1097–1116, 2003.
44. J. Oliver. Continuum modelling of strong discontinuities in solid mechanics using damage models. *Computational Mechanics*, 17:49–61, 1995.
45. J. Oliver. Modeling strong discontinuities in solid mechanics via strain softening constitutive equations. Part 1: Fundamentals. *International Journal for Numerical Methods in Engineering*, 39:3575–3600, 1996.
46. J. Oliver. Modeling strong discontinuities in solid mechanics via strain softening constitutive equations. part 2: numerical simulation. *International Journal for Numerical Methods in Engineering*, 39(21):3601–3623, 1996.
47. J. Oliver. On the discrete constitutive models induced by strong discontinuity kinematics and continuum constitutive equations. *International Journal of Solids an Structures*, 37:7207–7229, 2000.
48. J. Oliver, M. Cervera, and O. Manzoli. Strong discontinuities and continuum plasticity models: the strong discontinuity approach. *International Journal of Plasticity*, 15(3):319–351, 1999.
49. J. Oliver and A. Huespe. Continuum approach to material failure in strong discontinuity settings. *Computer Methods in Applied Mechanics and Engineering*, 193:3195–3220, 2004.
50. J. Oliver and A. Huespe. Theoretical and computational issues in modelling material failure in strong discontinuity scenarios. *Computer Methods in Applied Mechanics and Engineering*, 193:2987–3014, 2004.
51. J. Oliver, A. Huespe, M. Pulido, S. Blanco et al. New developments in computational material failure mechanics. In *Proceedings of 6th World Congress on Computational Mechanics*, Beijing, China, 2004.
52. J. Oliver, A. Huespe, M. Pulido, S. Blanco et al. Recent advances in computational modeling of material failure. In *Proceedings of ECCOMAS 2004*, 2004.
53. J. Oliver, A. Huespe, M. Pulido, and E. Samaniego. On the strong discontinuity approach in finite deformation settings. *International Journal for Numerical Methods in Engineering*, 56:1051–1082, 2003.
54. J. Oliver, A. Huespe, and E. Samaniego. A study on finite elements for capturing strong discontinuities. *International Journal for Numerical Methods in Engineering*, 56:2135–2161, 2003.
55. J. Oliver, A. Huespe, E. Samaniego, and E. Chaves. On strategies for tracking strong discontinuities in computational failure mechanics. In *Proceedings of 5th World Congress on Computational Mechanics*, Vienna, Austria, 2002.
56. J. Oliver, A.E. Huespe, E. Samaniego, and E.W.V. Chaves. Continuum approach to the numerical simulation of material failure in concrete. *International Journal for Numerical and Analytical Methods in Geomechanics*, 28(7–8):609–632, 2004.
57. M. Ortiz and A. Pandolfi. Finite-deformation irreversible cohesive elements for three-dimensional crack-propagation analysis. *International Journal for Numerical Methods in Engineering*, 44:1267–1282, 1999.

58. D. Owen, Y. Feng, E.S. Neto, M. Cottrell et al. The modelling of multi-fracturing solids and particulate media. *International Journal for Numerical Methods in Engineering*, 60(1):317–339, 2004.

59. A. Pandolfi and M. Ortiz. An efficient adaptive procedure for three-dimensional fragmentation simulations. *Engineering with Computer*, 18:148–159, 2002.

60. F.D. Pin, S. Idelsohn, E. Onate, and R. Aubry. The ALE/Lagrangian particle finite element method: A new approach to computation of free-surface flows and fluid-object interactions. *Computers and Fluids*, 36:27–38, 2007.

61. T. Rabczuk, P.M.A. Areias, and T. Belytschko. A simplified mesh-free method for shear bands with cohesive surfaces. *International Journal for Numerical Methods in Engineering*, 69:993–1021, 2006.

62. T. Rabczuk, S.P. Xiao, and M. Sauer. Coupling of mesh-free methods with finite elements: Basic concepts and test results. *Communications in Numerical Methods in Engineering*, 22:1031–1065, 2006.

63. J.N. Reddy. *An Introduction to Nonlinear Finite Element Analysis*. Oxford University Press, 2004.

64. S. Reese. Adaptive methods for continuous and discontinuous damage modeling in fracturing solids. PhD Thesis, University of Hannover, Institute of Mechanics and Computational Mechanics (IBNM), 2007.

65. M.C. Rivara. Algorithms for refining triangular grids suitable for adaptive and multigrid techniques. *International Journal for Numerical Methods in Engineering*, 20:745–756, 1984.

66. B. Schrefler, S. Secchi, and L. Simoni. Adaptive refinement techniques for cohesive fracture in multifield problems. In N.E. Wiberg and P. Diez (Eds.), *Proceedings ADMOS 2003*. CIMNE, Barcelona, 2003.

67. J. Simo and J. Ju. Strain- and stress-based continuum damage models – 1. Formulation. *International Journal of Solids and Structures*, 23(7):821–840, 1986.

68. J. Simo and J. Ju. Strain- and stress-based continuum damage models – 2. Computational aspects. *International Journal of Solids and Structures*, 23(7):841–869, 1986.

69. J. Simo and J. Oliver. A new approach to the analysis and simulation of strain softening in solids. In *Proceedings of Conference on Fracture and Damage in Quasibrittle Structures*, pp. 25–39, 1995.

70. J.C. Simo and T.J.R. Hughes. *Computational Inelasticity*. Interdisciplinary Applied Mathematics. Springer, 2000.

71. J.C. Simo, J. Oliver, and F. Armero. An analysis of strong discontinuities induced by strain-softening in rate-independent inelastic solids. *Computational Mechanics*, 12:277–296, 1993.

72. A. Simone, J. Remmers, and G. Wells. An interface element based on the partition of unity, 2001.

73. N. Sukumar, D.L. Chopp, N. Mos, and T. Belytschko. Modeling holes and inclusions by level sets in the extended finite-element method. *Computer Methods in Applied Mechanics and Engineering*, 190:6183–6200, 2001.

74. N. Sukumar, N. Mos, B. Moran, and T. Belytschko. Extended finite element method for three-dimensional crack modelling. *International Journal for Numerical Methods in Engineering*, 48(11):1549 – 1570, 2001.

75. P. Thoutireddy and M. Ortiz. A variational r-adaption and shape-optimization method for finite-deformation elasticity. *International Journal for Numerical Methods in Engineering*, 61:1–21, 2004.

76. P. Wriggers. *Computational Contact Mechanics*, 2nd edn. Springer, Berlin, 2006.

77. P. Wriggers. *Nonlinear Finite Element Methods*. Springer, Berlin, 2008.

78. X.P. Xu and A. Needleman. Numerical simulations of fast crack growth in brittle solids. *Journal of the Mechanics and Physics of Solids*, 42:1397–1434, 1994.

79. C. Zhao, B.E. Hobbs, A. Ord, P. Hornby et al. Particle simulation of spontaneous crack generation problems in large-scale quasi-static systems. *International Journal for Numerical Methods in Engineering*, 69:2302–2329, 2006.

80. G. Zi and T. Belytschko. New crack-tip elements for XFEM and applications to cohesive cracks. *International Journal for Numerical Methods in Engineering*, 57(15):2221–2240, 2003.
81. O. Zienkiewicz and J. Zhu. A simple error estimator and adaptive procedure for practical engineering analysis. *International Journal for Numerical Methods in Engineering*, 1987.
82. O.C. Zienkiewicz and R.L. Taylor. *The Finite Element Method*, volume 1: The Basics. Butterworth-Heinemann, Oxford, 2000.
83. O.C. Zienkiewicz and R.L. Taylor. *The Finite Element Method*, volume 2: Solid mechanics. Butterworth-Heinemann, Oxford, 2003.
84. T.I. Zohdi and P. Wriggers. *Introduction to Computational Micromechanics*, Lecture Notes in Applied and Computational Mechanics, Vol. 20, Springer, 2005.

New Strategies in Finite Element Analysis of Material Processing

B.-A. Behrens, A. Bouguecha, K.B. Sidhu, T. Matthias and I. Peshekhodov

Abstract Under the supervision of Professor Bernd-Arno Behrens, the head of the Institute of Metal Forming and Metal-Forming Machines (IFUM), three PhD projects were carried out during the span of the Research Training Group 615 (GRK 615). All the projects originated in the Department of Numerical Methods, which is headed by Dr. Anas Bouguecha, and were related to finite element modeling and simulation of materials processing. Thus, Kanwar Bir Sidhu developed and implemented remeshing and rezoning algorithms to cope with severe plastic deformation and crack propagation in thin sheet blanking. The subject of Ilya Peshekhodov was a multiscale approach in modeling of forming processes of metal sheets coated with thin polymer composite coatings. With the help of the fluid mechanics approach, Thorsten Matthias modeled bulk forming processes of aluminum alloys in a semi-solid state. These projects are presented below along with some exemplary results.

1 Finite Element Analysis of Blanking of Thin Metal Sheets

Blanking is the most widely used sheet metal cutting process in a broad range of mass production industries. Almost every sheet component that leaves the assembly line – either as a pre-formed piece or a finished part – undergoes blanking. During this process, the sheet is clamped between the blankholder and the die, and a predefined blank shape is punched out of the sheet with the cutting stamp (Fig. 1). The quality of the cut-edge profile depends on various factors such as sheet metal properties, blank shape and cutting tool geometry. Furthermore, the cut-edge profile is highly dependent on the clearance between the cutting tool and die. The radius of an unworn cutting stamp varies in a range between 10 to 20 μm [19]; the clear-

B.-A. Behrens · A. Bouguecha · K.B. Sidhu · T. Matthias · I. Peshekhodov
Institute of Metal Forming and Metal-Forming Machines (IFUM), Leibniz Universität Hannover,
An der Universität 2, 30823 Garbsen, Germany; e-mail: behrens@ifum.uni-hannover.de

E. Stephan & P. Wriggers (Eds.): Modelling, Simulation and Software Concepts, LNACM 57, pp. 117–131.

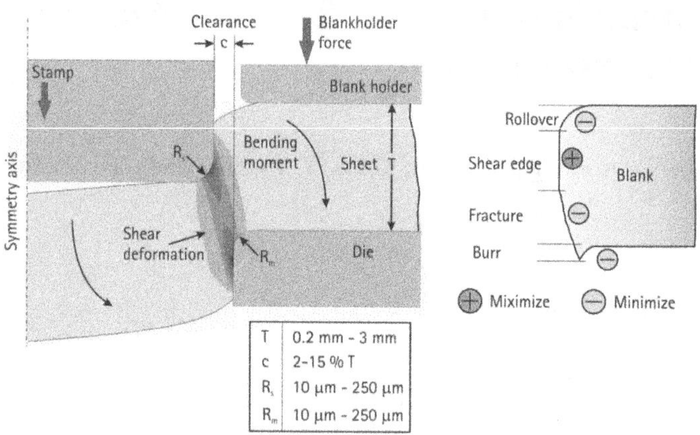

Fig. 1 Blanking and the resulting cut-edge profile.

ance may vary from 0.02 to 0.1 mm for a 1-mm thick sheet. During blanking, the sheet material is therefore subjected to large and highly localized shear deformation leading to ductile crack formation and ultimately to fracture.

The main objective of this project was to develop a finite element model to simulate blanking considering actual material and process parameters and to predict the cut-edge profile.

In the past two decades, there has been an exponential growth in the analysis of metal forming process using the Lagrangian finite element approach. The advantage of this approach is that it facilitates investigation on materials with history-dependent constitutive relations; its weakness is its inability to simulate large plastic deformation without remeshing. In order to avoid premature termination and to minimize the amount of computing time of a blanking simulation, it is essential to develop a remeshing technique. Remeshing is normally followed by mapping of the history variables from previous deformed mesh to the new mesh, which is also known as rezoning. In this work, an inverse distance method with an adaptive patch was developed in order to accurately rezone the history variables after a remeshing step.

The automatic remeshing technique based on an advancing front and quadtree methods, which are frequently used to cope with large plastic deformations, are not capable of generating an adaptive quadrilateral mesh as required for the simulation of blanking. In the present work, a remeshing algorithm based on [18] and [5] (Fig. 2) was developed and implemented into the MSC.Marc software.

During rezoning, the values of the nodal variables – such as temperature – are transferred from nodes of the previous mesh to the new mesh, and the values of the state variables at Gauss points, such as stress, strain and energy, are transferred from Gauss points of the previous mesh to Gauss points of the new mesh. In the common inverse distance method, the value of a history variable at a Gauss point, P, in a new mesh is calculated as a summation of the ratio of a history variable values at the Gauss points of the previous mesh located in a square patch around the new Gauss

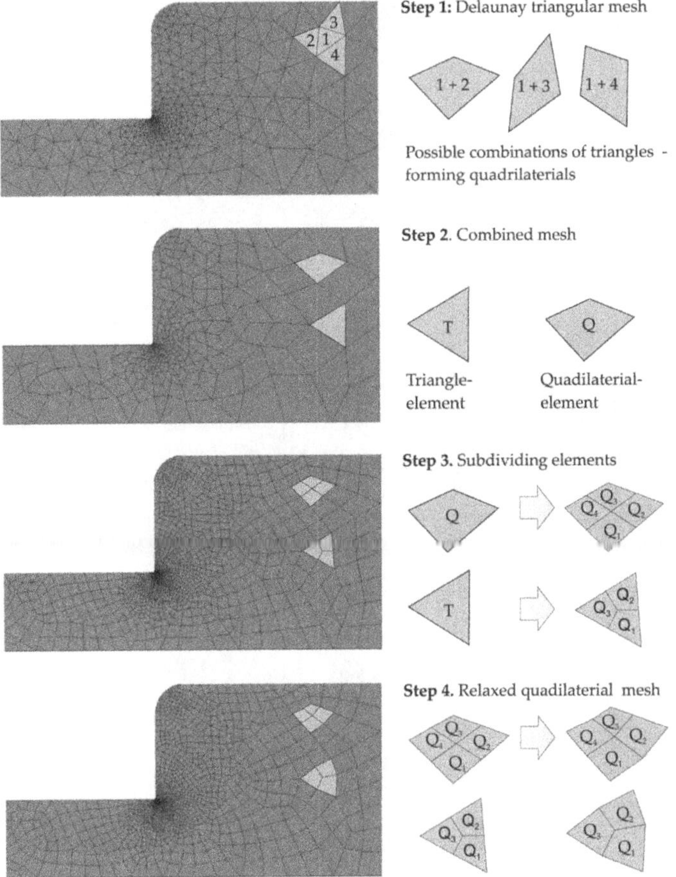

Fig. 2 Generation of the adaptive mesh [5, 18].

point divided by the square of distance, R_i, between the Gauss point P and the old Gauss point in this patch. In order to increase the state variables transfer speed, the size of the square patch was adapted in this project to the mesh density gradient. The size L of the square was thus not a constant value but determined by its position in the mesh as shown in Fig. 3.

The remeshing and rezoning codes were coupled with MSC.Marc. For that, a remeshing and rezoning code – QuadMesher – was developed. The coupling principle of the QuadMesher with MSC.Marc using FORTRAN subroutines is shown in Fig. 4.

The presented remeshing and rezoning algorithms help to deal with the severe plastic deformation. In order to incorporate the material separation process in the FE model, four different mesh separation techniques – element deletion, nodal release, modified nodal release, geometrical crack outline – were tested. In the element deletion approach, an element is removed from the mesh as soon as a predefined

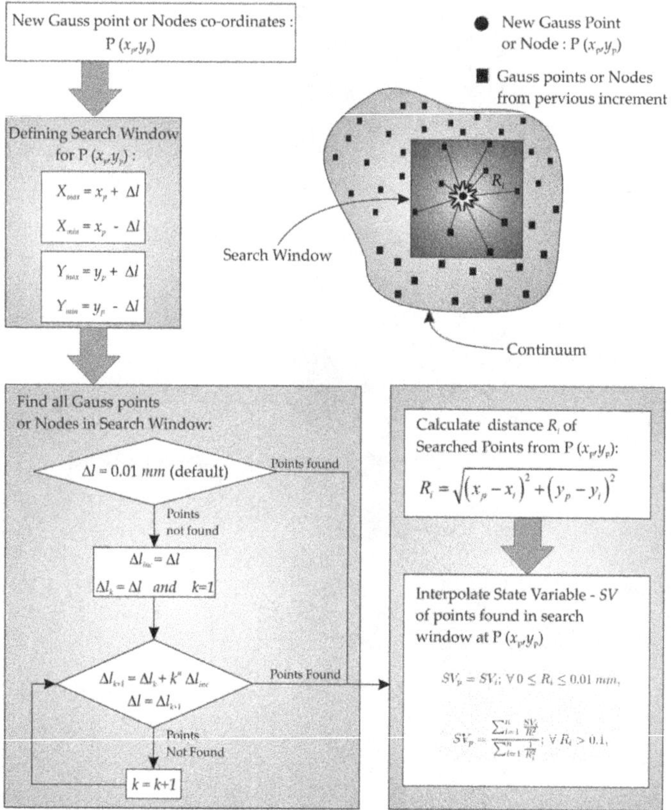

Fig. 3 Inverse distance method with an adaptive square patch.

critical fracture criterion value is achieved in it. The nodal release approach is a refined form of the element deletion, in which the crack can propagate along the element edges. Despite the simplicity of both these methods, the results are very mesh dependent, which consequently leads to unrealistic cut-edge profiles. In order to improve the results, a modification of the nodal release method was done by improving the local mesh around the crack tip during crack propagation. This mesh modification proved to lead to smooth fracture. In the final method – the geometrical crack outline – the crack propagation is realized by incrementally generating the new crack outline in the sheet followed by remeshing, the advantage being mesh independent. The results are presented in Fig. 5.

To simulate a blanking process, the isotropic constitutive model deep drawing sheet (DC06) of 1 mm thickness based on von Mises elasto-plastic model was taken. The flow curve of the sheet was derived from quasi-static uniaxial tensile tests. The blanking tools are modeled as rigid with quasi static stamp velocity of 0.1 mm/s; consequently, the thermo-mechanical effects are ignored which are normally observed due to relatively high stamp velocities. The fracture criterion implemented in

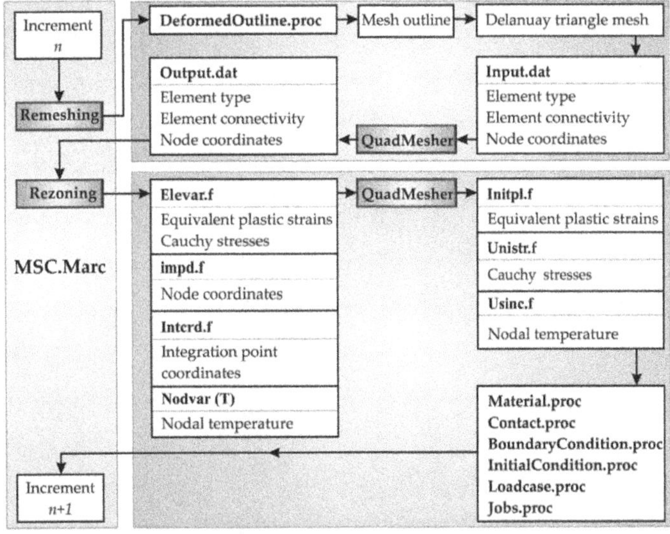

Fig. 4 Coupling of remeshing and rezoning code QuadMesher with MSC.Marc.

Table 1 Comparison of the cut-edge profile between simulation with the modified nodal release approach and experiment for DC06.

	Experiment	Simulation
Rollover	0.115 mm	0.127 mm
Shear edge	0.284 mm	0.299 mm

this research is based on integral of stress function provided by Oyane [15]:

$$\int_{\varepsilon_0}^{\varepsilon_n} \left(1 + 3.9 \frac{\sigma_H}{\bar{\sigma}}\right) d\varepsilon_p > C, \tag{1}$$

where $\sigma_H/\bar{\sigma}$ is the triaxiality, which is the hydrostatic pressure over the von Mises stress, C is a material and experiment set-up dependent constant. In this work $C = 7.7$. When the value of the integral gets larger then the threshold value then the crack is extended further. The rollover shape and sheared edge for the chosen sheet material can be predicted within the experimentally values for 10% clearances, however the burr length cannot be predicted correctly (Table 1).

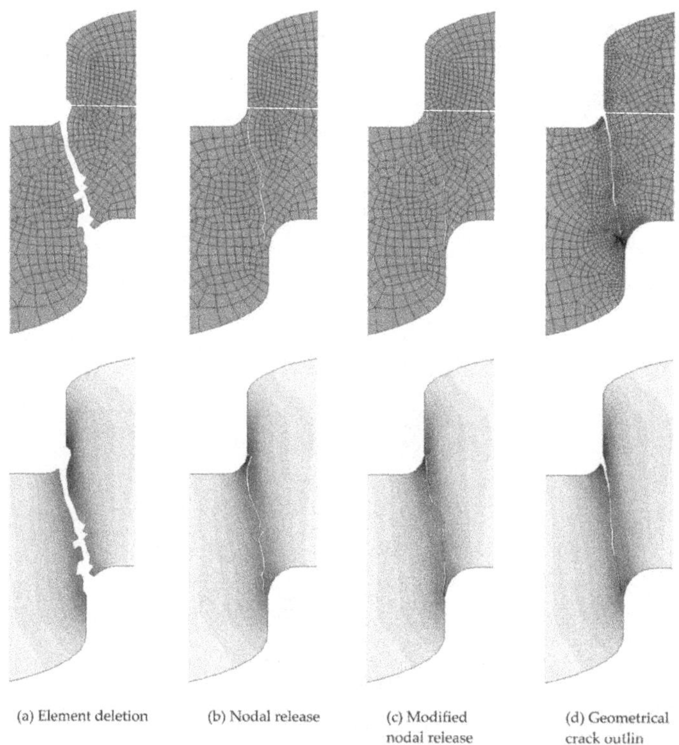

Fig. 5 Numerical analysis of the cut-edge profile using different discrete crack propagation method.

2 Finite Element Analysis of Forming of Polymer Coated Sheet Metal

The modern car body represents a multi-layer structure (Fig. 6). The first two organic layers – primer and intercoat – are applied onto the sheet via continuous coil coating before it is formed into a part [13]. This two-layer coating has to withstand sheet forming without being damaged. Damage in form of pores and cracks leads to reduction in corrosion resistance and aesthetics deterioration and is not acceptable.

There have been a number of researches dealing with an experimental investigation of damage in polymer coatings on metal substrates after sheet forming operations [6, 20]. Although these works provided a good understanding of the coating performance during forming, they all encompassed an extensive experiment program and yielded results for certain coatings and particular forming operations only; no relation between the coating microstructure or its mechanical properties and damage was established. Understanding this relation becomes however especially important when a multi-phase nature of most industrially produced coatings is considered. In fact, various pigments and fillers are usually added to the coat-

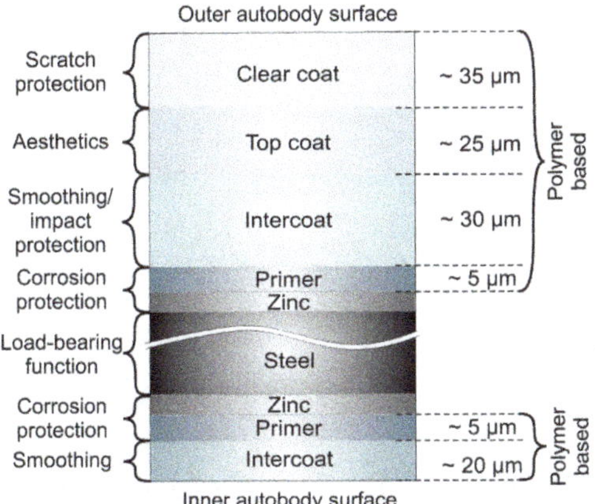

Fig. 6 Multi-layer structure of the steel car body.

ing resin to achieve the required corrosion protection, aesthetics, and mechanical properties [17].

To avoid a time-consuming experimental investigation of the coating condition after a forming process, finite element modeling can be employed to estimate the stress-strain state in the coating [1, 14] and subsequently relate it to damage [4]. However, small thickness of the polymer layers compared to those of the steel substrate leads to substantial problems in meshing and high computational costs when simulating industrial forming processes of coated sheets.

In the present work, a numerical study of coating damage is carried out on two scales. On the microscale, a representative volume of the primer and intercoat are investigated under the displacement constraints of the sheet and contact pressure exerted from the tool. On the macroscale, it is assumed that the coating solely defines the friction coefficient between the coated sheet and forming tools with no further influence on the sheet deformation. Hence, the polymer coating is not modeled on this scale and simulation of the forming process with the uncoated metal is carried out. Damage of the coating on the macroscale is estimated by means of a function which depends on sheet deformations in the sheet plane and tool contact pressure. This two-scale approach not only saves the computational costs of sheet forming simulations but also allows to investigate the influence of the coating microstructure on coating damage development during forming.

One-component polyurethanes (PUR) are widely used for primers and intercoats in car body construction and were therefore chosen for the present project. The coil coating system investigated consists of a PUR primer with a thickness of several micrometers and a PUR intercoat, which is approx. 20 μm thick. The pigments and fillers were determined with the help of the scanning electron microscopy (SEM) and energy dispersive X-ray spectroscopy (EDX) and are summarized in Table 2.

To study damage mechanisms in these multi-phase coatings induced by sheet forming, various forming processes were carried out. The coating condition was then analyzed with the SEM and EDX. The results of the analysis showed that dam-

Table 2 Pigments and fillers of the intercoat and primer studied.

Pigment	Approx. Shape	Mean Size	Main Function	Location
$BaSO_4$	Spheres	1 μm	Costs Reduction	Intercoat
C	Spheres	0.02 μm	Surface Covering	Intercoat
SiO_2	Spheres	0.1 μm	Rheology Control	Primer
TiO_2	Spheres	0.3 μm	Surface Covering	Primer
Zn	Flakes	1 μm	Corrosion Protection	Primer

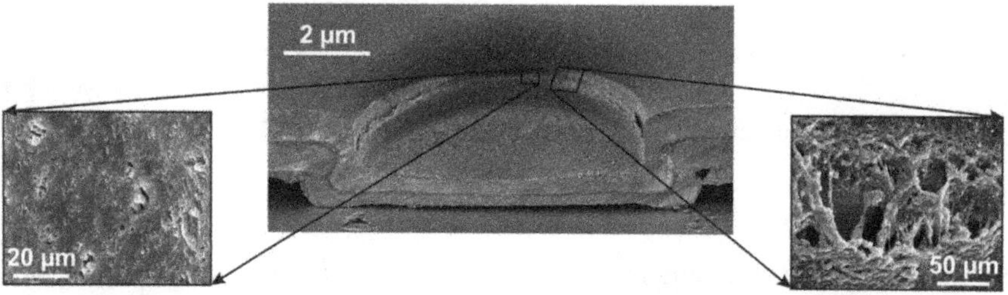

Fig. 7 Clinched DX56+Z275 sheets coated with PUR primer and PUR intercoat.

age starts at the interface between large $BaSO_4$ particles and PUR matrix and progresses with voids growth and their coalescence (Fig. 7).

The numerical approach requires in the first place a material model and corresponding material parameters that would accurately describe the coating behavior at large strains. In case of coatings, these parameters are unavailable and must be determined experimentally. One of widely applied methods to extract the mechanical properties of a thin film is nanoindentation, in which a finely shaped indentor usually made of diamond is driven into a sample to a depth of several micrometers at maximum. The load necessary to indent the sample as well as the indentation depth are continuously recorded, which gives a unique load-displacement curve. The research on nanoindentation of polyurethane-like materials is rather scarce. Oyen et al. [16] presented viscous-elastic-plastic models for time-dependent indentation behavior for several polymers including polyurethanes. Their model is developed in terms of the experimentally-observable extensive variables for the material – load, displacement, and time – rather than in terms of the intensive variables of stress, strain, and time. Until now, there is generally no analytical solution at hand to determine intrinsic mechanical properties of a polyurethane-like material from nanoindentation experiments.

In the present work, instrumented indentation was carried out with a Hysitron TriboIndenter^{TM} to determine the mechanical properties of the coating. The coating showed viscous-elastic-plastic behavior with plastic effects being attributed to damage accumulation. Relatively high surface roughness ($R_a = 0.1$ μm) scattered the results hindering an accurate material parameters estimation. The results of the experiment are presented in Fig. 7 with the approximated estimate of the Young's modulus of the coating system.

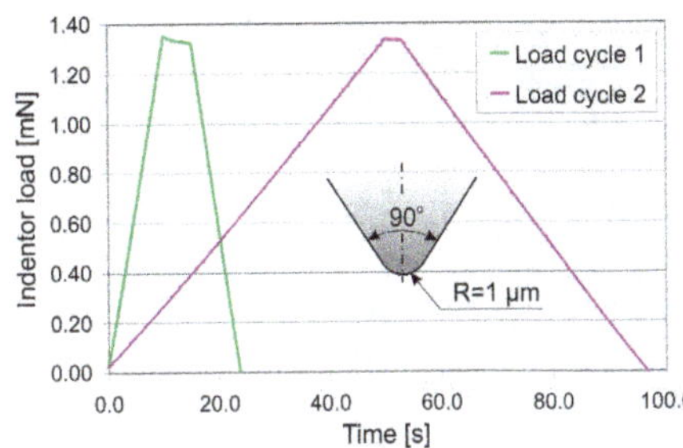

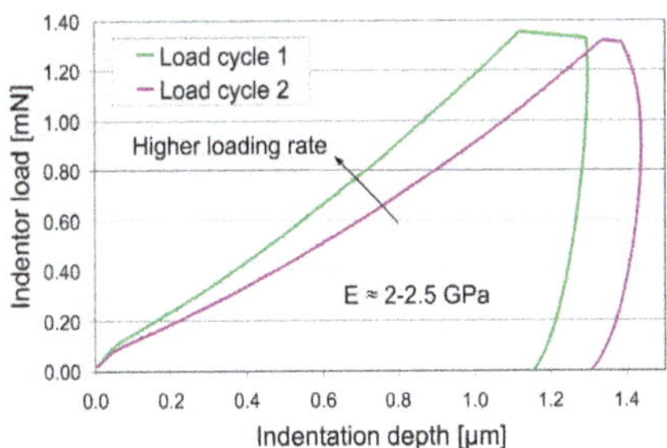

Fig. 8 Indentation of the coating consisting of the 5 μm primer and 20 μm intercoat.

Table 3 Mechanical properties of the main intercoat constituents.

Property	PUR Matrix	BaSO$_4$
Young's modulus, E [MPa]	250	60000
Poisson ratio, v [–]	0.43	0.32

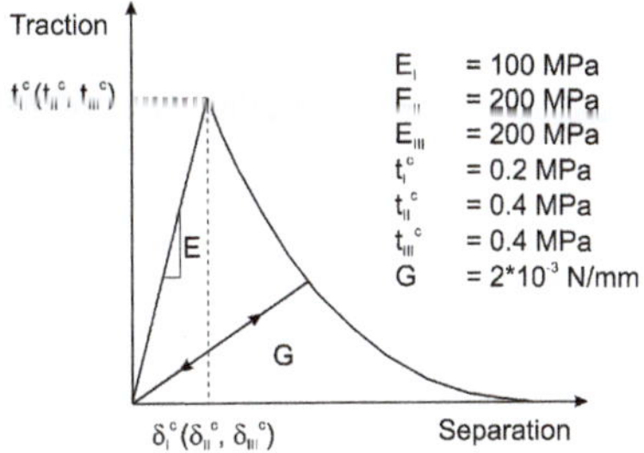

Fig. 9 Constitutive law for the interface between the PUR matrix and BaSO$_4$ particles.

On the microscale, damage was assumed to take place only at the interface between large BaSO$_4$ particles and PUR matrix with the smaller pigments contributing to polymer strengthening with no delamination from the matrix. The behavior of BaSO$_4$ was assumed to be linearly elastic; the Neo-Hookean formulation of the strain energy potential without viscous effects was chosen for the PUR matrix. The mechanical properties of the intercoat constituents are summarised in Table 3. The particle-matrix delamination between the phases of the intercoat was modeled with the help of an interfacial element with the constitutive behavior shown in Fig. 9.

In Fig. 10, scalar stiffness degradation along the single particle-matrix interface under applied displacement is presented. In the present work, overall coating damage D_{coating} is considered as a stiffness loss of BaSO$_4$-PUR interface averaged over the entire interface between the two phases.

In the future work, the mechanical properties of the primer and intercoat with and without BaSO$_4$ will be determined on free-standing films. An approach is to be elaborated for an estimation of the particle-matrix interface parameters. Coating damage will be numerically investigated on the microscale to determine the integrative coating damage variable as a function of principal sheet stretches and

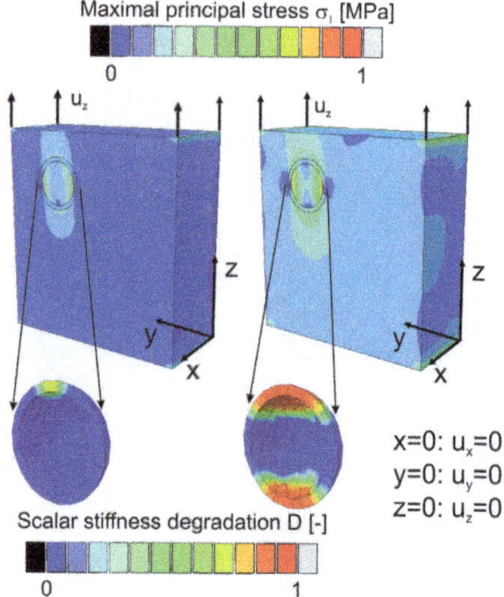

Fig. 10 Exemplary results of the interface degradation.

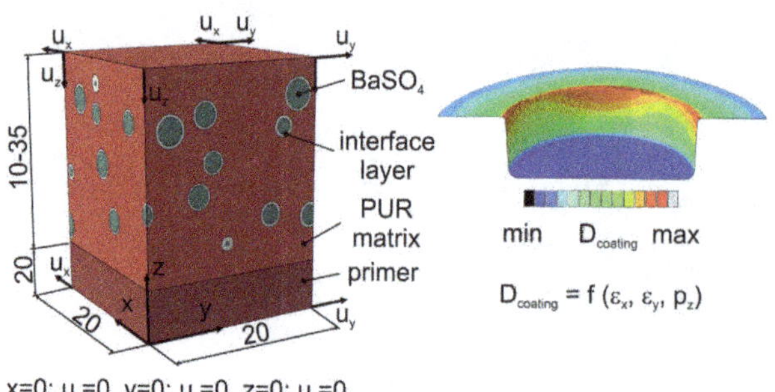

Fig. 11 Micromodel of the coating (left) and macromodel of the uncoated sheet (right).

tool contact pressure. Thus determined coating damage variable will be then implemented into a FE model of a deformation process with an uncoated sheet on the macroscale (Fig. 11).

3 Finite Element Analysis of Forming of Aluminium in a Semi-Solid State

Thixoforming is an innovative forming process under thixotropic conditions that combines the advantages of forging with those of casting. The forming of the material takes place in a semi-solid state, so that the temperature of the workpiece lies between the liquidus and the solidus lines. The process makes it possible to produce components with geometrical higher complexity. The complexity achieved is

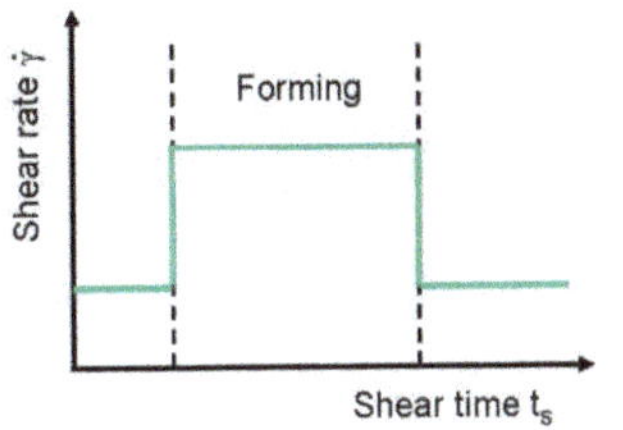
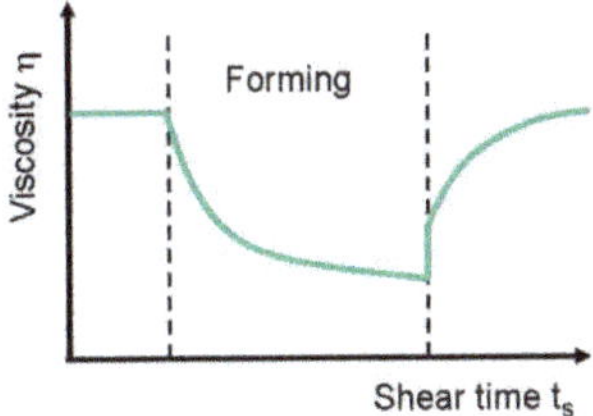

Fig. 12 Thixotropic phenomenon [8].

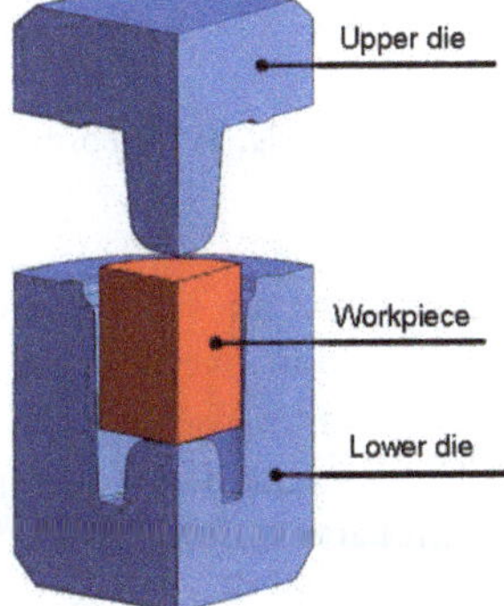

Fig. 13 CAD model of the process.

thereby higher than by means of forging and the mechanical properties of the produced part are better than after casting. If thixotropic material undergoes a shear load, it acts as thixotrope. Thixotropy describes a non-Newtonian fluid with time-dependent flow characteristics; its viscosity decreases under a shearing load during a shaping process, and after exposure of strain, its original viscosity can be restored (Fig. 12) [3, 7, 9].

The mathematical approach to the describtion of thixotropic flow characteristics of semi-solid aluminium depends on the solid fraction. At this point there are two fundamentally different approaches to modelling of the material's characteristics. For a solid fraction of 40–60%, an approach from fluid mechanics should be chosen. For higher solid fractions an approach from structure mechanics is preferred [10–12].

As Simulations-Software, `Ansys Polyflow` is to be used. The software is based on the finite element method and is well suited for modelling of non-Newtonian flow behaviour [21]. The model that is shown in Fig. 2 presents the thixoforging process of a cup and consists of the tools (upper and lower die) and workpiece. The upper die travels a distance of 100 mm at the speed of 50 mm/s. Thereby is the duration of the forming process 2 s long.

The tools are made of of the steel X37CrMoV5. This material is especially suited for high thermal and mechanical stess. Upper and lower die have a basic temperature of 500°C. The tool is modelled rigid. The workpiece is assigned the thixotropic material characteristics of the aluminium AlSi7Mg and has an initial temperature

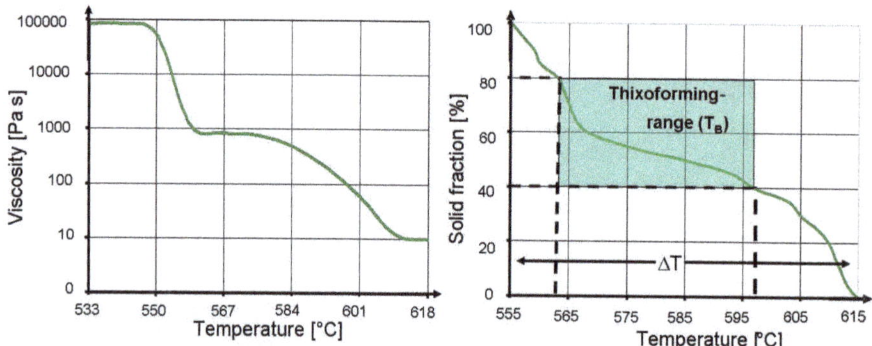

Fig. 14 Viscosity (left) and solid fraction (right) versus temperature.

of 585°C. As a result the material is in a semi-solid state. During the forming simulation, on all five numeric steps, an adaptive remeshing is carried out on those locations of the workpiece, where contact with the tool occurs.

The material of the workpiece used here is aluminium AlSi7Mg. The mechanical properties of this Al-alloy introduced in this section were taken from different literature [10–12]. In Fig. 14, the dependency of the viscosity on the temperature is shown. Starting at 550°C, the phase transformation from a solid to a semi-solid state takes place which leads to decrease of the viscosity. The solidus temperature lies at 555°C and the liquidus temperature at 615°C. Within this range (ΔT), the considered alloy posseses a solid as well as fluid phase. In the simulation, the workpiece has an initial temperature of 585°C, which corresponds to a solid content of approxiamately 50% (Fig. 14).

There are different approaches for describing the viscosity η of non-Newtonian fluids, e.g. Ostwald de Waele, Herschel–Bulkley and Carreau Yasuda [2, 3]. Figure 15 shows the implemented viscosity curves for the alloy studied. In this work, the Herschel–Bulkley law

$$F(\dot{\gamma}) = \begin{cases} \frac{\tau_0}{\dot{\gamma}} + k\left(\frac{\dot{\gamma}}{\dot{\gamma}_c}\right)^{n-1} & \dot{\gamma} > \dot{\gamma}_c \\ \frac{\tau_0\left(2 - \frac{\dot{\gamma}}{\dot{\gamma}_c}\right)}{\dot{\gamma}_c} + k\left[(2-n) + (n-1)\frac{\dot{\gamma}}{\dot{\gamma}_c}\right] & \dot{\gamma} \le \dot{\gamma}_c \end{cases} \tag{2}$$

is used, where τ_0 is the yield stress, $\dot{\gamma}_c$ is the critical shear rate, k is the consistency factor, and n is the power-law index. Furthermore, the approach has been expanded with the equation of William–Landel–Ferry (WLF)

$$\ln(H(T)) = \left(\frac{c_1(T_r - T_a)}{c_2 + T_r - T_a}\right) - \left(\frac{c_1(T - T_a)}{c_2 + T - T_a}\right) \tag{3}$$

so that the dependency of the viscosity on the temperature can be considered; where c_1 and c_2 are the WLF constants, T_r and T_a are reference temperatures. The entire

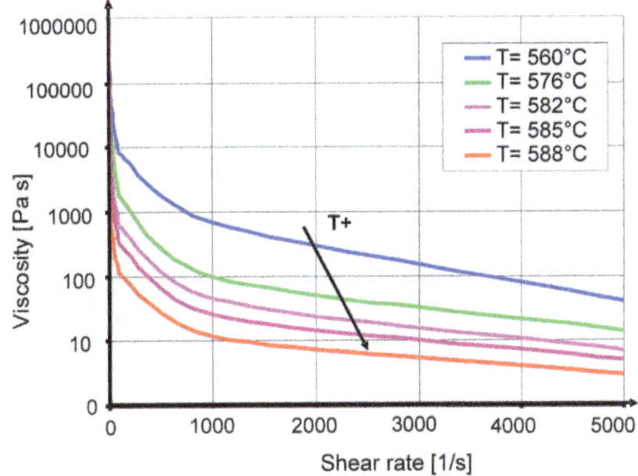

Fig. 15 Dependency of viscosity on the temperature and on the shear rate.

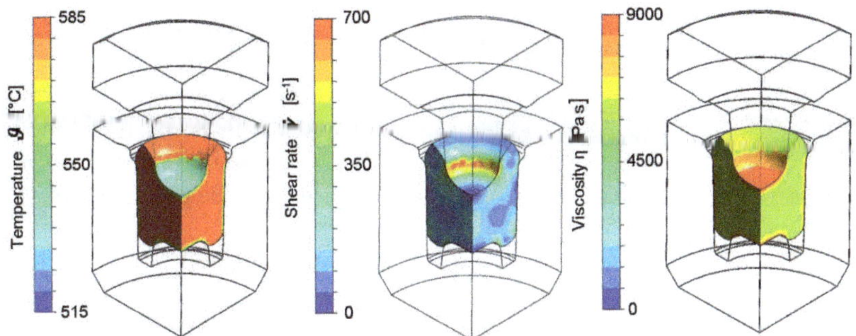

Fig. 16 Computational results after 50% of the forming.

formula to approximate the viscosity curves to describe the material behaviour of AlSi7Mg yields

$$\eta\left(\dot{\gamma}, T\right) = F\left(\dot{\gamma}\right) \cdot H\left(T\right). \tag{4}$$

During forming, the workpiece cools down when getting into contact with the surface of the tool; a solid phase forms. The shear rate describes the velocity of the fluid and is affected by neighbouring liquid layers and their distance. The layers are results from the different temperature fields. This phenomenon is especially pronounced on the edge of the workpiece where, as a result of the contact with the tool, a significantly strong cooling down takes place. It is to be observed that within the centre there is lower viscosity than on the edge. The viscosity distribution in the workpiece shows that the dependency of the viscosity on temperature and shear rate is successfully calculated (Fig. 16).

The solid fraction f_s of melt is directly linked to the actual temperature and can be calculated with the Scheil equation:

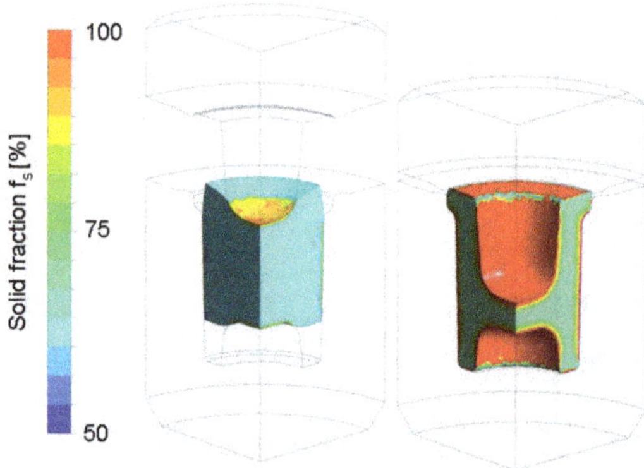

Fig. 17 Solid fraction distribution after 50 and 100% of the forming.

$$f_s = 1 - f_l = 1 - \left(\frac{T_M - T_L}{T_M - T}\right)^{1/(1-k)}, \tag{5}$$

where f_s and f_l are the solid and liquid fractions of the melt, T_M is the melting point of the pure metal (here pure aluminum), T_L is the liquidus temperature of the investigated alloy, T is the actual metal temperature, and k is the partition coefficient of the alloy.

The formula for the calculation of the partition coefficient k is

$$k = 1 - \left(\frac{m_l - L_m}{R - T_M}\right), \tag{6}$$

where m_l is the slope of the liquidus line in the hypo-eutectic alloy system, L_m is the latent heat of fusion of the pure metal, and R is the universal gas constant.

The solid fraction distribution makes clear that the contact between workpiece and tool leads to solid phase on the surface which is caused by the cooling (Fig. 17).

In the future, research will be extended to the thixoforging of steel. The material parameters for steel shall be determined with the help of experimental investigations. Furthermore, the physical parameters as for example Young modulus, heat conductivity, and heat capacity, should be determined as temperature dependent.

References

1. M.A.H. van der Aa. Wall ironing of polymer coated metal. PhD Thesis, Eindhoven University of Technology, 1999.
2. H.V. Atkinson. Modelling of semisolid processing of metallic alloys. *Progress in Materials Science* 50:314–412, 2005.
3. H.V. Atkinson. *Modelling of Semi-Solid Processing*. Shaker Verlag, Germany, 2008.
4. M.J. van den Bosch. Interfacial delamination in polymer coated metal sheet. PhD Thesis, Eindhoven University of Technology, 2007.

5. D. Brokken. Numerical modelling of ductile fracture in blanking. PhD Thesis, Eindhoven University of Technology, 1999.

6. F. Deflorian, L. Fedrizzi, and S. Rossi. Degradation of protective coatings after cupping test. *Corrosion* 55:1003–1111, 1999.

7. G. Hirt and R. Kopp. *Thixoforming – Semi-Solid Metal Processing*. Wiley-VCH, Germany, 2009.

8. J.M.M. Heussen. Untersuchungen zum Materialverhalten von Metallen im Bereich der Solidustemperatur, PhD Thesis, RWTH Aachen, 1996.

9. B. Hohn. *Beitrag zur numerischen Simulation des Thixoformings*. VDI Verlag, Germany, 2002.

10. Homepage of Fluent Benelux. http://www.polyflow.de, cited January 2010.

11. Y-L. Hsu and C-C. Yu. Computer simulation of casting process of aluminium wheels: A case study. *Journal of Engineering Manufacture* 220:203–211, 2006.

12. G. Messmer. Gestaltung von Werkzeugen für das Thixo-Schmieden von Aluminium und Messinglegierungen in automatisierten Schmiedezellen. PhD Thesis, University of Stuttgart, 2006.

13. B. Meuthen and A-S. Jandel. *Coil Coating* (Bandbeschichtung: Verfahren, Produkte und Märkte). Vieweg, Germany, 2008.

14. D.R.J. Owen, F.M. Andrade Pires, and M. Dutko. Computational strategies for polymer coated steel sheet forming simulations. In: J.M.A. Cesar de Sa and A.D. Santos (Eds.), *NUM-IFORM 07*, Portugal, 2007.

15. M. Oyane, T. Sato, K. Okimoto and S. Shima. Criteria for ductile fracture and their applications. *Mech. Work. Techn.* 4:65–81, 1980.

16. M.L. Oyen and R.F. Cook. Load-displacement behavior during sharp indentation of viscous-elastic-plastic materials. *Journal of Materials Research* 18:139–150, 2003.

17. D.Y. Perera. Effect of pigmentation on organic coating characteristics. *Progress in Organic Coatings* 50:247–262, 2004.

18. E. Rank, M. Schweingruber and M. Sommer. Adaptive mesh generation and transformation of triangular to quadrilateral meshes. *Communications on Numerical Methods in Engineering* 9:121–129, 1993.

19. E-F. Schaeper. Mehr Flexibilität und weniger Verschleißbeim Scherschneiden durch ein neues Werkzeugkonzept mit adaptivem Schneidspalt. PhD Thesis, Leibniz Universität Hannover, 2008.

20. R. Vayeda and J. Wang. Adhesion of coatings to sheet metal under plastic deformation. *International Journal of Adhesion and Adhesive* 18:480–492, 2007.

21. A. Wahlen. Processing of aluminium alloys in the semi-solid state. PhD Thesis, ETH Zürich, 2001.

Computational Techniques for Multiscale Analysis of Materials and Interfaces

Udo Nackenhorst, Dieter Kardas, Tobias Helmich, Christian Lenz and Wenzhe Shan

Abstract A series of four PhD projects worked out under the umbrella of the research training group GRK 615 are summarized in this contribution. The first is on the multiscale modeling of the mechanics of an atomic force microscope with special emphasis on the contact problem. At the relevant length scales atomic force interactions have been considered. The total device is modeled in a dimension adaptive manner using beam elements for the cantilever, solid elements for the tip and an atomic interaction approach for the contact problem. The second thesis is a straightforeward continuation of this research be setting up a powerful MD-FE coupling scheme especially for contact problems. Special emphasis has been led on the consistent coupling avoiding ghost forces by introducing dummy atoms and a boundary layer for the atomic domain. A second series is on the treatment of biomechanics of bones. For a better understanding of the biomechanical phenomena a computational multiscale environment has been implemented, where a cortical section with reinforcing osteons is modeled. The osteons itself are treated on a smaller length scale as laminar cross ply structures and the basic anisotropic properties of the layer are homogenized from the basic constituents, i.e. collagen matrix and hydroxyapatite crystals in dependency of the grade of mineralization. Based on a simple strain criterion detected at voids in between the layers of the osteons a closed control circuit has been realized to mimic the aging of bone. A micro-crack theory as basic origin for the cellular stimulation for bone remodeling has been realized in the last thesis. The strain driven evolution of interlaminar micro-cracks is simulated within an adaptively refined finite element framework. For studies on the released material integrity on the bone cells a sophisticated cell model in analogy to self-stabilizing tensegrity structures has been developed. By this model especially the amplification of stresses from the membrane into the nucleus is shown.

Udo Nackenhorst · Dieter Kardas · Tobias Helmich · Christian Lenz · Wenzhe Shan
Institut für Baumechanik und Numerische Mechanik, Gottfried Wilhelm Leibniz Universität Hannover, Appelstraße 9A, 30167 Hannover, Germany; e-mail: nackenhorst@ibnm.uni-hannover.de

E. Stephan & P. Wriggers (Eds.): Modelling, Simulation and Software Concepts, LNACM 57, pp. 133–167.
springerlink.com

1 Introduction

The development of computational techniques for a consistent approach of multiple length and time scales probably for the analysis of multi-physics problems is a subject of intensive research in computational mechanics. The needs for more sophisticated methods have been outlined for example in [83]. Basic problems which have to be overcome are on

- the limitations in spatial and temporal resolution caused from the discretization of partial differential field equations on the macroscopic continuum level, and
- bridging the gap between atomistic models and continuum models.

Increasing computational performance and simultaneously developed sound mathematical solution techniques nowadays provide a powerful platform for sophisticated modeling approach in engineering.

The DFG-research training group GRK 615 has been an ideal platform for the development of advanced modeling approaches and related computational techniques for the solution of multiscale problems with engineering applications. The interdisciplinary collaboration between mathematicians, computer scientists and engineers has been very fruitful for these developments. The role of applied mathematics has been on

- solver technology, and
- error estimates and adaptivity;

and the role of computer science on

- computational geometry, and
- data structures and data management

while engineers are working on modeling approaches and validation techniques.

In this report four PhD projects are summarized, which have been finished between 2005 and 2010 under the guidance of the first author. Two lines for engineering applications have been investigated, one has been on the investigation of contact problems and the other one on the investigation of biomechanical questions. In both applications sophisticated multiscale techniques are needed for an insight into the mechanisms observable in a macroscopic picture, but with their origin at much smaller length scales.

After a brief literature review on branches of multiscale methods in computational mechanics the first subject is on advanced contact mechanics. In the thesis by Helmich [43] a multiscale model for an atomic force microscope (AFM) has been developed. The total device has been modeled in a dimension adaptive manner. Special emphasis has been led on the contact model between the tip and the probe surface, where atomic forces play a special role. In continuation of this research Shan [92] developed a consistent and powerful adaptive MD-FE coupling method with special emphasis to contact problems. The performance of these techniques have been demonstrated on fully 3-D indentation problems.

A second subject of research has been on the biomechanics of bones with special focus on the biomechanical origin of bone remodeling. The mechanical demand of bone sensor cells (osteocytes) in their physiological environment are of interest. Lenz [59] developed a multiscale model for bone starting with a cortical section including initial osteons, scaling down to a single osteon model as lamellar crossply structure where the constitutive properties are homogenized from the basic ingredients, i.e. collagen and hydroxyapatite. Using a simple local strain criterion within a closed control circuit the creation of new osteons, their growth in hight and onging mineralization can be mimicked. This model has been refined in the thesis by Kardas [53] by the development of an adaptive micro-crack model. Basic assumption has been, that bone sensor cells detect the need for maintenance of bone tissue when a density of micro-cracks increases. In addition a sophisticated cell model has been developed and coupled with the bone tissue model. The cytoskeleton has been modeled as a tensegrity structure. It has been shown that by the cytoskeleton a strain amplification appears from the cell membrane to the nucleus.

2 Multiscale Methods in Computational Mechanics – A Brief Review

Multiscale methods in computational mechanics are mainly discussed in the context of modeling of material properties in order to replace phenomenological constitutive laws whose parameters are obtained from experiments at the so called macroscopic length scale by more physical related modeling approaches. Basic theories are dated back to 1960s, see e.g. [44], providing the general definitions of a representative volume element (RVE) and a periodic unit cell (PUC) and suitable boundary conditions for the homogenization procedure.

Early homogenization techniques for the computation of averaged mechanical properties of micro-heterogeneous materials have been developed within the framework of linear elasticity, Eshelby [28] for example provided analytical solutions for an elastic inclusion within a linear elastic full-space. For more dense particle distribution approximations like the Mori–Tanaka method [76], *self-consistent* scheme [45] or Hashin–Shtrikman bounds [40] have become popular. Prominent examples are also the Halpin–Tsai formulas for the calculation of macroscopic elastic properties of fiber reinforced materials, for a discussion the reader is referred to [37]. For further reading on these basic theories, often referred to as *micro-mechanics*, we refer to [78] and [80], a recent review article is published in [14].

With the availability of increasing computational power numerical techniques to tackle problems with fluctuations in the solution field at smaller length scales have been developed. Great impact had the *variational multiscale* method introduced by Hughes [46, 47], which has also been applied successfully for the resolution of vertices in large eddy simulations, e.g. [3, 36]. A concurrent development is called *partition of unity* method [73]. The general idea is to additively decompose the field into coarse and fine scale solutions and to enhance the finite element ansatz space with bubble functions of high order for example. In the same line the so called extended finite element method (XFEM) [24] is rated, where discontinuous enhancements are added to the finite element ansatz space to tackle cracks for example. Another research direction is on gradient enhanced theories [34], which have their origin obviously in earlier Cosserat theories [27].

A different line of research in computational multiscale mechanics has been opened by the FE^2 approach, see [54] for example. Here the basic idea is to resolve the micro-scale of heterogeneous materials by fine scale finite element models, which are solved at each integration point of the coarse scale finite element model. This for example enables the treatment of non-linear and inelastic material properties in a consistent manner. In contrast to those information passing approaches methods for direct coupling of fine scale finite element meshes with a macro-model have been suggested, the *Arlequin method* [1, 21] is one example.

Another trend is the coupling of atomistic simulations with macroscopic continuum models, for recent reviews the reader is referred to [35, 61, 84]. In this category also falls the Quasi-Continuum (QC) method [75, 95], where constitutive properties in the continuum domain are directly derived from atomic models. The direct coupling of Molecular Dynamics (MD) simulations and Finite Element Methods (FEM) is discussed in [74], we also refer to Section 3.2. Because MD simulations are based on constitutive models, e.g. Lennard–Jones potentials, Embedded Atom Method (EAM), etc., approaches for incorporating quantum mechanics have been proposed [10, 62].

3 A Multiscale Approach for Contact Interfaces

The mechanics of contact of solid bodies in the macroscopic picture is mathematically described by unilateral constraints, expressed by the Karush–Kuhn–Tucker (KKT) conditions

$$g_n \geq 0 \quad p \geq 0 \quad \text{and} \quad p \cdot g_n = 0 \tag{1}$$

where g_n is the normal gap function and p the normal contact pressure. Similarly, the tangential contact conditions are formulated like

$$\|\mathbf{s}\| \geq 0 \quad R \leq 0 \quad \text{and} \quad \|\mathbf{s}\| \cdot R = 0 \tag{2}$$

with the friction function $R = \|\tau\| - \tau_{\max}$. Here τ is the tangential shear traction in the contact interface and s is the slip-vector which expresses the relative tangential motion between contacting particles. The symbol $\tau_{\max}$ describes the constitutive behavior of friction, in the simplest case Coulombs law is used.

Within a computational framework the contact constraints can be treated like constrained optimization. Lagrange multiplier methods and penalty methods are quite familiar to enforce the kinematical constraints, the advantages and disadvantages within the framework of discretization methods have been discussed intensively and led to a broad class of alternative suggestions [57, 103]. Computational contact mechanics is still a branch of active research.

This mathematical framework expressed by eqns. (1) and (2) is non-smooth because the KKT conditions are not differentiable in a sufficient manner. This appears not obvious from a physical point of view, because frictionless normal contact is conservative and a potential energy function should exist. A conforming mathematical framework on that issue has been suggested in [19] by introducing a lower semi continuous functional approach.

The origin of the problems outlined before is an unsuitable description of physics by eqns. (1) and (2). The importance of smaller length scales in contact mechanics has been discussed intensively; for an overview, the reader is referred to [6, 87]. At smaller length scale contact of asperities which themselves carry hierarchical textures, often modeled by fractal description, are considered. As an example in Fig. 1 the spectral decomposition of a road surface is depicted, which has been investigated for rolling tire behavior [9]. With regard to the normal contact description the unilateral contact formulation can be simply scaled down to this smaller length scales. However, for the dynamic response of rolling tires for example, the bulk material behavior becomes important. In this context the hysteretic nature of rubber friction has been investigated intensively, e.g. [88], which have led to a basic understanding.

Within this down-scaling scenario one comes to a point, where continuum mechanics reaches the limits. When atomic structures are resolved, the physics of contact has to be described by interatomic forces rather than geometrical restraints. Macroscopic observable phenomena like adhesion are only explainable from this site. Motivated by the fact, that there are well investigated experimental technologies for studies on local interatomic contact behavior available, like atomic force microscopes for example, modeling approaches and computational techniques for the analysis of contact based on interatomic force interaction have been initiated. A dimension adaptive multiscale approach for the mechanical behavior of an AFM cantilever device has been developed, which will be described in the next subsection. Based on this experience a consistent and fully adaptive Molecular-Dynamics (MD) – Finite Element (FE) coupling procedure has been developed, which is described in Section 3.2 in more detail.

3.1 A Multiscale Model for AFM Operation

The basic working principle of an atomic force microscope (AFM) is sketched in Fig. 2, which is described in a simplified way as follows: A microscopic sharp tip is mounted at the end of a cantilever. When the tip approaches an object, atomic interaction forces (e.g. van der Waals forces) will deflect the cantilever beam. This deflection is measured by high resolution optical devices from which an image of the surface typology of the objects surface can be reconstructed. This idealized working principle will be tackled for the multiscale simulation approach outlined in this

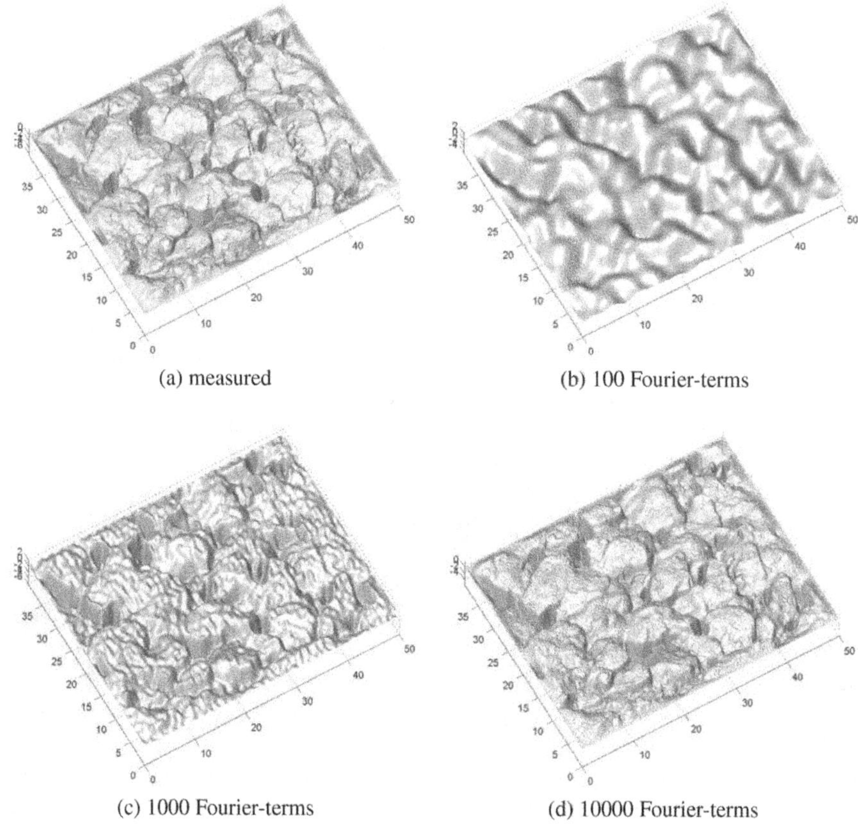

(a) measured

(b) 100 Fourier-terms

(c) 1000 Fourier-terms

(d) 10000 Fourier-terms

Fig. 1 Reconstruction of a measured road surface texture.

subsection. In addition to this quasi-static model assumption dynamic modes of operation are used in practice. Here the so called jump-to-contact mode will be studied.

Typical measures for these devices are cantilevers with a length of ≈ 500 μm, width of 40 μm and hight of 5 μm. The geometric dimensions of the tip are described by a hight of about 15 μm with a tip radius approaching few atoms diameters. To compute this system spanning length scales from Å to mm (i.e. 10^{-10} to 10^{-3} m) a dimension adaptive approach is suggested. Within a finite element framework the cantilever is discretized with 3-D Bernoulli–Euler beam elements, while the tip is modeled with 3-D hexagonal solid finite elements. For the interconnection a transition element has been implemented, which transforms the six degrees of freedom from one point at one end to three displacement degrees of freedom at four points at the other end. The general idea of this dimension adaptive approach mesh details for an AFM cantilever are depicted in Fig. 3.

Although we will concentrate on the mechanical contact in this section, it should be mentioned that at these small length-scales additional effects like electrostatic

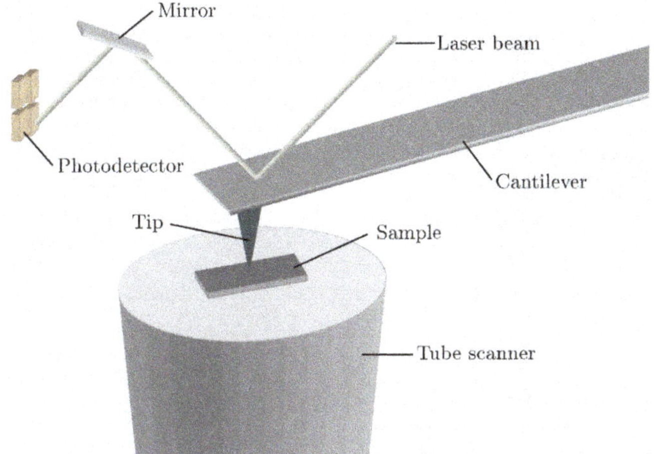

Fig. 2 Working principle of an AFM.

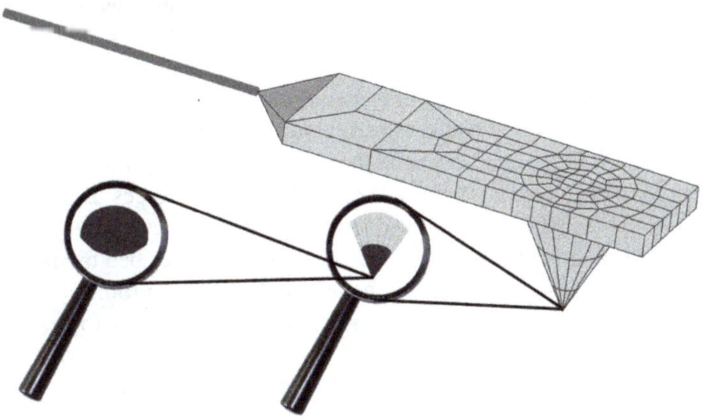

Fig. 3 Dimension adaptive modeling approach for an AFM device.

field interactions, etc., are apparent from which additional working modes of AFM have been developed. Computational techniques for the electrostatic field interaction on the mechanical deformation of the cantilever have been developed in a twin-project within the GRK 615 program [89]. Here only the general principle of this interdisciplinary collaboration is sketched by using Fig. 4.

Besides atomic interaction forces between the tip and the sample electrostatic forces are apparent, which also cause deflections of the cantilever. Separating these effects leads to concurrent measurement principles as described in the literature [6]. Here an approach for a coupled analysis is sketched. The electrostatic field interaction is described by the Maxwell equations, to be solved by a finite element approach

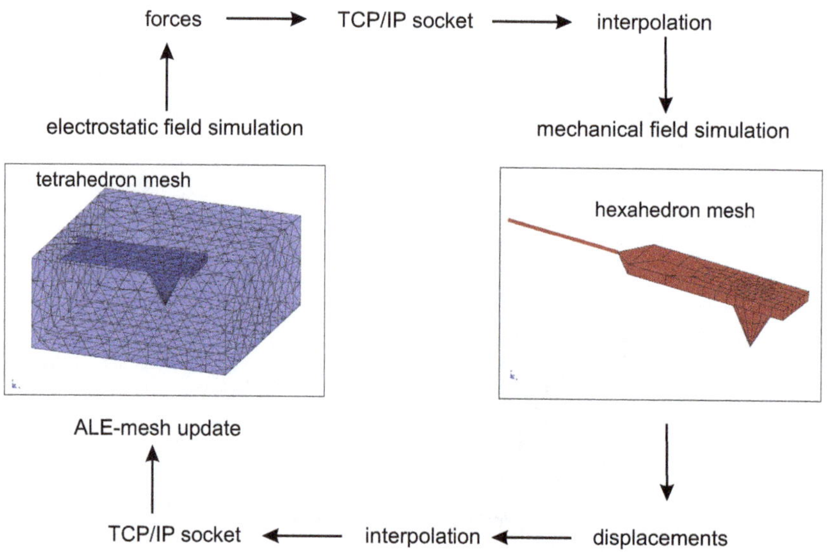

Fig. 4 Sketch of the staggered coupling scheme for electro-mechanic simulations.

with tetrahedral elements. The homogeneous electrostatic field is disturbed by the motion of the cantilever and the substrate surface. An ALE-technique is used to tackle the motion of the cantilever and the movement of the probe relative to the tip. From the geometric (contact) constellation electrostatic forces are derived, which result into mechanical deformations of the beam and changed boundary conditions for the electrostatic field. The pure mechanical part is described by the elastic cantilever carrying the tip for measuring the interatomic contact forces.

Between approaching atoms the presence of van der Waals forces is assumed, expressed by a potential

$$w(r) = -\frac{C_n}{r^n} + \frac{C_m}{r^m} \tag{3}$$

where C_i are constants which have to be evaluated experimentally for given materials and temperature; r is the distance of the nuclei and n, m are exponents, which are usually chosen as $n = 6$ and $m = 12$ for the repulsive and attractive part, respectively. It is emphasized, that eqn. (3) is an empirical expression derived for the interaction between two atoms. For suggestions to incorporate more detailed physics into this empirical formula, e.g. Keesom-interaction model, Debye-interaction model or dispersion effects introduced by London, the reader is referred to [43]. A generalized formulation for example, which incorporates different physical effects has been suggested by McLachlan [72]. However, here we will restrict ourself to the simple van der Waals-potential to incorporate atomic physics into a multiscale computational contact mechanics framework.

Because it has been intended to compute the overall working principle of an AFM within a structural and continuum mechanics framework, a prior step of homogenization has been performed. Instead of evaluating the interaction forces between atoms, interaction potentials between bodies have been evaluated following an approach introduced by Hamaker [38] and refined by Lifshitz [60] (for an overview, see [66, 85]). In general a analytical integration is performed over the atoms of the approaching bodies resulting into interaction potentials for specific geometries. It is mentionable, that the mathematical form (3) of the potentials does not change, i.e.

$$w(r) = -\frac{C_6}{|\mathbf{g}|^6} + \frac{C_{12}}{|\mathbf{g}|^{12}} , \tag{4}$$

where now $\mathbf{g}$ represents the distance vector between two points at the surfaces of the contacting bodies. But now the constants are evaluated for the interaction of bodies instead of atoms. In addition the exponents have been specified in correlation to the Lennard–Jones approach.

This assumptions enable for a straightforward implementation into a finite element environment, i.e. the interaction between two approaching bodies computed by numerical integration over surface elements similar like in established computational contact mechanics. Instead of computing the contact traction $\mathbf{t}_c$ from kinematic constraints, now the gradient fields of physical motivated potentials are used:

$$\mathbf{t}_c = \operatorname{grad} w(\mathbf{g}) . \tag{5}$$

This approach is non-local, in analogy to a boundary element method in general the interaction of each surface-point of the *slave-body* has to be evaluated for each surface-point of the *master-body*. Fortunately, due to the $|\mathbf{g}|^{-m}$ dependency, only the nearest neighbors have to be considered in practical computations.

The advantage in comparison with classical computational contact mechanics is that variational equalities in contrast to variational inequalities have to be tackled. The algorithmic *active set concept* is not needed any more. Nevertheless, because of the highly non-linear potential an incremental iterative scheme, e.g. the Newton–Raphson Method, is needed. In addition, physical instability (in case of negative gradients of the potential) has to be controlled when the attractive part gets predominant (e.g. *jump to contact*).

Under quasistatic assumptions a coupled finite element system of the form

$$\begin{bmatrix} \mathbf{K}_1 + \mathbf{K}_{c11} & -\mathbf{K}_{c12} \\ -\mathbf{K}_{c21} & \mathbf{K}_2 + \mathbf{K}_{c22} \end{bmatrix} \begin{bmatrix} \Delta\mathbf{u}_1 \\ \Delta\mathbf{u}_2 \end{bmatrix} - \begin{bmatrix} \mathbf{f}_1 - \mathbf{f}_{c1} \\ \mathbf{f}_2 + \mathbf{f}_{c2} \end{bmatrix} , \tag{6}$$

is derived, where $\mathbf{K}_i$ and $\mathbf{f}_i$ are the stiffness matrices and applied external forces acting at the two bodies. The contact contributions are obtained from the integration of the contact tractions and their linearization. In detail, the contact forces are obtained from

$$\mathbf{f}_{c1} = \int_{\Gamma_1} \mathbf{N}_1^T \mathbf{t}_c d\Gamma \tag{7}$$

$$\mathbf{f}_{c2} = \int_{\Gamma_2} \mathbf{N}_2^T \mathbf{t}_c d\Gamma \tag{8}$$

and the related tangent matrices are expressed as

$$\mathbf{K}_{c11} = \int_{\Gamma_1} \mathbf{N}_1^T \Delta \mathbf{t}_c \mathbf{N}_1 d\Gamma \tag{9}$$

$$\mathbf{K}_{c12} = \int_{\Gamma_1} \mathbf{N}_1^T \Delta \mathbf{t}_c \mathbf{N}_2 d\Gamma \tag{10}$$

$$\mathbf{K}_{c21} = \int_{\Gamma_2} \mathbf{N}_2^T \Delta \mathbf{t}_c \mathbf{N}_1 d\Gamma \tag{11}$$

$$\mathbf{K}_{c22} = \int_{\Gamma_2} \mathbf{N}_2^T \Delta \mathbf{t}_c \mathbf{N}_2 d\Gamma \ . \tag{12}$$

Herein $\mathbf{N}_i$ denote the matrices of the finite element ansatz functions, $\mathbf{t}_c$ is the contact traction vector field computed from the contact potential as

$$\mathbf{t}_c = \nabla w(\mathbf{g}) = \left(\frac{6C_6}{(g_1^2 + g_2^2 + g_3^2)^4} - \frac{12C_{12}}{(g_1^2 + g_2^2 + g_3^2)^7} \right) \begin{bmatrix} g_1 \\ g_2 \\ g_3 \end{bmatrix} \tag{13}$$

where g_i are the cartesian coordinates of the distance vector between two points of the contacting surfaces. The linearization $\Delta \mathbf{t}_c$ with respect to the displacement fields is computed straightforward. It is clearly seen that the contact traction and the stiffness contributions are dominated by the nearest distance points and decrease rapidly for neighboring points. This gives rise to reduce the computational effort by integration only between the nearest points of interaction.

Systematic convergence studies regarding the finite element discretization and the related integration scheme for the evaluation of the contact forces have been reported in [43]. While by the finite element discretization the description of the geometry and the elastic deformation of the contacting bodies is controlled, the evaluation of the contact forces is controlled by the order of the numerical integration scheme. Gauss-Quadrature has been used to compute the related integrals and it has been figured out that converged solutions are obtained for orders, where the distance between the Gauss-points is less than 1Å, which means in the magnitude of the distance of atoms within a crystal lattice.

A computational example is sketched in Fig. 5, an idealized tip of an AFM approaches a non-plane surface. A conical geometry of the tip with a spherical tip radius of 10 nm has been assumed. The spherical tip surface has been discretized with bilinear finite elements of mean element length of about 0.35 nm. By controlled displacements the tip approaches the surface, while the elastic deformation of both bodies appeared negligible.

The contact potential for three contact distances are depicted in Fig. 6, due to the sloped surface geometry an asymmetry can be observed. The corresponding finite element contact forces are depicted in Fig. 7, from which it is clearly seen that a strong concentration with decreasing distance onto single nodes of the finite element mesh happens. Therefore, the accuracy of this approach for the overall performance of an AFM might also depend on the spatial resolution of the finite element mesh.

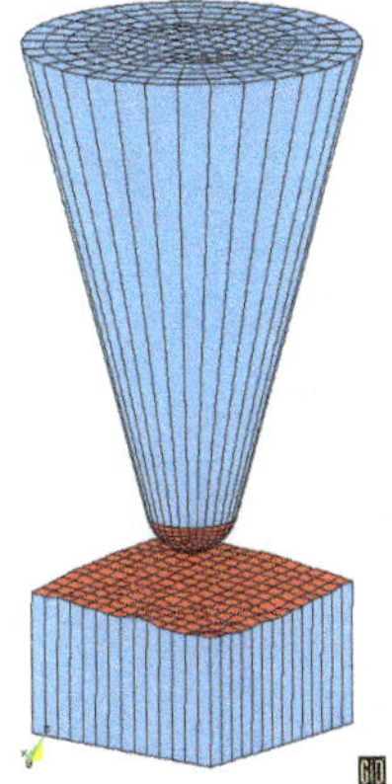

Fig. 5 Finite element model of an AFM-tip approaching a non-smooth probe.

For dynamic operation modes the contact model has been integrated into a transient dynamics simulation environment, where the Newmark scheme has been chosen for the temporal discretization. The dimension adaptive model as depicted in Fig. 3 has been used for the simulation of the jump-to-contact phenomenon. In Fig. 8 the time-displacement curve from this finite element simulation is compared with the solution of a simple 1-D dynamic oscillator [6]. It is clearly shown that the 1-D model is not capable to mimic the jump-to-contact phenomenon, because it is simply reflected. In contrast, the dimension adaptive finite element model sticks in contact, oscillating with small decreasing amplitudes.

3.2 A Consistent MD-FE Coupling Approach

Motivated from the experience that the contact description by simple pair potential interaction outlined in the previous subsection is neither satisfactory from the physical point of view nor efficient in the computational sense, a consistent molecular-dynamics (MD) coupling with finite element methods (FEM) has been developed [92, 93]. The idea behind is to model the contact interaction by an atomic interaction approach, while the deformable behavior of the bulk material is described as continuum, discretized with finite elements. A fully adaptive scheme has been implemented, where based on a local error estimation first an h-refinement of the finite element domain is performed and in a second step, when a critical size of the elements is reached, the FE domain is switched into an atomic domain.

A quasi-static zero Kelvin description has been chosen for the atomic region for simplicity and interatomic force potentials like Lennard–Jones (LJ) potential or Embedded Atom Method (EAM) have been assumed. Because the interaction is not limited to the nearest neighborhood the atomic domain is also referred to as non-local domain, whereas for a simple continuum only local deformation states

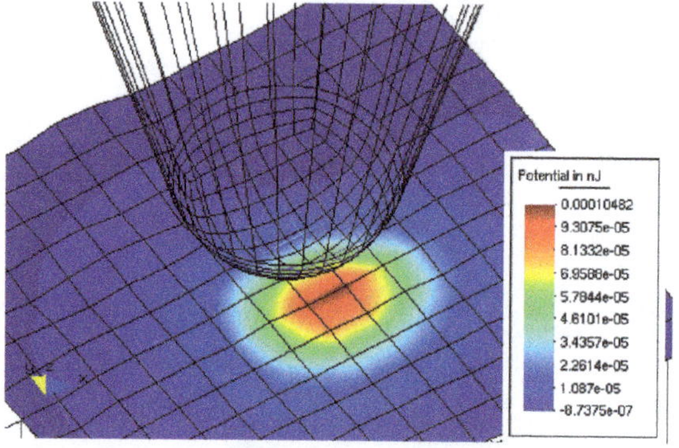

(a) Step 26, Distance ≈ 4.88 nm

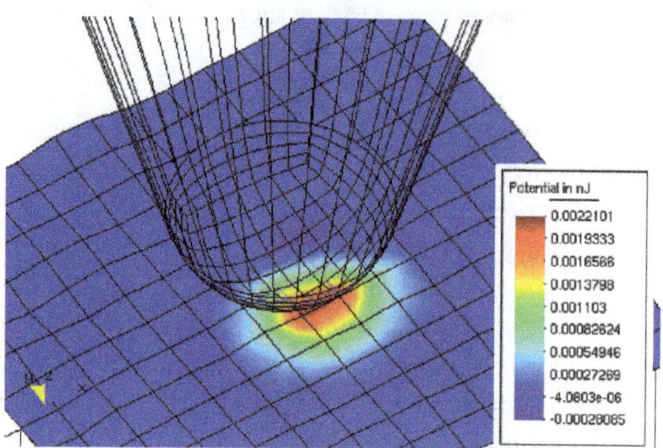

(b) Step 52, Distance ≈ 2.28 nm

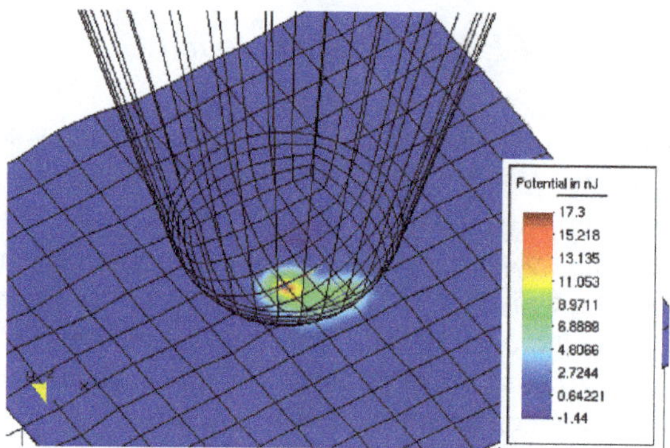

(c) Step 72, Distance ≈ 0.26 nm

Fig. 6 Contact potential for an approaching tip.

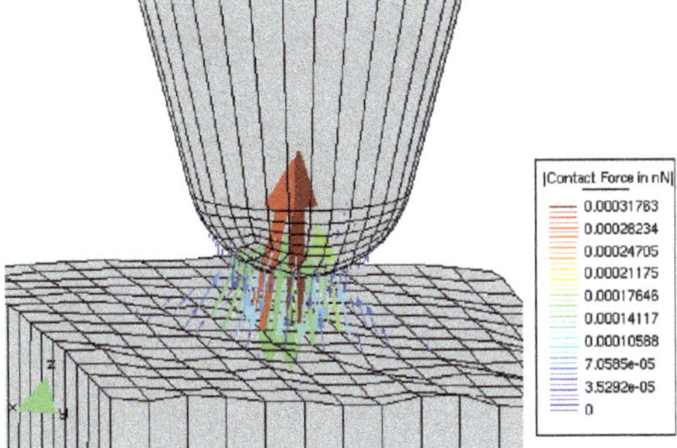

(a) Step 26, Distance ≈ 4.88 nm (scaling factor $5 \cdot 10^4$)

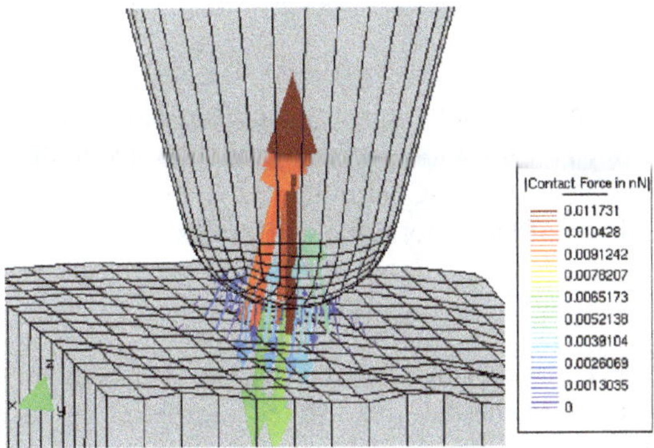

(b) Step 52, Distance ≈ 2.28 nm (scaling factor $2 \cdot 10^3$)

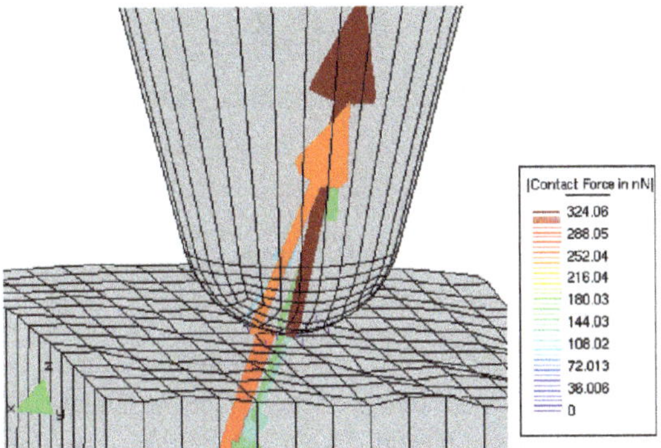

(c) Step 72, Distance ≈ 0.26 nm (scaling factor $1 \cdot 10^{-1}$)

Fig. 7 Contact forces computed for the approaching tip.

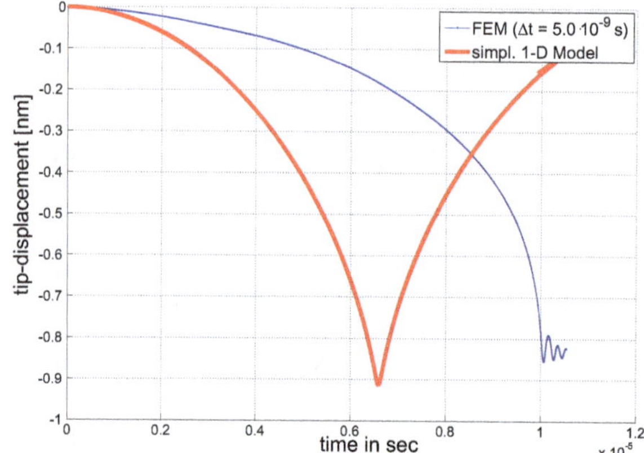

Fig. 8 Simulation of jump-to-contact mode.

are considered. For computational efficiency the decreasing far-field interaction is limited by a user defined cut-off radius, which limits the size of the neighbor-list.

Basic assumption for the Quasi-Continuum (QC) approach is the Cauchy–Born (CB) rule, by which an affine deformation of atoms related to the deformation gradient $\mathbf{F}$,

$$\mathbf{r} = \mathbf{F} \cdot \mathbf{R} \, , \tag{14}$$

is assumed, where $\mathbf{R}$ is the distance between atoms in the undeformed reference lattice and $\mathbf{r}$ the distance in the deformed state. In order to derive a strain energy density function for the continuum model equation (14) is rewritten in terms of the right Cauchy–Green tensor $\mathbf{C}$,

$$\mathbf{r}^2 = \mathbf{R}^T \cdot \mathbf{C} \cdot \mathbf{R} \, . \tag{15}$$

From the EAM-potential for FCC-aluminum lattice structure for example, a strain energy function is derived as

$$W(\mathbf{C}) = \frac{1}{V_a} \left[U \left(\sum_j \rho(|\mathbf{r}_j(\mathbf{C})|) \right) + \frac{1}{2} \sum_j \Phi(|\mathbf{r}_j(\mathbf{C})|) \right] \, , \tag{16}$$

where j is the index for neighbors of the representative core. The first part represents the embedding energy depending on the electron density ρ and the second part represents the atomic pair potential. The usage of the strain energy function as constitute law in the finite element domain is referred to concurrent lattice homogenization. For a single crystal aluminum the elastic constants listed in Table 1 have been obtained ab initio, which are in good agreement with experimentally obtained results.

To underline the overall consistency of this approach the stress- and strain fields are computed in the non-local region. The deformation gradient in the atomistic do-

Table 1 Elastic constants obtained at equilibrium status for the representative lattice oriented in $\langle 100 \rangle$ directions, in comparison with experimental results published in [26, 52].

	This work	[26]	[52]
C_{11} (GPa)	117.7	118.1	114.3
C_{12} (GPa)	62.2	62.3	61.9
C_{44} (GPa)	32.25	36.7	31.6

main is recovered by a least squares approach from the Cauchy–Born rule (14); for details, the reader is referred to [93]. The virial stress appears to be the most common approach to compute the stress state in the non-local domain [29], for a detailed derivation of viral stress, see [63]. The Cauchy stress σ_i for an atom position i is related to the interatomic forces $\mathbf{f}_{ij}$ and the distance $\mathbf{r}_{ij}$ to the neighboring atoms by

$$\sigma_i = \frac{1}{v_a} \sum_{j \neq i} \mathbf{f}_{ij} \otimes \mathbf{r}_{ij} \ . \tag{17}$$

From a straightforward approach within the QC-framework, [92], one obtains

$$\sigma_i = \frac{1}{2 v_a} \sum_{j \neq i} \mathbf{f}_{ij} \otimes \mathbf{r}_{ij} \ . \tag{18}$$

Here the sampling volume is twice as large as in the common description. However, this is in agreement with the more robust approach proposed by Hardy [39], which has been discussed in [106].

The equivalence between the QC-finite element method and the direct MD-approach has been outlined so far neglecting free boundaries. At surfaces the atomistic model relaxes due to missing neighbor elements, which can be incorporated into the QC-approach by a boundary layer potential,

$$\Pi^{bc} = \underbrace{\int_{\Omega_{\text{int}}} W(\mathbf{C}) \mathrm{d}V}_{\Pi^{bc}_{\text{FE}}} + \underbrace{\sum_{i=1}^{N_{bc}} \Pi_i}_{\Pi^{bc}_{\text{MD}}} \ . \tag{19}$$

The lattice strain energy $W(\mathbf{C})$ is computed for the interior domain with volume fraction Ω_{int}, N_{bc} is the number of boundary atoms inside an element and Π_i is the potential energy of the boundary atom i. The general concept is illustrated in Fig. 9. With this modification a perfect match of QC finite element results and MD solution has been obtained for 3-D examples with free surfaces under homogeneous stress conditions [92].

Furthermore, this concept has been proven as an accurate and efficient approach for coupling MD and FE domains. The existence of non-physical ghost-forces by the direct nodal-based MD-FE coupling has been discussed intensively in the literature,

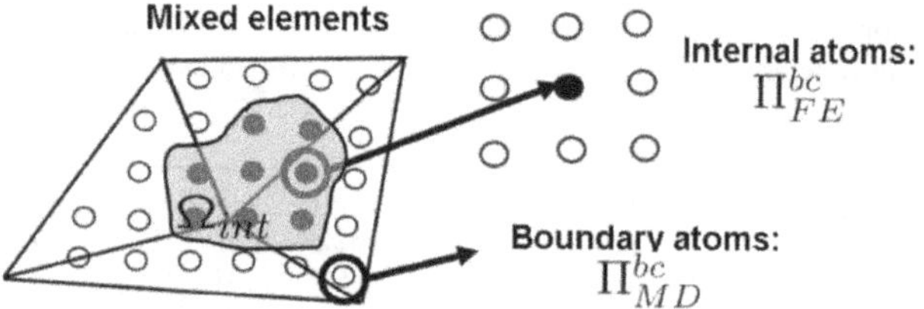

Fig. 9 Illustration of interface elements containing both internal atoms and boundary atoms.

see [23,75] for example. The mathematical origin as well as the locality of this effect has been illustrated in [92] on a 1-D example. In addition the node based coupling concept, introduced by [95], leads to further effort in adaptive mesh refinement, because newly generated FE-nodes have to be placed at atoms positions within the lattice. To circumvent this disadvantages, an energy based coupling approach has been suggested in [93]. The general idea is to introduce so called dummy-atoms at lattice position within the finite elements at the coupling boundary.

For efficient computations a two stage adaptive scheme has been developed. In a first stage the size of the finite elements in the continuum domain is reduced based on a deformation gradient error criterion,

$$\|\mathbf{e}_F\|^2 = \sum_{i=1}^{N_{\text{el}}} \underbrace{\int_{\Omega_i} (\mathbf{F} - \tilde{\mathbf{F}})^T \mathbf{C} (\mathbf{F} - \tilde{\mathbf{F}}) d\Omega}_{\|\mathbf{e}_F\|_i^2} \tag{20}$$

where N_{el} is the number of local QC elements, $\tilde{\mathbf{F}}$ is the deformation gradient directly evaluated at the Gaussian points, $\mathbf{C}$ is a positive diagonal scaling matrix and $\mathbf{F}$ is the interpolated deformation gradient obtained from a superconvergent projection scheme. In a second step, when the size of a finite element gets smaller that the reference lattice, the QC domain is converted into a MD domain. A formal criterion for the critical size is given as

$$\min(h_i) < a \, R_{\text{ref}}, \tag{21}$$

where h_i is the characteristic element length, R_{ref} is the radius of the reference lattice and $a \geq 1$ is a user defined parameter to be chosen for balancing computational performance and accuracy.

The performance of the adaptive scheme has been demonstrated from the simulation of an indentation test as sketched in Fig. 10. The indentation cone is assumed to be rigid and the process is computed by controlled displacements. A fully atomistic model has been computed as reference solution. The initial coupled QC model consists of a atomic top layer with 662 non-local repatoms and a finite element model with 14 nodes.

The effect of local/non-local conversion is depicted in Fig. 11. The L_2-norm of the maximal displacement error is plotted against the unknowns. It is clearly

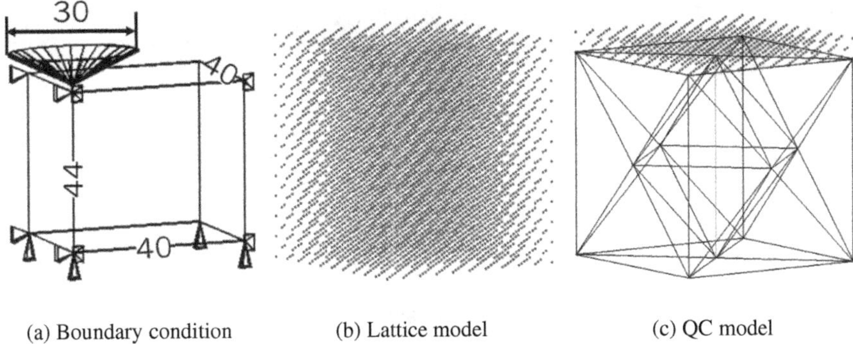

(a) Boundary condition   (b) Lattice model   (c) QC model

Fig. 10 Models used for accuracy test for QC model. The indentation depth in (a) has been set to 0.4 (dimensionless).

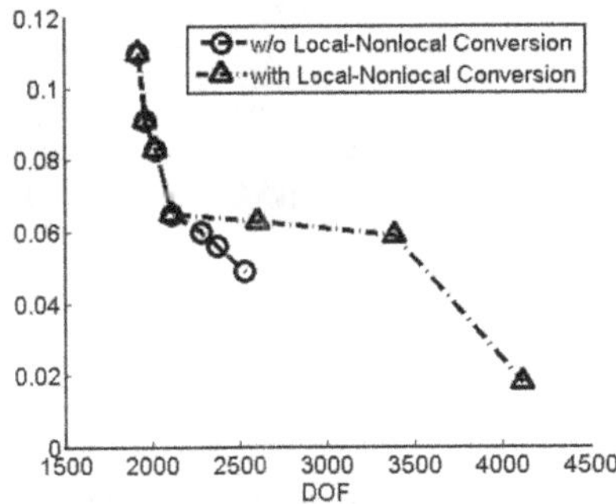

Fig. 11 Comparison of the maximum displacement error of QC models with and without local/non-local conversion.

seen that after three refinement steps there is no mentionable improvement without the conversion of local finite elements into non-local MD domain. A few steps are needed to convert all the necessary elements until a further reduction of the error is obtained.

The evolution of the non-local (MD) domain for this example is depicted in Fig. 12.

With special emphasis to contact analysis a second indentation example has been analyzed, the principal setup and the initial model are shown in Fig. 13. Now the spherical indentor is modeled as perfect lattice structure, the top-layer of the probe again is modeled as non-local MD model with three layers of atoms while the main body is discretized with local QC-finite elements. As reference solution again a fully atomistic model containing about 37,000 atoms has been used.

The total indentation of 3 μm has been computed within 30 incremental steps. The finally obtained displacement field and the adaptively refined model are shown in Fig. 14. The final model consists of about 4,000 nodes.

The computational effort of the adaptive QC model growth linearly with the number of unknowns, as depicted in Fig. 15. The computation of the adaptive QC model

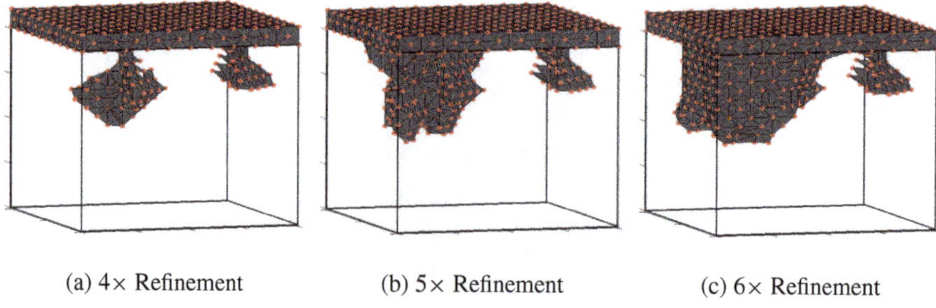

(a) $4\times$ Refinement   (b) $5\times$ Refinement   (c) $6\times$ Refinement

Fig. 12 Evolvement of non-local QC region during the refinement process.

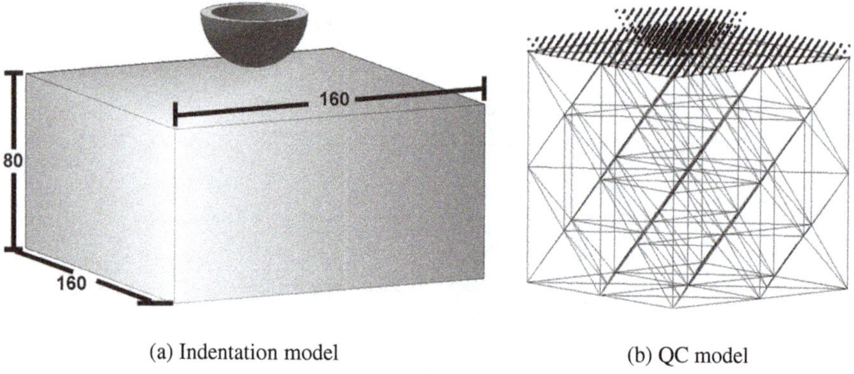

(a) Indentation model   (b) QC model

Fig. 13 Indentation model: (a) full model (b) one fourth of the coupled QC model with symmetric boundary condition in the $x-y$ direction.

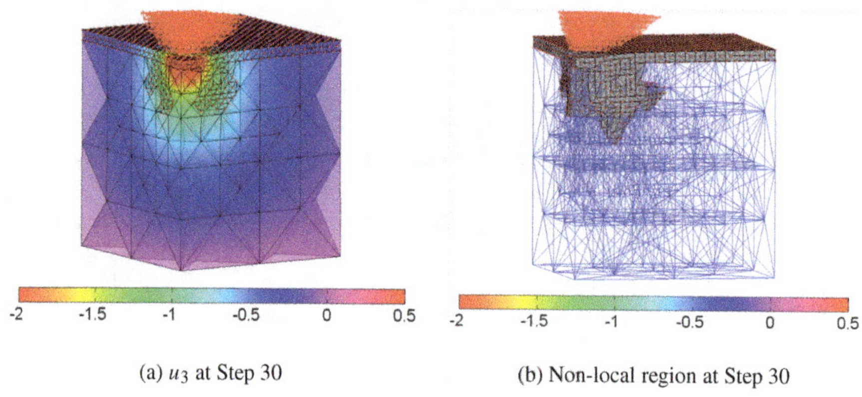

(a) u_3 at Step 30   (b) Non-local region at Step 30

Fig. 14 Indentation process simulated by adaptive QC-method.

lasts 40 minutes on a standard desktop PC (Intel 3 GHz CPU, 3 GB RAM), using Matlab 2008b and Windows XP 32bit environment. In comparison, the computation of the reference solution with a full lattice model took about 15 hours on a

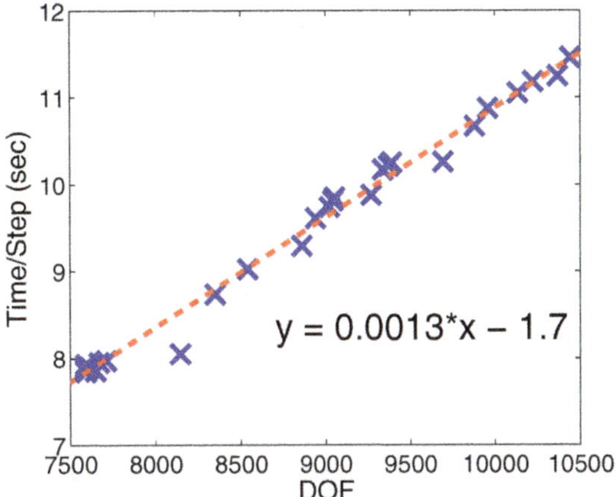

Fig. 15 Computational cost per iteration step of the adaptive QC model for the indentation test.

64bit UNIX (2.66 GHz CPU, 32 GB RAM) machine, which underlines the progress obtained by the outlined research.

Based on the presented computational techniques next steps on more complicated contact problems are straightforward, with emphasis to nano-technology on high performance surface treatment for example. Basic work will be dedicated on sophisticated interatomic surface potentials for alloys and special surface dotation.

4 Biomechanics of Bones

Bones are living organs with the ability to adapt themselves to changing mechanical demand, already proposed in the often cited booklet from Wolff [102] 120 years ago. This knowledge has been applied in clinical practice by Pauwels, who successfully treated pseudo-arthrosis using mechanical stimulation [86]. Computational mechanics came into play with early attempts on bone remodeling prediction, pure phenomenological approaches based on continuum damage mechanics have been suggested to explain aseptic loosening of artificial hip-joint implants for example [2, 18, 100]. Early problems regarding the stability of numerical algorithm are solved, see [51,99] for example, and nowadays stable and reliable computational methods to predict bone remodeling caused by changed mechanical environment are available based on the concepts of continuum constitutive theory and computational inelasticity. In the following a brief outline on the phenomenological continuum approach on bone modeling and remodeling will be presented, where we restrict ourselves to a first order approach. Special attention will be led on the formulation of the boundary conditions. We will continue with a multiscale modeling approach in order to get more insight into the biological origin of the remodeling process and continue with a micro-crack model where growing micro-cracks in cortical bone are assumed as mechanical source for remodeling activities. Finally a cell model will be presented.

4.1 Continuum Approach

A first order approach of this biomechanical simulation is outlined as follows. In a first step the mechanical equilibrium

$$\operatorname{div} \boldsymbol{\sigma} = \mathbf{0} \tag{22}$$

is solved, where $\boldsymbol{\sigma}$ represents the Cauchy stress tensor in a small strain environment. Permanent (time averaged) local strain out of a physiological tolerable level causes biological reactions leading to changes in bone constitution. Within the continuum framework the bone constitution is described by the bone mineral density (BMD), which is a radiographically measurable quantity. In the simplest case the mechanical stimulus is described by the local strain energy density ψ, see for example [2, 100]. By this the constitutive model assumptions are described by the statement of a biomechanical target function

$$\mathcal{F} = \psi - \psi_{\mathrm{bt}} = 0 \,, \tag{23}$$

where ψ_{bt} represents the (long term averaged) biological comfortable target value and $\psi = \psi(\mathbf{F}, \varrho)$ represents the strain energy density in dependence of the local deformation, expressed symbolically by the deformation gradient $\mathbf{F}$ and BMD ϱ, to be interpreted as internal variable. Within a thermodynamic consistent constitutive framework a evolution rule for the bone mineral density is derived as

$$\dot{\varrho} = \lambda \frac{\partial \mathcal{F}}{\partial \varrho} \,, \tag{24}$$

where λ is a Lagrangian computed using well established implicit Euler schemes, e.g. [104]. So far missing is the back-coupling to the mechanical model, which is expressed by the dependency of the mechanical properties from the BMD. With the assumption of linear elastic mechanical behavior, from the constitutive theory a relationship between Young's modulus E and BMD

$$E = E_0 \left(\frac{\varrho}{\varrho_0} \right)^2 \tag{25}$$

is argued, see [55] for details, which is supported by a statistical analysis reported in [91].

It is emphasized, that already by this *first order approach* the typical osseous structures in bone observable in X-rays are computable from scratch, i.e. starting with a homogeneous BMD-distribution and suitable boundary conditions [79]. Studies in comparison with clinical experience for quite different bone implants underline the qualitative predictive behavior of this phenomenological approach, see [30, 64, 65] for example. From that it is concluded, that mechanical stimulation is of first order meaning for bone remodeling processes and computational mechanics is suitable for principle predictions, e.g. for the development of biomechanical compatible implant design.

The resolution of the outlined continuum mechanics approach is comparable with radiographic measurements. Thus, it enables for a qualitative description only. In order to explain the bio-physiological mechanism in more detail, one has to go out to observe smaller length scales, which will be outlined below.

Remark: In comparison to the very simple phenomenological modeling approach sketched before, more sophisticated constitutive models for stress adaptive bone remodeling simulation have been suggested. This started with two phase modeling approaches, already introduced in [16] and recently investigated by, for example, Ebinger et al. [25]. Further activities have been reported on the description of the mechanical anisotropy in the continuum framework, e.g. [51, 55, 107]. More sophisticated constitutive models for single phase material have been reported in, for example, [22, 56]. However, it appears hard to validate these models in vivo, e.g. from radiological measurement

4.1.1 Boundary Conditions

For the remodeling scheme described before, a boundary value problem for the mechanical equilibrium conditions has to be solved. The related boundary conditions are described by the muscle forces and joint load, which are in general not measurable. Measurements with instrumented hip-joint prostheses are reported in [42], resulting hip-joint forces for different motions are public in the orthoload database (www.orthoload.com). An approach for the computation of related muscle forces has been suggested in [97]. However, these are short term data recorded for a gait-cycle for example while bone remodeling processes take place over month and years. Thus, these data have to be averaged over daily activity.

An alternative approach has been suggested in [64], here Neumann conditions have been computed by an inverse optimization technique from radiographic data. Starting with the geometry reconstruction from CT-images the interior Hounsfield units are translated into associated BMD-information and mapped to the generated finite element model. With genetic algorithms the *static equivalent* muscle forces and joint loads are computed such that the computed BMD-distribution fits best with the image data.

4.2 Multiscale Modeling of Cortical Bone Tissue

A consistent concept for the phenomenological description of bone remodeling phenomena has been sketched before. However, the biological origin of these processes

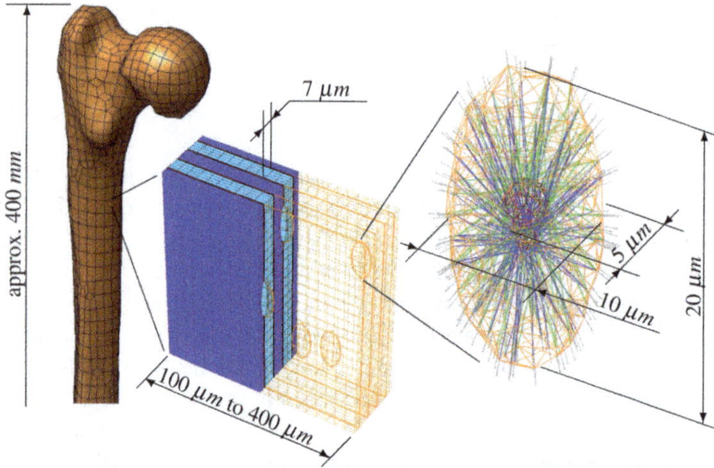

Fig. 16 Length scales of the multiscale approach. Left: macro-scale (human femur, approx. 400 mm), center: micro-scale (lamellar bone, 100 to 400 μm), right: cell-scale (osteocyte cytoskeleton, 5 μm × 10 μm × 20 μm). In the micro-model, light and dark blue color denote different collagen fiber orientations. The ellipsoidal cavities represent lacunae.

is described unsatisfactory, the question is by what is bone remodeling initiated, or more precisely what is the mechanical stimulus for the bone cells. There is a well developed knowledge on the cellular biology and the cell communication, for an overview the reader is referred to [17] for example. For bone remodeling mainly three different types of cells are responsible, *osteoclasts* resorb damaged bone tissue, *ostoblast* build new bone and *osteocytes* are the sensor-cells by which the remodeling process is initiated. The osteocytes are placed between the laminar bone, see Fig. 16, and the questions is how they are mechanically stimulated to signal demand for maintenance. There are quite different theories derived from in vitro experiments with cell cultures, outlined in more detail in Section 4.3. Goal of this section is to provide a computational framework to simulate the physiological environment of bone cells.

The mechanical environment for osteocytes which are buried in cortical bone has to be described as hierarchical structure [90]. A multiscale computational framework is set up for the mechanical environment as shown in Fig. 17; for details, see [59]. At the largest scale a section of cortical bone is discretized as amorphous matrix including cylindrical reinforcements, by which the osteons are mimicked. The boundary conditions for the section is derived from the finite element analysis of a femur model (see Fig. 16), which has been analyzed for static equivalent loading conditions, see Section 4.1.1. The *homogenized* constitutive properties of the osteons at this level are computed from the next smaller modeling scale, at which isolated osteons are modeled as cross-ply laminated structures. The Dirichlet boundary conditions are derived from the cortical section approach. The constitutive properties of each layer are obtained from the basic composite ingredients, which is

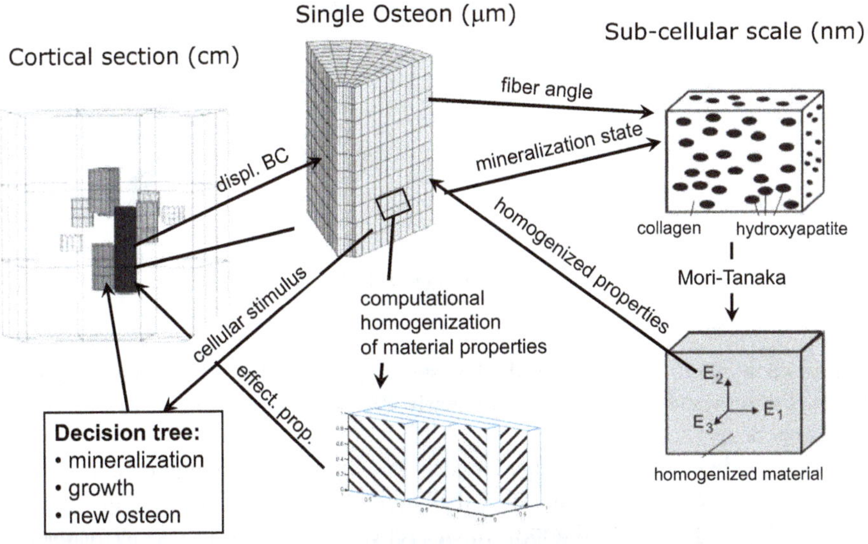

Fig. 17 Computational multiscale model for osteonal development in cortical bone.

modeled as homogeneous matrix of collagen with ellipsoidal inclusions (mean axis diameter of about 3 × 20 nm) of hydroxylapatite crystals based on elastic inclusion theory and Eshelby solutions. The homogenized transversal isotropic properties for the individual layers are computed by Mori–Tanaka method in dependency of the grade of mineralization. The extracellular matrix produced by osteoblasts consists of collagen mainly, while the mineralization proceeds later on. Fully mineralized bone consists of about 65% minerals. This approach enables for the computation of the elastic properties of newly build bone tissue and its maturing at basic tissue level.

A quite simple bioregulatory closed circuit loop has been set up to simulate the creation of new osteons, their growth in length and their mineralization in time. As mechanical stimulus the averaged interlaminar shear stress computed within detailed osteon models has been assumed. In addition soft elements have been placed at the interface between two lamellae randomly to mimic the osteocytes. Based on this heuristical criterion within a time stepping scheme the need for maintenance (creation of new osteons), their growth in hight and their mineralization in time has been implemented. A sequence of bone maturing simulated by this approach is depicted in Fig. 18. Starting with five initial osteons in 35 incremental steps in total about 200 new osteons have been created, each of them growing in hight and mineralizing with time. For the cortical section model the challenge of efficient mesh generation is obvious, which has been solved automatically using efficient Delauny techniques. A geometry based error estimation has been used to control the remesh-

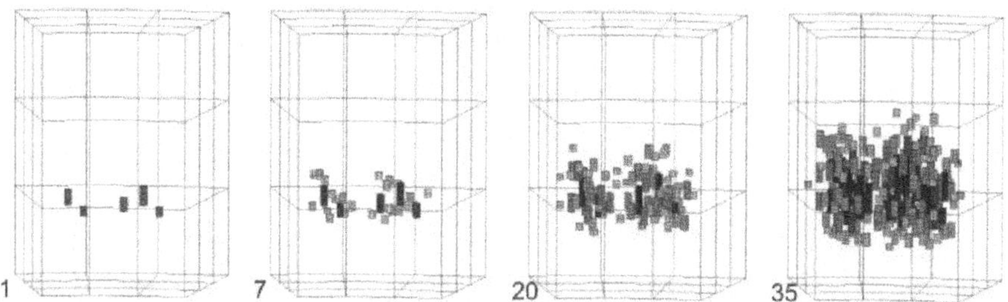

Fig. 18 Development of osteons inside a section of cortical bone.

ing for each incremental step. In the last step depicted in Fig. 18 the finite element mesh of the cortical section consists of nearly 400,000 linear tetrahedrons with about 70,000 nodes.

These studies have to be judged as demonstration for the computability of the multiscale modeling approach, rather than a bio-mechanic consistent approach for cellular stimulation. Related improvements for that will be outlined in the next subsection. To provide an impression on the computational effort for these multiscale investigations, it is emphasized that all computations have been performed on a single desktop computer. For example, the solution of the final step has been computed within 3.5 hours, which includes the computation of 200 isolated osteons and their mineral grade dependent constitutive material properties, the mesh regeneration for the cortical section and the solution of the linear system for the cortical section with more than 200,000 unknowns.

4.3 An Adaptive Computational Approach for Interlaminar Micro-Cracks

As outlined before, there is a controversy discussion on the mechanical stimulation of bone cells in the literature. One line is based on fluid shear assumptions, e.g. [4, 71, 77, 101, 105], which have their origin in laboratory experiments with cell cultures in non-physiological environment. Another line is on bone tissue deformation [7, 20, 70, 81]. A plausible theory that micro-cracks in cortical bone tissue could be the reason for maintenance has already be stated by Frost [31, 32] and underlined by experiments, e.g. [5, 12]. This idea has been picked up by different research groups, for example [13, 69, 82]. There are two different explanations why micro-cracks stimulate bone cells for remodeling activities. The first is on direct injury of the cellular network, [11, 67, 68], the second is on strain amplification due to loss of material integrity [7, 71, 81].

Micro-cracks in cortical bone can be interpreted as interlaminar bondage failure [41, 58, 96]. For the simulation of micro-cracks in cortical bone anisotropic debondage model has been chosen based on the damage criterion introduced by

Brewer and Legace [8]. The damage criterion for an interlaminar layer reads

$$\left(\frac{\langle\sigma_{11}\rangle}{\tilde{\sigma}_{11}}\right)^2 + \left(\frac{\sigma_{12}}{\tilde{\sigma}_{12}}\right)^2 + \left(\frac{\sigma_{13}}{\tilde{\sigma}_{13}}\right)^2 = 1 \,. \tag{26}$$

Herein σ_{12} and σ_{13} are the in-plane shear stresses and σ_{11} is the normal stress perpendicular to the interlaminar layer, while $\tilde{\sigma}_{ij}$ are the related threshold values for delamination and $\langle\bullet\rangle$ is the McCaulay bracket. For linear elastic material response equation (26) can be rewritten by an equivalent strain based failure criterion,

$$\mathcal{F}(\boldsymbol{\varepsilon}) = \varepsilon_v - \tilde{\varepsilon} = 0, \tag{27}$$

where $\tilde{\varepsilon}$ describes the strain threshold for damage initiation. The equivalent strain measure is written as

$$\varepsilon_v = \frac{\sqrt{\boldsymbol{\varepsilon}^{\mathrm{T}}\, \mathbf{C}^{\mathrm{T}}\, \mathbb{P}\, \mathbf{C}\, \boldsymbol{\varepsilon}}}{2\, \mathbb{C}_{1212}} \tag{28}$$

where $\mathbb{P}$ is a projection operator with the non-zero entries

$$\bar{\mathbb{P}}_{11} = \left(\frac{\tilde{\sigma}_{12}}{\tilde{\sigma}_{11}}\right)^2, \quad \bar{\mathbb{P}}_{44} = 1 \quad \text{and} \quad \bar{\mathbb{P}}_{55} = \left(\frac{\tilde{\sigma}_{12}}{\tilde{\sigma}_{13}}\right)^2. \tag{29}$$

The damaged constitutive tensor for an interlaminar layer reads

$$^{d}\mathbf{C} = \begin{bmatrix} h\,\mathbb{C}_{1111} & \mathbb{C}_{1122} & \mathbb{C}_{1133} & 0 & 0 & 0 \\ \mathbb{C}_{2211} & \mathbb{C}_{2222} & \mathbb{C}_{2233} & 0 & 0 & 0 \\ \mathbb{C}_{3311} & \mathbb{C}_{3322} & \mathbb{C}_{3333} & 0 & 0 & 0 \\ 0 & 0 & 0 & g\,\mathbb{C}_{1212} & 0 & 0 \\ 0 & 0 & 0 & 0 & g\,\mathbb{C}_{1313} & 0 \\ 0 & 0 & 0 & 0 & 0 & \mathbb{C}_{2323} \end{bmatrix}. \tag{30}$$

where $g = 1-d$ with a scalar damage variable $d \in [0, 1]$ and $h = 1 - d\langle\sigma_{11}\rangle/|\sigma_{11}|$. Following to [33] a phenomenological damage evolution rule is used:

$$d\,(\gamma) = 1 - \left(\frac{^{i}\gamma}{\gamma}\right)^{\beta} \left(\frac{^{u}\gamma - \gamma}{^{u}\gamma - \,^{i}\gamma}\right)^{\alpha}, \tag{31}$$

where

$$\gamma = \max\left(\varepsilon_v\right), \tag{32}$$

$^{i}\gamma$ is the threshold for damage initiation and $^{u}\gamma$ is a value that describes total failure. By the parameters α and β the progressive damage can be controlled.

This approach has been implemented in combination with an adaptive mesh refinement for the crack-tip resolution. Additional measures for efficient and reliable computations, like viscous regularization and non-local formulation as well as intensive numerical tests are described in detail in [53].

A computed sequence of growing cracks is shown in Fig. 19 where the comparison between an unrefined mesh and a refined mesh is displayed. Results of the stress-norm

$$|\sigma| = \sqrt{\varepsilon^{\mathrm{T}} \mathbf{C}^{\mathrm{T}} \mathbb{P} \mathbf{C} \, \varepsilon} \tag{33}$$

that indicates damage initiation and the damage variable d are shown for different load increments λ between 0.2 and 1.0. At first glance it can be observed that the crack contour and the stresses in the refined mesh are more smooth than those in the unrefined mesh. Since the distribution of the stress norm is smooth around the crack shape in the coarse mesh, the refinement is only performed in the region of the crack tip. In the left column of Fig. 19 different refinement levels are displayed. Orange-colored elements are unrefined, green-colored elements have a refinement level of one and blue elements have a refinement level of two.

4.4 Mechanical Model for Bone Cells

With the preparations described in the former sections now the environment for the simulation of osteocytes placed within the lacunae between the osteonal lamellae is prepared. Detailed models of bone cells are created based on the major components of osteocytes with respect to mechanical aspects, see Fig. 20. These components are:

- integrins that are responsible for the adhesion of osteocytes to the bone matrix,
- the nucleus which is supposed to play a major role in the mechanosensory process,
- the centrosome which contains γ-tubulin for the creation of microtubules,
- microfilaments that are mostly observable under the cell membrane,
- microtubules that connect the centrosome with the cell-membrane,
- microfilaments that connect the nucleus with the cell-membrane.

In vivo these cell components exhibit eigenstresses.

In [50] it is shown that microfilaments and intermediate filaments are tensioned while microtubules are compressed. It is assumed these eigenstresses lead to a self-stabilizing structure similar to tensegrity structures [15, 48–50, 94, 98]. Tensegrity structures are self-stabilizing truss systems. The characteristic feature of these structures is the stable shape which results from initial stresses of the bar elements. Tension and compression hold the balance in equilibrium, whereas the elements have torqueless connections.

In order to consider eigenstresses in the computation of cytoskeleton of osteocytes, a third fictious and stress-free configuration (denoted by B_f) is added to the ordinary continuum mechanical description for large deformations where B_0 is the initial configuration and B_t is the current configuration, see Fig. 21.

The deformation gradients $\mathbf{F}_0$ and $\tilde{\mathbf{F}}$ describe the geometrical mapping from the stress-free configuration to the initial and current configuration. Hence, $\tilde{\mathbf{F}}$ can be computed by a multiplicative split

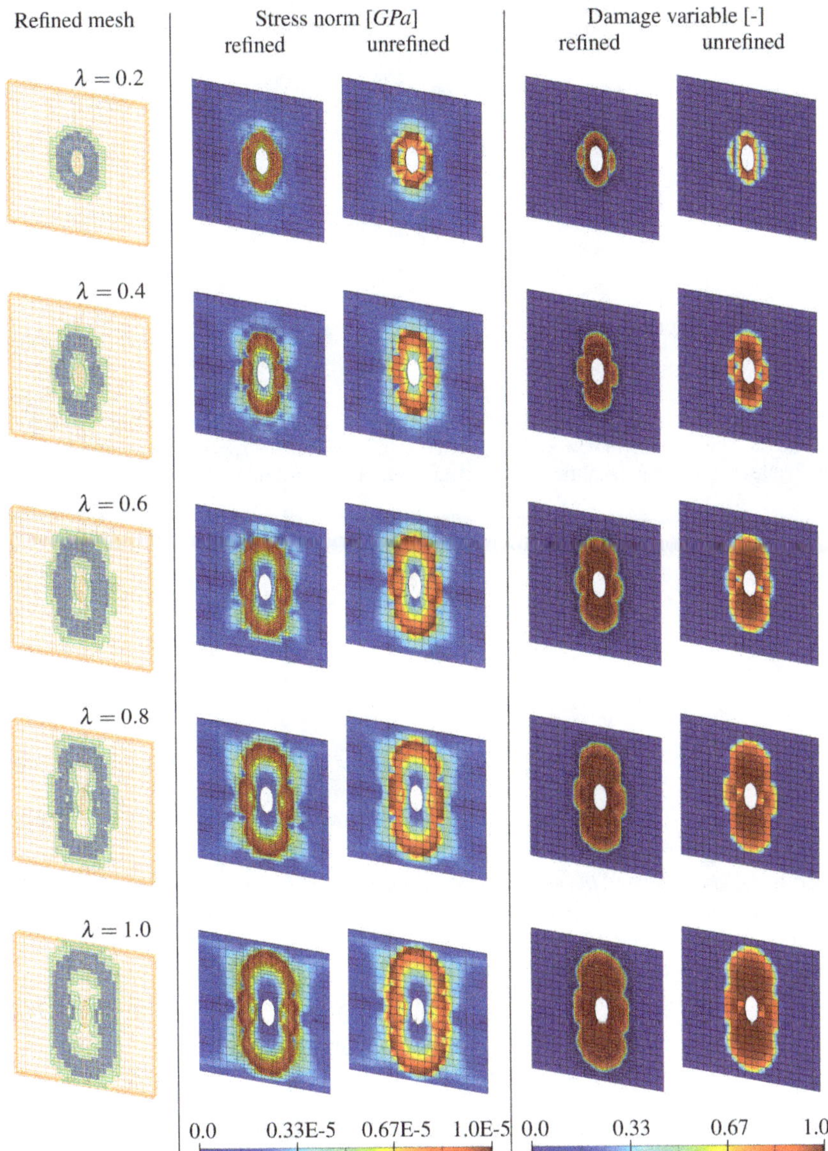

Fig. 19 Crack growth with adaptive mesh refinement at the crack tip.

$$\tilde{\mathbf{F}} = \mathbf{F}\,\mathbf{F}_0 = \frac{\partial \mathbf{x}}{\partial \mathbf{X}}\,\frac{\partial \mathbf{X}}{\partial \tilde{\mathbf{X}}} = \frac{\partial \mathbf{x}}{\partial \tilde{\mathbf{X}}}\,. \tag{34}$$

Herein, $\tilde{\mathbf{X}}$ denotes the position vector in the stress-free configuration. This leads to the internal part of the weak form of the balance of linear momentum as

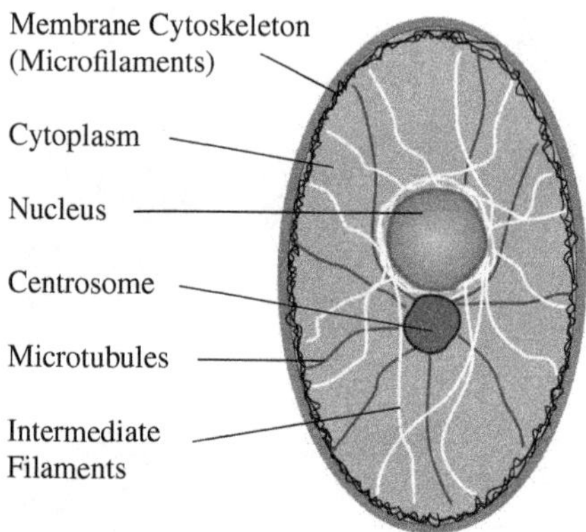

Fig. 20 Illustration of cytoskeleton components. Microfilaments build up the membrane-cytoskeleton. This protein framework is connected with the cell nucleus, via intermediate filaments. Microtubules connect the membrane-cytoskeleton with the centrosome.

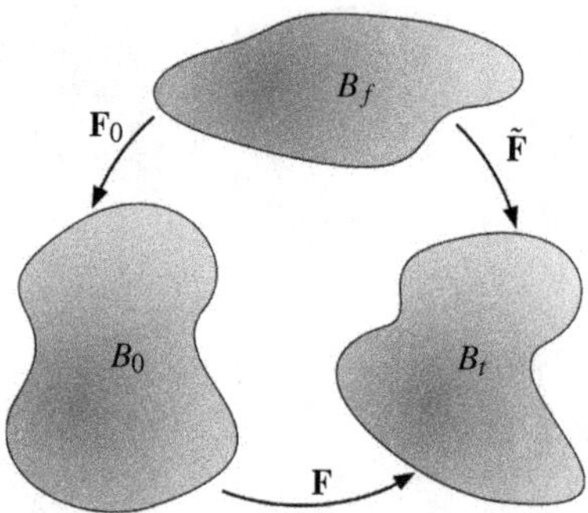

Fig. 21 Initial (B_0), current (B_t) and stress-free configuration (B_f) of a continuum. $\mathbf{F}_0$ maps infinitesimal line elements from the stress-free configuration to the initial configuration and $\tilde{\mathbf{F}}$ maps line elements from the stress-free configuration to the current configuration.

$$G^{\text{int}} := \int_{B_0} \frac{1}{J_0} \left(\tilde{\mathbf{F}} \, \tilde{\mathbf{S}} \, \mathbf{F}_0^{\mathrm{T}} \right) : \text{Grad}(\delta\mathbf{u}) \, dV. \tag{35}$$

J_0 is the Jacobian of $\mathbf{F}_0$ and $\tilde{\mathbf{S}}$ is the second Piola–Kirchhoff stress tensor in the stress-free configuration. In order to idealize the cytoskeleton network as truss framework, finally the internal force vector for bar elements results in

$$F^{\text{int}} = \frac{1}{\tilde{L}} \, \tilde{\mathbb{C}} \, \tilde{E} \, A \, \mathbf{Q} \,, \tag{36}$$

with the element length $\tilde{L}$, the Young's modulus $\tilde{\mathbb{C}}$, the Green–Lagrange strain $\tilde{E}$ and the cross section A. $\mathbf{Q}$ is a vector which describes the transformation into a 3-D coordinate system. The tilde symbols in (36) denote that the variables are present

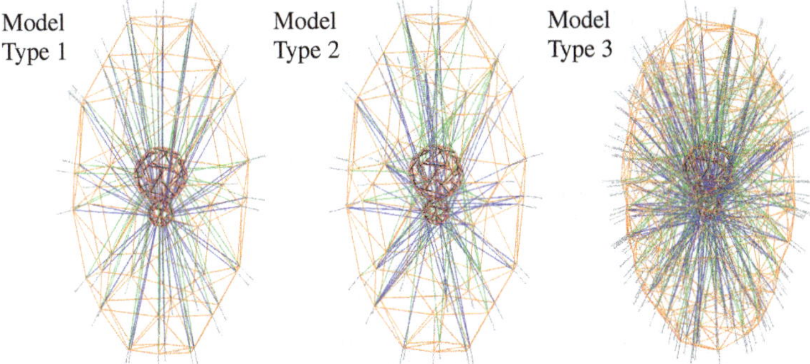

Fig. 22 Osteocyte cytoskeleton model approaches. Each model consists of microfilaments (orange), microtubules (blue), intermediate filaments (green), integrins (gray), the nucleus (upper sphere) and the centrosome (lower sphere).

in the stress-free configuration. The discretized system is defined by the tangential element stiffness matrix written as

$$\mathbf{K}_T = \frac{1}{\tilde{L}^3}\,\tilde{\mathbb{C}}\,A\,\mathbf{Q}\,\mathbf{Q}^{\mathrm{T}} + \frac{1}{\tilde{L}}\,\tilde{\mathbb{C}}\,\tilde{E}\,A \begin{bmatrix} \mathbf{I}_3 & -\mathbf{I}_3 \\ -\mathbf{I}_3 & \mathbf{I}_3 \end{bmatrix}. \tag{37}$$

Here $\mathbf{I}_3$ denotes the 3×3 identity tensor.

According to the cell structure shown in Fig. 20, the computational models of the cytoskeleton are shown in Fig. 22. These model types differ in the distribution of the cytoskeletal components where type 1 has a regular assembly. Type 3 has all protein fibers (microfilaments, microtubules and intermediate filaments) arranged randomly. In type 2 only the microfilaments are regular distributed. These different model types are studied and compared in detail in [53].

Results of the multiscale computation where the cells are embedded in their physiological environment (see Fig. 16) have shown that the randomized distributions of the cytoskeleton components has a significant effect on the mechanical loading of the cell nucleus. It can be assumed that the deformation of the extracellular bone matrix is directly transferred to the nucleus. In [53] it has been shown that the strain at the extracellular matrix is amplified by the cytoskeleton such that nucleus is strained much more. This supports the idea that the nucleus is the target of mechanosensation.

5 Conclusions

Model hierarchy in engineering mechanics starts with a pencil sketch on a sheet of paper, for example a free body diagram of a simple supported beam to compute support reactions and interior forces. Based on these results the stress of material is analyzed, e.g. the combination of bending normal stress and shear stress within a I-shape beam. As far as the boundary conditions at this level are well defined and the modeling approach is validated via suitable experiments, we can try to set up models for the explanation of the physical behavior on the next smaller length and time scale, e.g. to simulate the bonding of composites, etc. A broad variety of computational techniques for the treatment of this intuitive multiscale approach have been developed in the past. Nowadays we are on a way that computational mechanics meets computational material science.

In this contribution computational multiscale techniques have been developed for the more sophisticated modeling of contact problems. In the macroscopic picture contact is modeled by unilateral constraints which leads to a non-smooth mathematical description and related challenges for the computations. Motivated from the application of atomic force microscopes a sophisticated contact model based on atomic force interactions has been developed and implemented into a finite element system. The total device has been modeled in a dimension adaptive manner and simulated under static and dynamic conditions. In a second step the contact model has been transferred in to a consistently coupled MD-FE modeling approach. A powerful and fully adaptive 3-D implementation has been proven for sophisticated contact analysis.

An other line of engineering problems for which multiscale methods are needed has been outlined for the biomechanics of bones. Here the mechanics of bone cells in their physiological environment is of primal interest. A multiscale model for a cortical section of bone as a portion of the human femur has been developed. The computability of this model within a closed control circuit for growth and adaption has been proven. Because it is assumed that the need for maintenance of bone tissue is initiated by micro-cracks, an adaptive micro-crack model has been integrated. Furthermore a sophisticated cell model has been developed where the cytoskeleton is modeled as a tensegrity like structure. It has been shown, that there is a strain amplification from the local tissue strain to the nucleus of the cell.

Acknowledgement

The research presented in this chapter has been funded by the German Research Foundation under the Grant GRK 615. The authors acknowledge the financial support for this fruitful research training.

References

1. P.T. Bauman, J.T. Oden, and S. Prudhomme. Adaptive multiscale modeling of polymeric materials with Arlequin coupling and goals algorithms. *Computer Methods in Applied Mechanics and Engineering*, 198(5-8):799–818, 2009.
2. G.S. Baupre, T.E. Orr, and D.R. Carter. An approach for time-dependent bone modeling and remodeling-applications: A preliminary remodeling simulation. *Journal of Orthopaedic Research*, 8:662–670, 1990.

3. Y. Bazilevs, V.M. Calo, J.A. Cottrell, T.J.R. Hughes, A. Reali, and G. Scovazzi. Variational multiscale residual-based turbulence modeling for large eddy simulation of incompressible flows. *Comput. Methods Appl. Mech. Engrg.*, 197:173–201, 2007.

4. T. Beno, Y.-J. Yoon, S. C. Cowin, and S. P. Fritton. Estimation of bone permeability using accurate microstructural measurements. *Journal of Biomechanics*, 39:2378–2387, 2006.

5. V. Bentolila, T.M. Boyce, D.P. Fyhrie, R. Drumb, T.M. Skerry, and M.B. Schaffler. Intracortical remodeling in adult rat long bones after fatigue loading. *Bone*, 23:275–281, 1998.

6. B. Bhushan. *Springer Handbook of Nanotechnology*. Springer-Verlag, New York, 2004.

7. A.R. Bonivtch, L.F. Bonewald, and D.P. Nicolella. Tissue strain amplification at the osteocyte lacuna: A microstructural finite element analysis. *Journal of Biomechanics*, 40:2199–2206, 2007.

8. J.C. Brewer and P.A. Lagace. Quadratic stress criterion for initiation of delamination. *Quadratic Stress Criterion for Initiation of Delamination*, 22:1141–1155, 1988.

9. M. Brinkmeier and U. Nackenhorst. An approach for large-scale gyroscopic eigenvalue problems with application to high-frequency response of rolling tires. *Computational Mechanics*, 41(4):503–515, 2008.

10. J.Q. Broughton, F.F. Abraham, N. Bernstein, and E. Kaxiras. Concurrent coupling of length scales: Methodology and application. *Physical Review B – Condensed Matter and Materials Physics*, 60(4):2391–2403, 1999.

11. E.H. Burger and J. Klein-Nulend. Mechanotransduction in bone – Role of the lacunocanalicular network. *FASEB Journal*, 13:S101–S112, 1999.

12. D.B. Burr. Targeted and nontargeted remodeling. *Bone*, 30:2–4, 2002.

13. D.B. Burr, R.B. Martin, M.B. Schaffler, and E.L. Radin. Bone remodeling in response to in-vivo fatigue microdamage. *Journal of Biomechanics*, 18:189–200, 1985.

14. N. Charalambakis. Homogenization techniques and micromechanics. A survey and perspectives. *Applied Mechanics Review*, 63, 2010.

15. C.S. Chen and D.E. Ingber. Tensegrity und Mechanoregulation: Vom Skelett zum Zytoskelett. *Osteopathische Medizin, Zeitschrift für ganzheitliche Heilverfahren*, 9:4–17, 2008.

16. S.C. Cowin. Bone poroelasticity. *Journal of Biomechanics*, 32(3):217–238, 1999.

17. S.C. Cowin (Ed.). *Bone Mechanics Handbook*. CRC Press, 2001.

18. S.C. Cowin and D.H. Hegedus. Bone remodeling I: Theory of adaptive elasticity. *Journal of Elasticity*, 6:313–326, 1976.

19. A. Curnier. Unilateral contact mechanical modelling. In P. Panagiotopoulos and P. Wriggers (Eds.), *CISM Lecture Series*. Springer, 1999.

20. D.D. Deligianni and C.A. Apostolopoulos. Multilevel finite element modeling for the prediction of local cellular deformation in bone. *Biomechanics and Modeling in Mechanobiology*, 7:151–159, 2008.

21. H.B. Dhia and G. Rateau. The Arlequin method as a flexible engineering design tool. *International Journal of Numerical Methods in Engineering*, 62:1442–1492, 2005.

22. M. Doblare and J.M. García. Anisotropic bone remodelling model based on a continuum damage-repair theory. *Journal of Biomechanics*, 35(1):1–17, 2002.

23. M. Dobson, R. Elliott, M. Luskin, and E. Tadmor. A multilattice quasicontinuum for phase transforming materials: Cascading cauchy born kinematics. *Journal of Computer-Aided Materials Design*, 14:219–237, 2007.

24. J. Dolbow, N. Moes, and T. Belytschko. Discontinuous enrichment in finite elements with a partition of unity method. *Finite Elements in Analysis and Design*, 36:235–260, 2000.

25. T. Ebinger, S. Diebels, and H. Steeb. Numerical homogenization techniques applied to growth and remodelling phenomena. *Computational Mechanics*, 39(6):815–830, 2007.

26. F. Ercolessi and J. Adams. Interatomic potentials from first-principles calculations: the force-matching method. *Europhysics Letters*, 26:583–588, 1994.

27. A. C. Erringen. *Microcontinuum Field Theories*. Springer, 1998.

28. J.D. Eshelby. The determination of the elastic field of an ellipsoidal inclusion, and related problems. *Proceedings of the Royal Society of London. Series A, Mathematical and Physical Sciences*, 241:376–396, 1957.

29. J. Fish, M. Nuggehally, M. Shephard, C. Picu, S. Badia, M. Park, and M. Gunzburger. Concurrent AtC coupling based on a blend of the continuum stress and the atomistic force. *Computer Methods in Applied Mechanics and Engineering*, 196:4548–4560, 2007.

30. T. Floerkemeier, A. Lutz, U. Nackenhorst, F. Thorey, H. Waizy, H. Windhagen, and G. von Lewinski. Core decompression and osteonecrosis intervention rod in osteonecrosis of the femoral head: Clinical outcome and finite element analysis. *International Orthopaedics*, DOI 10.1007/s00264-010-1138-x, 2010.

31. H.M. Frost. Presence of microscopic cracks in vivo in bone. *Bull Henry Ford Hosp*, 8:25–35, 1960.

32. H.M. Frost. Bone microdamage: Factors that impair its repair. In H.K. Uhthoff (Ed.), *Current Concepts in Bone Fragility*, pages 123–148, Springer, 1985.

33. M.G.D. Geers. Experimental analysis and computational modelling of damage and fracture. PhD Thesis, Technische Universiteit Eindhoven, 1997.

34. M.G.D. Geers, V.G. Kouznetsova, and W.A.M. Brekelmans. Multi-scale computational homogenization: Trends and challenges. *Journal of Computational and Applied Mathematics*, DOI 10.1016/j.cam.2009.08.77, 2009.

35. N.M. Ghoniem, E.P. Busso, N. Kioussis, and H. Huang. Multiscale modelling of nanomechanics and micromechanics: An overview. *Philosophical Magazime*, 83(31–34):3475–3528, 2003.

36. V. Gravemeier. Scale-separating operators for variational multiscale large eddy simulation of turbulent flows. *Journal of Computational Physics archive*, 212(2):400–435, 2006.

37. J.C. Halpin and J.L. Kardos. The Halpin–Tsai equations: A review. *Polymer Engineering and Science*, 16(5):344–352, 1976.

38. H.C. Hamaker. The London–van der Waals attraction between spherical particles. *Physica*, 4:1058–1072, 1937.

39. R. Hardy. Formulas for determing local properties in molecular-dynamics simulations: Shock waves. *Journal of Chemical Physics*, 76:622–628, 1982.

40. Z. Hashin and S. Shtrikman. Note on a variational approach to the theory of composite elastic materials. *J. Franklin Inst.*, 271:336–341, 1961.

41. J.G. Hazenberg, M. Freeley, E. Foran, T.C. Lee, and D. Taylor. Microdamage: A cell transducing mechanism based on ruptured osteocyte processes. *Journal of Biomechanics*, 39:2096–2103, 2006.

42. M.O. Heller, G. Bergmann, G. Deuretzbacher, L. Dürselen, M. Pohl, L. Claes, N.P. Haas, and G.N. Duda. Musculo-skeletal loading conditions at the hip during walking and stair climbing. *Journal of Biomechanics*, 34(7):883–893, 2001.

43. T. Helmich. Elektromechanisch gekoppelte Kontaktmodellierung auf Mikroebene. PhD Thesis, Universität Hannover, 2007.

44. R. Hill. Elastic properties of reinforced solids: Some theoretical principles. *J. Mech. Phys. Solids*, 11:357–372, 1963.

45. R. Hill. A self-consistent mechanics of composite materials. *J. Mech. Phys. Sol.*, 13:213–222, 1965.

46. T. Hughes. Multiscale phenomena: Green's functions, the dirichlet-to-neumann formulation, subgrid scale models, bubbles and the origin of stablized methods. *Computer Methods in Applied Mechanics and Engineering*, 127:387–401, 1995.

47. T. Hughes, G. Feijo, L. Mazzei, and J. Quincy. The variational multiscale method – A paradigm for computational mechanics. *Computer Methods in Applied Mechanics and Engineering*, 166:3–24, 1998.

48. D.E. Ingber. Cellular tensegrity: Defining new rules of biological design that govern the cytoskeleton. *Journal of Cell Science*, 104:613–627, 1993.

49. D.E. Ingber. Tensgrity: The architectual basis of cellular mechanotransduction. *Annual Review of Physiology*, 59:575–599, 1997.

50. D.E. Ingber. Tensegrity I. Cell structure and hierarchical systems biology. *Journal of Cell Science*, 116:1157–1173, 2003.

51. C.R. Jacobs, J.C. Simo, G.S. Beaupre, and D.C. Carter. Adaptive bone remodeling incorporating simultaneous density and anisotropy considerations. *Journal of Biomechanics*, 6:603–613, 1997.

52. G. Kamm and G. Alers. Low-temperature elastic moduli of aluminium. *Journal of Applied Physics*, 35:327–330, 1964.

53. D. Kardas. A multiscale computational approach for microcrack evolution in cortical bone and related mechanical stimulation of bone cells. PhD Thesis, Institut für Baumechanik und Numerische Mechanik, Leibniz Universität Hannover, 2010.

54. V. Kouznetsova, W.A.M. Brekelmans, and F.P.T. Baaijens. Approach to micro-macro modeling of heterogeneous materials. *Computational Mechanics*, 27(1):37–48, 2001.

55. N. Krstin, U. Nackenhorst, and R. Lammering. Zur konsitutiven Beschreibung des anisotropen beanspruchungsadaptiven Knochenumbaus. *Technische Mechanik*, 20(1):31–40, 2000.

56. E. Kuhl, A. Menzel, and P. Steinmann. Computational modeling of growth. A critical review, a classification of concepts and two new consistent approaches. *Computational Mechanics*, 32(1–2):71–88, 2003.

57. T. Laursen. *Computaional Contact and Impact Mechanics*. Springer, 2003.

58. H. Leng, X. Wang, R.D. Ross, G.L. Niebur, and R.K. Roeder. Micro-computed tomography of fatigue microdamage in cortical bone using a barium sulfate contrast agent. *Journal of the Mechanical Behavior of Biomedical Materials*, 1:68–75, 2008.

59. C. Lenz. Numerical micro-meso modelling of mechanosensation driven osteonal remodelling in cortical bone. PhD Thesis, Institut für Baumechanik und Numerische Mechanik, Universität Hannover, 2005.

60. E.M. Lifshitz. The theory of molecular attractive forces between solids. *Soviet Physics JETP (Engl. Transl.)*, 2:73–83, 1956.

61. B. Liu, Y. Huang, H. Jiang, S. Qu, and K.C. Hwang. The atomic-scale finite element method. *Computational Methods in Applied Mechanics and Engineering*, 193:1849–1864, 2004.

62. G. Lu, E.B. Tadmor, and E. Kaxiras. From electrons to finite elements: A concurrent multiscale approach for metals. *Physical Review B – Condensed Matter and Materials Physics*, 73(2):1–4, 2006.

63. J. Lutsko. Stress and elastic constants in anisotropic solids: Molecular dynamics techniques. *Journal of Applied Physics*, 64:1152–1154, 1988.

64. A. Lutz and U. Nackenhorst. Numerical investigations on the biomechanical compatibility of hip-joint endoprostheses. *Archive of Applied Mechanics*, 80(5):503–512, 2010.

65. A. Lutz, U. Nackenhorst, G. von Lewinski, H. Windhagen, and T. Floerkemeier. Numerical studies on alternative therapies for femoral head necrosis – A finite element approach and clinical experience. *Biomechanics and Modeling in Mechanobiology*, DOI 10.1007/s10237-010-0261-3, 2010.

66. J. Mahanty and B.W. Ninham. *Dispersion Forces*. Academic Press, 1976.

67. G. Marotti, M. Ferretti, M.A. Muglia, C. Palumbo, and S. Palazzini. A quantitative evaluation of osteoblast-osteocyte relationships on growing endosteal surface of rabbit tibiae. *Bone*, 13:363–368, 1992.

68. R.B. Martin. Toward a unifying theory of bone remodeling. *Bone*, 26:1–6, 2000.

69. R.B. Martin, D.B. Burr, and N.A. Sharkey. *Skeletal Tissue Mechanics*. Springer, 1998.

70. B.R. McCreadie and S.J. Hollister. Strain concentrations surrounding an ellipsoid model of lacunae and osteocytes. *Computer Methods in Biomechanics and Biomedical Engineering*, 1:61–68, 1997.

71. J.G. McGarry, J. Klein-Nulend, M.G. Mullender, and P.J. Prendergast. A comparsion of strain and fluid shear stress in stimulating bone cell responses – A computational and experimental study. *The FASEB Journal*, 19:482–484, 2005.

72. A.D. McLachlan. Three-body dispersion forces. *Molecular Physics*, 6(4):423–427, 1963.

73. J.M. Melenk and I. Babuska. The partition of unity finite element method: Basic theory and applications. *Computer Methods in Applied Mechanics and Engineering*, 139(1–4):289–314, 1996.

74. R. Miller and E. Tadmor. A unified framework and performance benchmark of fourteen multsicale/continuum coupling methods. *Modelling and Simulation in Materials Science and Engineering*, 17:1–51, 2009.

75. R.E. Miller and E.B. Tadmor. The quasicontinuum method: Overview, applications and current directions. *J. Comp.-Aid. Mat. Design*, 9:203–239, 2002.

76. T. Mori and K. Tanaka. Average stress in matrix and average elastic energy of materials with misfitting inclusions. *Acta Metallurgica*, 21:571–574, 1973.

77. M. Mullender, A.J. El Haj, Y. Yang, M.A. van Duin, E.H. Burger, and J. Klein-Nulend. Mechanotransduction of bone cells in vitro: Mechanobiology of bone tissue. *Medical & Biological Engineering & Computing*, 42:14–21, 2004.

78. T. Mura. *Micromechanics of Defects in Solids*. Springer, 1987.

79. U. Nackenhorst. Biomechanics of bones: Modeling and computation of bone remodeling. In *Handbook of Biomineralization*, chapter 3, pages 35–48, Wiley VCH, 2007.

80. S. Nemat-Nasser and M. Hori. *Micromechanics: Overall Properties of Heterogeneous Solids*. North-Holland, Amsterdam, 1993.

81. D.P. Nicolella, D.E. Moravits, A.M. Gale, L.F. Bonewald, and J. Lankford. Osteocyte lacunae tissue strain in cortical bone. *Journal of Biomechanics*, 39:1735–1743, 2006.

82. T.L. Norman and Z. Wang. Microdamage of human cortical bone: Incidence and morphology in long bones. *Bone*, 20:375–379, 1997.

83. J.T. Oden, T. Belytschko, I. Babuska, and T.J.R. Hughes. Research directions in computational mechanics. *Computer Methods in Applied Mechanics and Engineering*, 192(7–8):913–922, 2003.

84. H.S. Park and W.K. Liu. An introduction and tutorial on multiple-scale analysis in solids. *Computer Methods in Applied Mechanics and Engineering*, 193:1733–1772, 2004.

85. V.A. Parsegian. Long-range physical forces in the biological milieu. *Annual Review of Biophysics and Bioengineering*, 2:221–255, 1973.

86. F. Pauwels. *Gesammelte Abhandlungen zur Funktionalen Anatomie des Bewegungsapparates*. Springer, 1965.

87. B.N.J. Persson. *Sliding Friction: Physical Principles and Applications (Nanoscience and Technology)*. Springer, 2000.

88. B.N.J. Persson. Theory of rubber friction and contact mechanics. *Journal of Chemical Physics*, 115(8):3840–3861, 2001.

89. T. Preisner, M. Greiff, U.B. Bala, and W. Mathis. Numerical computation of magnetic fields applied to magnetic force microscopy. *COMPEL*, 28(1):120–129, 2009.

90. J.Y. Rho, L. Kuhn-Spearing, and P. Zioupos. Mechanical properties and the hierarchical structure of bone. *Medical Engineering and Physics*, 20:92–102, 1998.

91. J.C. Rice, S.C. Cowin, and J.A. Bowman. On the dependence of elasticity and strength of concellous bone apparent density. *Journal of Biomechanics*, 21:155–168, 1988.

92. W. Shan. Multiscale coupling based on the quasicontinuum framework, with application to contact problems. PhD Thesis, Leibniz University Hannover, 2009.

93. W. Shan and U. Nackenhorst. An adaptive FE-MD model coupling approach. *Computational Mechanics*, 46(4):577–596, 2010.

94. D. Stamenovic, J. Fredberg, N. Wang, J.P. Butler, and D.E. Ingber. A microstructural approach to cytoskeletal mechanics based on tensegrity. *Journal of Theoretical Biology*, 181:125–136, 1996.

95. E. Tadmor, M. Ortiz, and R. Phillips. Quasicontinuum analysis of defects in solids. *Philosophical Magazine A*, 73:1529–1563, 1996.

96. D. Taylor. Fracture and repair of bone: A multiscale problem. *Journal of Material Science*, 42:8911–8918, 2007.

97. W.R. Taylor, M.O. Heller, G. Bergmann, and G.N. Duda. Tibio-femoral loading during human gait and stair climbing. *Journal of Orthopaedic Research*, 22(3):625–632, 2004.

98. N. Wang, K. Naruse, D. Stamenovic, J.J. Fredberg, S.M. Mijailovich, I. M. Tolic-Nørrelykke, T. Polte, R. Mannix, and D.E. Ingber. Mechanical behavior in living cells consistent with the tensegrity model. *Proceedings of the National Academy of Sciences of the United States of America*, 98:7765–7770, 2001.

99. H. Weinans, R. Huiskes, and H.J. Grootenboer. The behavior of adaptive bone remodeling simulation models. *Journal of Biomechanics*, 25:1425–1441, 1992.
100. H. Weinans, R. Huiskes, and H.J. Grootenboer. Effects of fit and bonding characteristics of femoral stems on adaptive bone remodeling. *Journal of Biomechanical Engineering*, 116:393–400, 1994.
101. S. Weinbaum, S.C. Cowin, and Y. Zeng. A model for the excitation of osteocytes by mechanical loading-induced bone fluid shear stresses. *Journal of Biomechanics*, 27:339–360, 1994.
102. J. Wolff. *Das Gesetz der Transformation der Knochen*, 1982.
103. P. Wriggers. *Computational Contact Mechanics*. Springer, 2006.
104. P. Wriggers. *Nonlinear Finite Element Methods*. Springer, Berlin, 2008.
105. L. You, S.C. Cowin, M.B. Schaffler, and S. Weinbaum. A model for strain amplification in the actin cytoskeleton of osteocytes due to fluid drag on pericellular matrix. *Journal of Biomechanics*, 34:1375–1386, 2001.
106. J. Zimmerman, E. Webb III, J. Hoyt, R. Jones, P. Klein, and D. Bammann. Calculation of stress in atomistic simulation. *Modelling and Simulation in Materials Science and Engineering*, 12:S319–S332, 2004.
107. P.K. Zysset and A. Cournier. An alternative model for anisotropic elasticity based on fabric tensors. *Mechanics of Materials*, 21(4):243–250, 1995.

Numerical Modelling and Simulation of Atomic Force Microscopes

Wolfgang Mathis, Thomas Preisner and Uzzal B. Bala

Abstract Electrostatic force microscopes (EFM) and magnetic force microscopes (MFM) are very important tools for the investigation of electric and magnetic properties at the nanometer scale. Basically these measurement instruments are atomic force microscopes (AFM) which operate in a non-contact mode. In this operating mode if some requirements for the measurement setup are fulfilled, the electrostatic and the magnetic force, respectively, become the main interaction between the sharp tip and a sample surface. In this chapter we discuss concepts for modelling EFMs and MFMs and consider some numerical aspects of solving the descriptive equations of these models. Moreover some numerical results are presented.

1 Atomic Force Microscopes

Significant progress in nanotechnology has been observed over the last twenty years. This progress has also been influenced by the development of new high resolution measurement instruments. Due to the rapid miniaturization of integrated devices into the mesoscopic regime and the increasing interest in very small structures, these instruments have become very important. An interesting example is the AFM. Based on the design of the scanning tunneling microscope (STM), the first AFM was developed in 1986 by Binnig and his coworkers in collaboration between IBM and Stanford university (see e.g. [1]). Since then a new era of topographical imaging, as well as for measuring force-separation interactions between a probe and substrate began. The ability of AFMs to scan surfaces with nearly atomic resolution and their versatility make them one of the most important measurement devices in nanotechnology. Since their functionality depend on the interaction between the sample and the AFM tip different types of AFMs are available.

Wolfgang Mathis · Thomas Preisner · Uzzal B. Bala
Institut für Theoretische Elektrotechnik, Leibniz Universität Hannover, Appelstr. 9A,
30167 Hannover, Germany; e-mail: {mathis, preisner}@tet.uni-hannover.de

E. Stephan & P. Wriggers (Eds.): Modelling, Simulation and Software Concepts, LNACM 57, pp. 169–180.
springerlink.com　　　　　　　　　　　　　　　　© Springer-Verlag Berlin Heidelberg 2011

Basically AFMs consist of a cantilever where a sharp tip is placed underneath. If the sample under investigation holds a charge distribution and the distance between the conducting AFM tip and the sample is kept sufficient large then all other interaction forces except the electrostatic force can be neglected. This special working mode of the AFM is known as EFM which can be used for scanning electric fields with nearly atomic resolution. The EFM has many materials-related applications including measuring the surface potential or contact potential, detecting charges on surfaces or nanocrystals, etc. Another concept is the MFM where the AFM tip is coated with magnetic material and the force between the tip and the magnetic surface is measured. In each case concerning a numerical model these types of AFMs consist of a mechanical and an electrical or magnetic part and therefore the mechanical force has to be calculated in a very accurate manner from the electrical or magnetic field.

In this chapter we consider 2D and 3D physical modelling of EFM and MFM and some numerical methods are proposed to calculate the electric or magnetic field with high accuracy and very efficiently. Based on the electromagnetic fields the mechanical force density at the cantilever has to be calculated where different concepts are discussed and compared with respect to global as well as local forces. Due to the origin of these mechanical forces we denote them in the following as electromagnetic forces. Finally these approaches will be illustrated by means of some examples.

2 Physical Foundations of AFMs

For the EFM and MFM, respectively, the dominant interaction force would be the electromagnetic force between the biased atomically sharp tip and the sample, that is the electrostatic force for EFMs and the magnetic force for MFMs. In addition van der Waals forces between the tip and the sample are always present. However van der Waals forces and the electromagnetic forces have different dominant regions of attraction since van der Waals forces are proportional to $1/r^6$ whereas the electromagnetic force is proportional to $1/r^2$. Thus when the tip is close to the sample van der Waals forces are dominant. If the tip is moved away from the sample the electromagnetic force is dominant. The scanning process of EFMs/MFMs is usually done in two steps. First the topography of the sample is done by tapping scanning mode which is also known as intermittent contact (IC) mode. In this case van der Waals forces play a significant role. Second using this topographical information a constant tip-sample distance is maintained while scanning, where the electromagnetic force is dominant, a technique which is known as lift scanning. In this technique it is assumed that the influence of all short-range forces can be neglected and only the electromagnetic force plays the vital role for surface imaging. To detect electrostatic forces in EFMs between the tip and the material surface a voltage is applied between the cantilever tip and the sample. The cantilever oscillates near its resonance frequency which changes in response to any additional force gradient. A diode

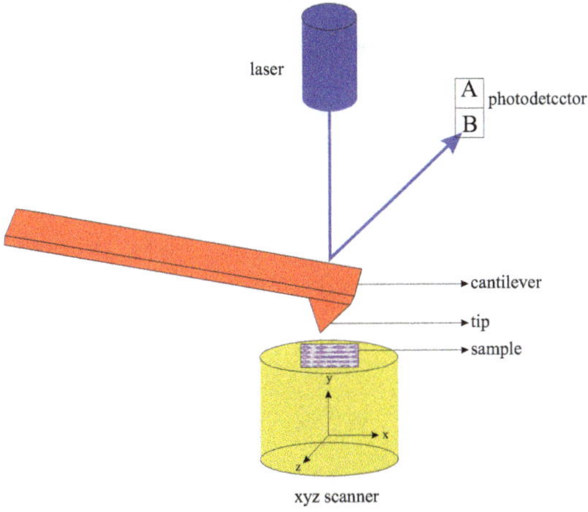

Fig. 1 Experimental setup of the EFM and MFMs [14].

laser is focused on the back of the reflective cantilever and the reflected light is col-
lected by a position sensitive detector (PSD). This usually consists of two (or four)
closely spaced photodiodes. Any angular displacement of the cantilever results in
collecting more light in one of the photodiodes and a higher output voltage is generated in this diode. This voltage then plots the local intensity of the electrostatic field
of the sample. In a similar way a MFM is working where forces between the tip and
the surface are measured by means of a tip that is coated with a magnetic material. The experimental setup of EFMs and MFMs is shown in Fig. 1. Some typical
geometric parameters of EFMs and MFMs, i.e., the length of the cantilever is about
200–300 µm, the height of the tip is nearly 5–20 µm and the end of the tip is usually
less than 10 nm. So for modelling and simulating EFMs and MFMs, different physical aspects must be taken into consideration. In the consequence we are confronted
from the numerical point of view with a multi-scale problem. Therefore the application of advanced numerical methods is necessary. As the cantilever frequently
changes its position during scanning, the coupled mechanical and electromagnetic
behavior have to be taken into account. This can be achieved by dividing the model
into an electromagnetic and a mechanical part; see Fig. 2. The interaction between
them can conveniently be modelled by using a staggered simulation approach where
the mechanical and electromagnetic model equations are solved successively. The
results from one part are used as input data for the next part such that we end up
with an iterative approach. Otherwise a combined physical model can be developed
as well [31]. For developing a model of EFMs and MFMs different effects have to
be considered. For example long distance interaction, charge distribution and possible non-linearity of the material properties, singularity etc. In order to take into
consideration these effects the simulation region is divided into several subregions.

In the following we consider modelling equations of EFMs and MFMs and discuss some numerical results.

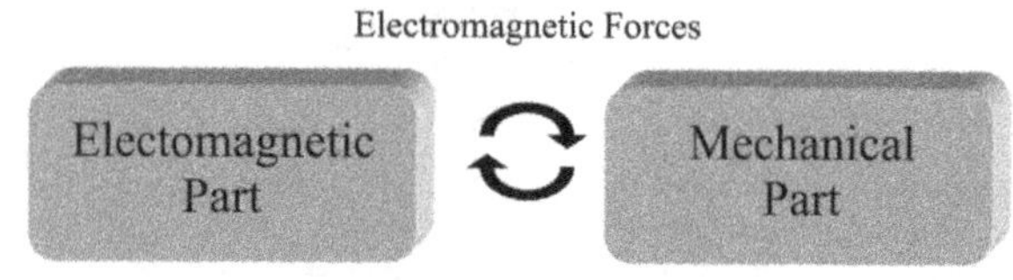

Fig. 2 Electromagnetic and mechanical parts of EFM/MFM models.

2.1 Mechanical Foundations

In the mechanical part the deflection of the cantilever and its tip, respectively, due to the influence of the electromagnetic force is calculated. The forces which are calculated at each point of the cantilever in the electromagnetic part will be used as input for mechanical part. The governing equations and the boundary conditions of the mechanical part are

$$\mathrm{div}(\sigma) + \rho_M \mathbf{f} = 0 \tag{1}$$

$$\sigma = \lambda \left(\nabla \cdot \mathbf{u} \right) \mathbf{I} + \mu \left(\nabla \mathbf{u} + \nabla \mathbf{u}^{tr} \right) \tag{2}$$

$$\mathbf{u} = \mathbf{u}_0 \ \text{on} \ \Gamma_u$$
$$\sigma \mathbf{n} = \mathbf{f_e} \ \ \text{on} \ \Gamma_e \tag{3}$$

$$\Gamma_e = \Gamma_e(\mathbf{u})$$

where eq. (1) is the local form of conservation of linear momentum for the static case where σ is the Cauchy stress tensor, ρ_M the mass density, and $\mathbf{f}$ the inner force, as well as eq. (2) is the constitutive equation (material law) for a linear isotropic material where $\mathbf{u}$ is the displacement vector, λ and μ are the Lamé coefficients and $\mathbf{I}$ is the identity tensor and eq. (3) are boundary conditions for displacement and force (normal component) respectively. By utilizing the above equations, boundary conditions and the forces acting at each point of the cantilever, deflection will be calculated at each point of the cantilever as shown in Fig. 3. As a result the cantilever will move to a new position. This deflection will be used in the electromagnetic part of the model to modify the geometry. Considering this new position the calculations of the electromagnetic part will start again. Therefore, an efficient modelling of the coupling between electromagnetic and mechanical parts is necessary.

3 Electrostatic Force Microscopes (EFM)

The EFM is used to investigate the electrical properties of a material surface. It is a non-contact AFM sensitive to the variations in the potential difference between

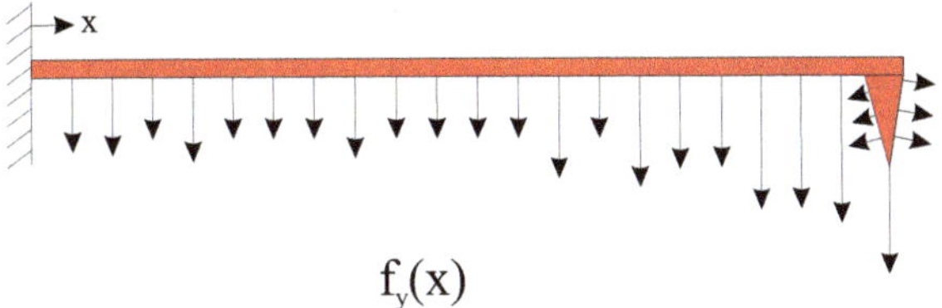

Fig. 3 Beam model of an AFM cantilever [14].

the sharp tip and the sample. As already mentioned in Section 1 the electric field is the dominant interaction between the AFM tip and a surface. Although the van der Waals force is always present between the tip and the sample it can be omitted with respect to the distance between tip and sample. During the scanning process of EFM, the sharp tip oscillates near its resonance frequency, and the phase and frequency shift is proportional to its electrostatic force gradient. Most force gradient comes from the end of the sharp tip.

Since the scanning velocity of such an AFM is rather low the electric field $\mathbf{E}$ can be approximated using the negative gradient of the electrostatic potential φ, that is $\mathbf{E} = -\nabla \varphi$. The potential φ can be calculated in regions holding a volume charge ρ by solving the Poisson equation

$$\nabla \cdot (\varepsilon \nabla \varphi) = -\rho \tag{5}$$

taking into account given boundary conditions.

Since the EFM geometry is rather complex a numerical approach is needed for solving the electrostatic problem. It seems to be a standard problem in numerical electromagnetic field simulation such that, e.g., the finite element method (FEM) can be applied. However the tip is very sharp – comparing to the other geometric objects of the EFM – such that singularities of the electrical field arise at the edges and corners. Furthermore because of the scanning process of an AFM its geometry changes in time with respect to the rough surface. Moreover, for a suitable interpretation of the measurement data the distance between the tip and the surface has to be constant during the electrical measurement such that the cantilever has to be changed its position in dependence of the surface topology. Therefore advanced numerical techniques have to be applied.

With respect to the field singularities special methods exist in order to solve the 2D and 3D Poisson equation (5) with high accuracy in regions around edges and corners (see e.g. [21]). This method is called augmented FEM. However at least in 3D problems an adaptive grid of the 3D region can be an efficient alternative approach.

The above mentioned time-varying geometry of the EFM model has to be taken into account in the electrostatic simulation. Therefore the FEM mesh has to be adapted to the changing geometry in each time step. This can be done by running the mesh generator again, but in order to save calculation time, another approach is used in our work. The mesh in the electrostatic part of the model is treated as a massless elastic net which is deformed by the changing boundaries. In this arbitrary Lagrangian Eulerian (ALE) approach a vector Laplace equation

$$\Delta \mathbf{v} = \mathbf{0} \tag{6}$$

is solved for the mesh deformation $\mathbf{v}$ by using a linear FEM, where the movement of sample and probe is brought in as Dirichlet boundary condition $\mathbf{v} = \mathbf{v}_0$. At the boarder between FEM and BEM regions the normal component of $\mathbf{v}$ is fixed while the tangential components $\mathbf{v}_t$ are kept loose and therefore treated as Neumann boundary condition. Some more details about using the ALE concept in EFM modelling can be found in [22].

The coupling between FEM and the boundary element method (BEM) can be used as very powerful numerical simulation method for the modelling and simulation of MEMS. Since only the boundary of the FEM volume needs to be discretised in the BEM, it reduces the complexity of the problem by one dimension which saves time for the discretisation and in the same way some computation time. Another important point is that FEM always requires a bounded domain whereas BEM can work with models which has an unbounded exterior domain. This allows the BEM to deal with problems which have unbounded open geometry like in potential problems with long-range potentials, e.g. in electrostatics. With respect to the FEM/BEM coupling the reader is reffered to, e.g., [11]. Further details about the formulation of the FEM-BEM equations for EFMs and their numerical implementation are included in the monograph of Bala [14] and further publications [15, 16].

Electrostatic forces are used as an input for the mechanical calculation of the cantilever deflection. Since the electrostatic field is singular near corners or edges, the results obtained by numerical calculation will lack accuracy in those regions. Therefore special numerical techniques for robust and accurate calculations of the force on pieces of the cantilever surface including the tip are needed. This can be done by using the Maxwell stress tensor

$$\mathbf{T}_e = \begin{pmatrix} \varepsilon(E_x^2 - \frac{1}{2}\|\mathbf{E}\|^2) & \varepsilon E_x E_y & \varepsilon E_x E_z \\ \varepsilon E_x E_y & \varepsilon(E_y^2 - \frac{1}{2}\|\mathbf{E}\|^2) & \varepsilon E_y E_z \\ \varepsilon E_x E_z & \varepsilon E_y E_z & \varepsilon(E_z^2 - \frac{1}{2}\|\mathbf{E}\|^2) \end{pmatrix} \tag{7}$$

where the force $\mathbf{f}_e$ can be obtained by evaluating the following surface integral of the Maxwell's stress tensor

$$\mathbf{f_e} = \iint_A \mathbf{T_e} d\mathbf{A} \tag{8}$$

where A denotes a piece of the cantilever surface. Note that higher accuracy of the force calculation can be obtained if A is modified in a certain sense; see [18] for more details.

For a certain timestep numerical results of the electrical potential and the corresponding electrical field of a EFM configuration are shown in Fig. 4. We observe the highest electrical field (color "red" to "yellow") is around the end of the tip of the cantilever and therefore the maximum mechanical force appears at this point.

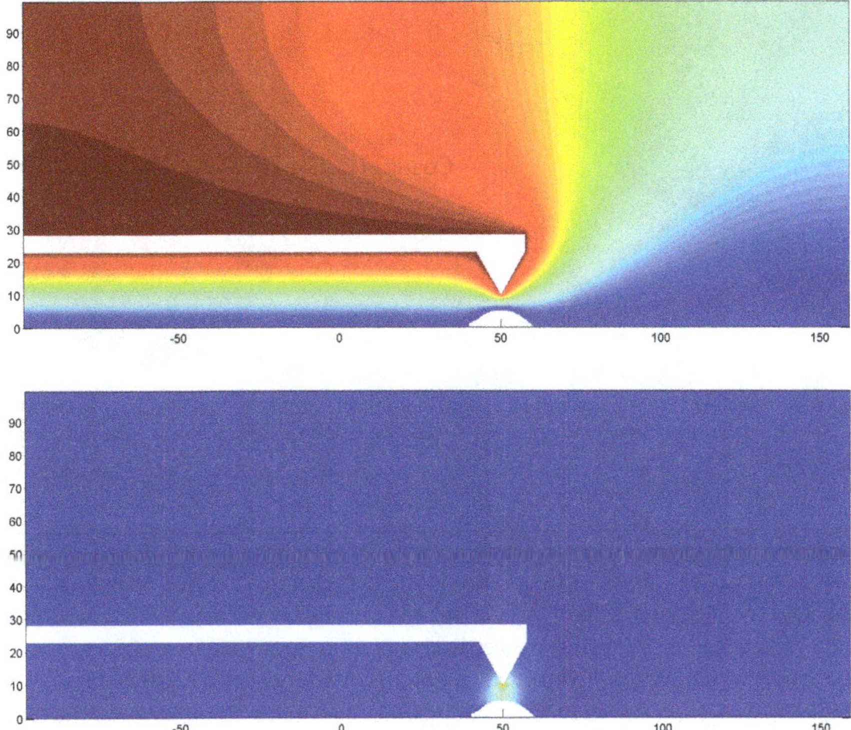

Fig. 4 Calculated potential and electric field using augmented FEM, linear FEM and BEM of a certain timestep where the white region (roughness of the surface material) is located under the EFM tip [14].

4 Magnetic Force Microscopes (MFM)

In order to simulate a MFM scanning process, the magnetic interaction between tip and sample surface has to be considered. As it was also assumed for the EFM, the scanning velocity is rather low. This assumption offers the opportunity to describe the problem in a magneto-static sense. For current free regions a magnetic scalar potential approach is used, where the magnetic field strength $\mathbf{H}$ can be calculated by $\mathbf{H} = -\nabla\phi$ and the magnetic scalar potential ϕ should solve the Poisson equation

$$\nabla \cdot (\mu \nabla \phi) = \mu_0 \nabla \cdot \mathbf{M}, \tag{9}$$

where $\mathbf{M}$ is the material magnetization of the magnetic domains and μ is the permeability. If the sample consists of electric currents the curlcurl equation is solved here

$$\nabla \times \frac{1}{\mu}\nabla \times \mathbf{A} = \nabla \times \frac{\mu_0}{\mu}\mathbf{M} + \mathbf{J}, \tag{10}$$

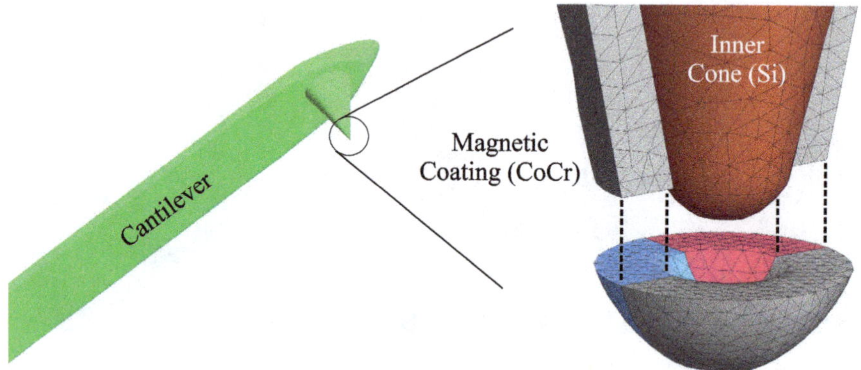

Fig. 5 An example for the cantilever and coated tip model. On the left side the large cantilever ($\approx 200\ \mu$m) is shown, whereas on the right side the magnetic coated tip and the end of the tip are demonstrated [24].

where **A** is the magnetic vector potential and **J** is the current density. The magnetic induction **B** can be obtained by solving eq. (10) and applying the curl-operator to the magnetic vector potential; $\mathbf{B} = \nabla \times \mathbf{A}$. Due to the complex geometry of the investigated problem and the large differences in size (10 nm tip radius versus 200 μm cantilever length, see Fig. 5) the magnetic part of the MFM model is separated into two subdomains. As it is shown in [24], for the first one (cantilever, tip and sample material) the Finite Element Method (FEM) is used to solve (9) and (10), respectively, whereas in the second domain the Boundary Element Method (BEM) is implemented which allows the field evaluation in the outer region. With regard to a magneto-mechanical model a computation of occurring magnetic forces is necessary. Several studies are dealing with the relation between field calculation inaccuracies and the error propagation obtaining the occurring forces [9, 10]. Thus, force calculations should be handled with care and remain a challenging research topic. Many different methods have been developed, but a calculation technique with a sufficiently high accuracy valid for any possible experimental configuration is still missing. As seen from results of former studies [3, 4], the solution is rather dependent on the problem under investigation. For this reason several force calculation methods are implemented and compared to each other, namely the Maxwell Stress Tensor (MST), the Virtual Work (VW1) principle, the Equivalent Magnetic Currents and Equivalent Magnetic Charges method (EMS1, EMS2). Furthermore, based on an idea in [5], a further virtual work method (VW2) is implemented [26] which also considers the intrinsic energy of the magnetized body. Concerning the total force acting on an investigated body all methods yield approximately the same results, but in the case of permanent magnetic materials the force distributions strongly differ from each other [8, 24, 25]. This fact also occurred by modelling the magnetic tip coating. With respect to the tip configuration shown in Fig. 5 and an alternating magnetized sample surface the five mentioned force calculations methods lead to nearly the same total force (Fig. 6a), but the occurring force distributions on the colored half

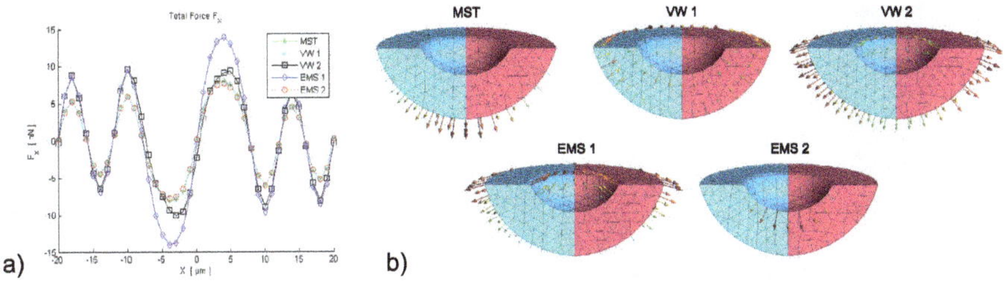

Fig. 6 Comparison of several magnetic force calculation methods [24]. (a) Total force, (b) local forces on the half of the magnetic coated end of the tip.

of the end of the tip, demonstrated in Fig. 6b (normal force components are shown), are totally different. Related to the physical behavior at the end of the tip, it seems that the VW2 approach should be appropriate, and thus leading to the best results, due to a more precise force model in contrast to the other methods. Furthermore, another investigation of a configuration with two magnetized cubes with an existing analytical force solution supports this assumption. For the magneto-mechanical simulation of MFM's the BEM/FEM approach for the magnetic behavior and an FEM approach for the structural analysis are coupled in an iterative manner. Furthermore, in order to simulate a whole scanning process the moving-material method [2,7,13] as well as the moving-mesh technique [6,7,12] are implemented and combined. While the first method is applied to change the material magnetization at each time step, the second one modifies the finite elements of the magnetic part to be also able to handle mechanical degrees of freedom in the case of small displacements. These approaches allow avoiding a remeshing procedure. A result of the developed weak coupled model is shown in Fig. 7. Thereby, alternating magnetic domains in the sample material are assumed, that lead in the illustrated case to an attractive force which acts on the cantilever tip and leads to a deflection Δu_y.

5 Conclusion

In this chapter we presented concepts for hybrid numerical models for the simulation of EFMs as well as MFMs but these models can be also applied to other MEMS devices. The developed hybrid model consists of a mechanical and an electromagnetic part that can to be solved numerically in an iterative manner. The resulting electromagnetic forces calculated on the basis of the numerical approximation of the electromagnetic field represent the input data to the mechanical part. On the other hand the results of the mechanical part determine the geometry of the electromagnetic part. The numerical methods include an augmented finite element method, the finite element method and the boundary element method. Obviously for the calculation of electromagnetic forces an electromagnetic field has to be calculated with high accuracy. In order to study parasitic effects of the EFM/MFM the forces should

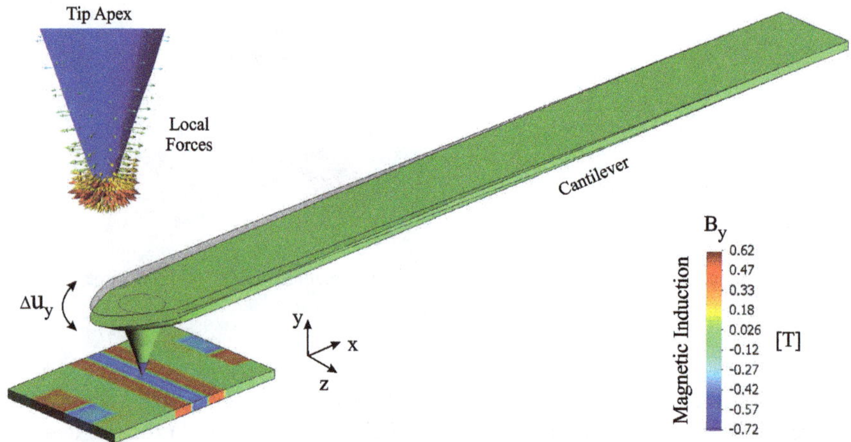

Fig. 7 Magneto-mechanical model of the MFM with a deflected cantilever due to magnetic interactions between tip and sample surface. The Virtual Work approach (VW2) was used for the total and local force evaluation.

be calculated at each point on the cantilever including the tip. In EFMs and MFMs field singularities and the multi-scale aspects are some of the problems in modelling and numerical simulation. Furthermore the accurate calculation of the force density is another essential problem. The numerical formulation considered here can be used in 2D as well as 3D cases. More details about our EFM models are described in [14–23] whereas the MFM models and contributions on magnetic field and force calculation methods can be found in [24–30].

Acknowledgements

At first we would like to thank our former colleagues M. Sc. I. Basol, Dr.-Ing. M. Greiff, Dipl.-Hydrol. Josefine Freitag and former student Dipl.-Ing. D. Mente for their close collaboration and many discussions over the years. Special thanks to our cooperation partners within the post graduate programme DFG GRK 615: Prof. Dr.-Ing. U. Nackenhorst, and Dr.-Ing. T. Helmich, as well as Prof. Dr. rer. nat. E. P. Stephan, Prof. Dr. rer. nat. M. Maischak. Furthermore we thank Prof. Dr.-Ing. A. Kost, Univ. Prof. Dipl.-Ing. Dr. U. Langer, Linz (Austria) and Prof. Dr.-Ing. P. Wriggers for many helpful discussions. Finally we would like to thank the German Research Society (DFG) for their financial support within the DFG GRK 615.

References

1. S. Morita, R. Wiesendanger, and E. Meyer. *Noncontract Atomic Force Microscopy*. Springer-Verlag, Berlin/Heidelberg, 2002.
2. B. Bendjima, K. Srairi, and M. Feliachi. A coupling model for analysing dynamical behaviours of an electromagnetic forming system. *IEEE Trans. Magn.*, 33:1638–1641, 1997.
3. Y. Chun and J. Lee. Comparison of magnetic levitation force between a permanent magnet and a high temperature superconductor using different force calculation methods. *Physica C*, 372:1491–1494, 2002.
4. L.H. De Medeiros, G. Reyne, and G. Meunier. Comparison of global force calculations on permanent magnets. *IEEE Trans. Magn.*, 34:3560–3563, 1998.
5. L.H. De Medeiros, G. Reyne, and G. Meunier. Distribution of electromagnetic force in permanent magnets. *IEEE Trans. Magn.*, 34:3012–3015, 1998.
6. B. Gaspalou, F. Colarmartino, C. Marchland, and Z. Ren. Simulation of an electromagnetic actuator by a coupled magnetomechanical modelling. *COMPEL*, 14:203–206, 1995.
7. M. Kaltenbacher. *Numerical Simulations of Mechatronic Sensors and Actuators*. Springer-Verlag, Berlin/Heidelberg, 2007.
8. D-H. Kim, D.A. Lowther, and J.K. Sykulski. Efficient global and local force calculations based on continuum sensitivity analysis. *IEEE Trans. Magn.*, 43:1177–1180, 2007.
9. S. McFee and D.A. Lowther. Towards accurate and consistent force calculation in finite element based computational magnetostatics. *IEEE Trans. Magn.*, 5:3771–3773, 1987.
10. W. Müller. Comparison of different methods of force calculation. *IEEE Trans. Magn.*, 26:1058–1061, 1990.
11. E.P. Stephan. Coupling of boundary element methods and finite element methods. In: *Encyclopedia of Computational Mechanics*, Vol. 1, Chapter 13, pp. 375–412, 2004.
12. N. Sadowski, Y. Lefevre, M. Lajoie-Mazenc, and J. Cros. Finite element torque calculation in electrical machines while considering the movement. *IEEE Trans. Magn.*, 28:1410–1413, 1992.
13. K. Srairi, M. Feliachi, and Z. Ren. Electromagnetic actuator behavior analysis using finite-element and papmetrization methods. *IEEE Trans. Magn.*, 31:3497–3499, 1995.
14. U.B. Bala. *Hybrid Numerical Modelling and Simulation of Electrostatic Force Microscope*. Dissertation, VDE Verlag, Berlin, 2009.
15. U.B. Bala, M. Greiff, T. Preisner, and W. Mathis. Numerical modelling of electrostatic force microscopes considering charge and dielectric constant. *COMPEL*, 28(1):109–119, 2009.
16. U.B. Bala, M. Greiff, and W. Mathis. Hybrid numerical simulation of electrostatic force microscopes considering charge distribution. *Piers Online*, 3(3):300–304, 2007.
17. I. Basol, M. Greiff, and W. Mathis. A staggered ALE approach for coupled electromechanical systems with application to a scanning probe microscope. In: *10th GMM Workshop: Methoden und Werkzeuge für den Entwurf von Mikrosystemen*, Cottbus, 2004.
18. M. Greiff. Entwicklung eines Simulationsprogramms für mikroelektromechanische Systeme (MEMS) mit Anwendung auf ein elektrostatisches Kraftmikroskop (EFM). Dissertation, http://www.tib.uni-hannover.de, 2010.
19. M. Greiff, U.B. Bala, and W. Mathis. A dynamic simulation approach for electrostatic force microscopy on inhomogeneous sample material. In: *Proceedings Piers 2007*, Prague, pp. 282–286, 2007.
20. M. Greiff, U.B. Bala, J. Freitag, and W. Mathis. Numerical calculation of electrostatic forces in micro electro mechanical systems (MEMS). *Przeglad Elektrotechniczny*, 83(11), 18–20, 2007.
21. M. Greiff, M. Maischak, U.B. Bala, W. Mathis, and E.P. Stephan. Numerical computation of electric fields including singularities in Micro Electro Mechanical Systems (MEMS). *Przeglad Elektrotechniczny*, 83(11):47–49, 2007.
22. M. Greiff, U.B. Bala, and W. Mathis. A staggered ALE approach for coupled electromechanical systems. In: A.M. Anile, G. Ali, and G. Mascali (Eds.), *Scientific Computing in Electrical Engineering SCEE 2006*, Mathematics in Industry, Vol. 9, pp. 33–38. Springer, Heidelberg, 2006.

23. M. Greiff, U.B. Bala, and W. Mathis. Hybrid numerical simulation of micro electro mechanical systems. *Piers Online*, 2(3):270–274, 2006.
24. T. Preisner and W. Mathis. A three dimensional FEM–BEM approach for the simulation of magnetic force microscopes. *Piers Online*, 5:272–277, 2009.
25. T. Preisner and W. Mathis. Berechnung der auftretenden lokalen Kräfte auf der magnetischen Beschichtung eines magnetischen Rasterkraftmikroskops. *Advances in Radio Science*, 7:37–41, 2009.
26. T. Preisner and W. Mathis. Magnetic force calculations applied to magnetic force microscopy. In: J. Roos and L.R.J. Costa (Eds.), *Scientific Computing in Electrical Engineering SCEE 2008*, Mathematics in Industry, Vol. 14, pp. 101–109. Springer, Heidelberg, 2010.
27. T. Preisner and W. Mathis. A comparison of different magnetic force distributions with respect to mechanical deformations using a hybrid calculation method. In: *Proceedings International Symposium on Electromagnetic Theory - EMTS2010*, Berlin, Germany, 16–19 August 2010, accepted.
28. T. Preisner, M. Greiff, U.B. Bala, and W. Mathis. Numerical computation of magnetic fields applied to magnetic force microscopy. *COMPEL*, 28(1):120–129, 2009.
29. M. Barke, T. Preisner, and W. Mathis. Effiziente Verfahren zur Parallelisierung von EMV-Simulationen für den Entwurfsprozess integrierter Schaltungen. *Advances in Radio Science*, 7:101–105, 2009.
30. T. Preisner, Ch. Bolzmacher, A. Gerber, K. Bauer, E. Quandt, and W. Mathis. Numerical modeling of a micrometer scaled actuator considering different force calculation methods. In: *Proceedings of the 15th International Symposium on Theoretical Electrical Engineering - ISTET09*, Lübeck, Germany, 22–24 June, pp. 159–163, 2009.
31. J. Freitag and W. Mathis. Numerical modelling of nonlinear electromechanical coupling of an atomic force microscope with finite element method. *Advances in Radio Science*, 2010, in press.

Finite Element and Boundary Element Approaches to Transmission and Contact Problems in Elasticity

Ernst P. Stephan

Abstract Under the leadership of Professor Stephan, four PhD projects were investigated during the running period of the GRK615 on the field of error controlled finite element/boundary element methods, namely on adaptive FE/BE couplings for transmission problems in elasticity (S. Oestmann) and for fluid-structure interaction (C. Domínguez) and on contact problems with friction (A. Chernov) as well as with delamination (L. Nesemann).
The Galerkin coupling formulations in Sections 1 and 4 are based on the fundamental paper by Costabel and Stephan [7], whereas the least squares coupling approach – applicable to mixed formulations and avoiding locking – extends the results of Maischak and Stephan [15]. The presented a posteriori error estimates for the h-version coupling extends those from Carstensen and Stephan [2, 3]. Contact problems with friction are considered with non-matching grids on the contact boundary and the foundation in Section 3. With mortar techniques, formulations are derived for a variational inequality on the contact boundary with the Poincaré–Steklov operator (Dirichlet-to-Neumann map), and the hp-version is analyzed and implemented for better numerical simulation [5]; penalty formulations are investigated in [4]. For adhesion problems (delamination) leading to a hemivariational inequality due to the set-valued adhesion law, existence and uniqueness results for the solution, representing the displacement, are given in Section 2 from [16]. Here, the bundle-Newton algorithm is used as a solver for the discrete finite element system and numerical simulations are given with a heuristic refinement algorithm, based on a residual-type error estimator.

Ernst P. Stephan
Institute for Applied Mathematics, Leibniz University, Welfengarten 1, 30167 Hannover, Germany;
e-mail: stephan@ifam.uni-hannover.de

E. Stephan & P. Wriggers (Eds.): Modelling, Simulation and Software Concepts, LNACM 57, pp. 181–199.
springerlink.com   © Springer-Verlag Berlin Heidelberg 2011

1 Adaptive FE/BE Coupling for Elasticity Problems

First, we present a Galerkin FE/BE coupling for elasticity problems, together with adaptive refinements with error estimators of residual and hierarchical type. We consider a transmission problem with non-linear elastic material in a bounded region $\Omega \subset \mathbb{R}^3$ and linear elastic material in the unbounded exterior region $\Omega^c := \mathbb{R}^3 \backslash \overline{\Omega}$. For given volume force $\mathbf{F} \in \mathbf{L}^2(\Omega)$ and $\mathbf{u}_0 \in \mathbf{H}^{1/2}(\Gamma)$ and $\mathbf{t}_0 \in \mathbf{H}^{-1/2}(\Gamma)$ on the transmission boundary $\Gamma = \overline{\Omega} \cap \overline{\Omega^c}$, find $\mathbf{u}_1 \in \mathbf{H}^1(\Omega)$, $\mathbf{u}_2 \in \mathbf{H}^1_{\mathrm{loc}}(\Omega)$ such that (with $\mathbf{n}$ denoting the exterior normal to Γ)

$$-\mathrm{div}\,\sigma(\mathbf{u}) = \mathbf{F}, \quad \mathrm{in}\ \ \Omega, \tag{1}$$

$$-\mu_2 \Delta \mathbf{u}_2 - (\lambda_2 + \mu_2)\,\mathrm{grad}\,\mathrm{div}\,\mathbf{u}_2 = 0, \quad \mathrm{in}\ \ \Omega^c, \tag{2}$$

$$\mathbf{u}_1 = \mathbf{u}_2 + \mathbf{u}_0, \quad \sigma(\mathbf{u}_1) \cdot \mathbf{n} = T(\partial., \mathbf{n})\mathbf{u}_2 + \mathbf{t}_0, \quad \mathrm{on}\ \ \Gamma, \tag{3}$$

$$\mathbf{u}_2 = \mathcal{O}(|x|^{-1}), \quad (|x| \longrightarrow \infty), \tag{4}$$

with stress σ and strain ε

$$\sigma(\mathbf{u}_1) = \left(k - \frac{2}{3}\mu(\gamma(\mathbf{u}))\right) \mathrm{tr}\,\varepsilon(\mathbf{u})I + 2\mu(\gamma(\mathbf{u}))\varepsilon(\mathbf{u}),$$

and normal traction

$$T(\partial_x, \mathbf{n})\mathbf{u}_2 = \lambda_2 \mathbf{n}\,\mathrm{div}\,\mathbf{u}_2 + 2\mu_2 \frac{\partial \mathbf{u}_2}{\partial \mathbf{n}} + \mu_2(\mathbf{n} \times \mathrm{curl}\,\mathbf{u}_2),$$

where the non-linear function $\gamma(\mathbf{u})$ is chosen such that existence of solution is guaranteed [19]. For further transmission problems in FE/BE couplings, see the survey article [20].

In the exterior domain the solution $\mathbf{u}_2$ is given by the Somigliana representation formula

$$\mathbf{u}_2(x) = \int_\Gamma \{T(\partial_y, \mathbf{n})\Gamma(x-y)^T \mathbf{u}_2(y) - \Gamma(x-y)T(\partial_y, \mathbf{n})\mathbf{u}_2(y)\}ds_y, \quad x \in \Omega^c, \tag{5}$$

with the Kelvin-Matrix ($i, j = 1, 2, 3$)

$$\Gamma_{ij}(x - y) =$$

$$\frac{1}{8\pi\mu_2(\lambda_2 + 2\mu_2)} \left((\lambda_2 + 3\mu_2)\frac{\delta_{ij}}{|x - y|} + (\lambda_2 + \mu_2)\frac{(x_i - y_i)(x_j - y_j)}{|x - y|^3}\right). \tag{6}$$

Using the representation formula, applying the boundary traction and the jump relations for the integral operator, for the limit $x \longrightarrow \Gamma$ we obtain the symmetric coupling formulation where $\phi := T(\partial_x, \mathbf{n})\mathbf{u}_2 = \sigma(\mathbf{u}_1) \cdot \mathbf{n} - \mathbf{t}_0$.

For given $\mathbf{F}, \mathbf{u}_0, \mathbf{t}_0$ as above, find $\mathbf{u} \in \mathbf{H}^1(\Omega)$, $\phi \in \mathbf{H}^{-1/2}(\Gamma)$:

$$-\mathrm{div}\,\sigma(\mathbf{u}) = \mathbf{F}, \quad \mathrm{in}\ \ \Omega, \tag{7}$$

$$2\boldsymbol{\phi} = -W(\mathbf{u} - \mathbf{u}_0) + (I - K')\boldsymbol{\phi}, \quad \text{on} \quad \Gamma, \tag{8}$$

$$0 = (I - K)(\mathbf{u} - \mathbf{u}_0) + V\boldsymbol{\phi}, \quad \text{on} \quad \Gamma, \tag{9}$$

Here we use the integral operators for $x \in \Gamma$ [12]

$$V\boldsymbol{\phi}(x) = \int_{\Gamma} \Gamma(x - y)\boldsymbol{\phi}(y)ds_y, \quad \text{single layer potential}, \tag{10}$$

$$K\mathbf{v}(x) = \int_{\Gamma} T(\partial_y, \mathbf{n})\Gamma(x - y)^T \mathbf{v}(y)ds_y, \quad \text{double layer potential}, \tag{11}$$

$$K'\boldsymbol{\phi}(x) = T(\partial_x, \mathbf{n}) \int_{\Gamma} \Gamma(x - y)^T \boldsymbol{\phi}(y)ds_y, \quad \text{adjoint double layer potential}, \tag{12}$$

$$W\mathbf{v}(x) = -T(\partial_x, \mathbf{n}) \int_{\Gamma} T(\partial_y, \mathbf{n})\Gamma(x - y)^T \mathbf{v}(y)ds_y, \quad \text{hypersingular operator}. \tag{13}$$

For a variational form we multiply the equilibrium condition (7) for the inner problem by a function $\mathbf{v} \in \mathbf{H}^1(\Omega)$, integrate over Ω, and obtain

$$\int_{\Omega} \sigma(\mathbf{u}) : \varepsilon(\mathbf{u})dx - \int_{\Gamma} \sigma(\mathbf{u}) \cdot \mathbf{n}\mathbf{v}ds = \int_{\Omega} \mathbf{F}\mathbf{v}dx.$$

Next, we multiply (8), (9) by $\mathbf{v}|_{\Gamma} \in \mathbf{H}^{1/2}(\Gamma)$, $\boldsymbol{\psi} \in \mathbf{H}^{-1/2}(\Gamma)$ and integrate over Γ and obtain the variational formulation of (7)–(9):
   Find $(\mathbf{u}, \boldsymbol{\phi}) \in \mathbf{H}^1(\Omega) \times \mathbf{H}^{-1/2}(\Gamma) =: \mathcal{H}$, with

$$B((\mathbf{u}, \boldsymbol{\phi}), (\mathbf{v}, \boldsymbol{\psi})) := \int_{\Omega} \sigma(\mathbf{u}_1) : \varepsilon(\mathbf{v})dx$$

$$+ \frac{1}{2}\langle W\mathbf{u}|_{\Gamma} + (K' - I)\boldsymbol{\phi}, \mathbf{v}|_{\Gamma}\rangle + \frac{1}{2}\langle \boldsymbol{\psi}, V\boldsymbol{\phi} + (I - K)\mathbf{u}|_{\Gamma}\rangle$$

$$= \int_{\Omega} \mathbf{F}\mathbf{v}dx + \frac{1}{2}\langle \boldsymbol{\psi}, (I - K)\mathbf{u}_0\rangle + \left\langle \mathbf{t}_0 + \frac{1}{2}W\mathbf{u}_0, \mathbf{v}|_{\Gamma}\right\rangle$$

$$=: L(\mathbf{v}, \boldsymbol{\psi}), \tag{14}$$

for all $(\mathbf{v}, \boldsymbol{\psi}) \in \mathcal{H}$, where $\langle \boldsymbol{\phi}, \boldsymbol{\psi}\rangle := \int_{\Gamma} \boldsymbol{\phi}\boldsymbol{\psi}ds$.
   We see that (7)–(9) and (14) are equivalent, i.e. if $(\mathbf{u}_1, \mathbf{u}_2)$ solves (1)–(4) then $(\mathbf{u}, \boldsymbol{\phi})$ solves (14) and vice versa (cf. [7]).
   Next, we introduce the FE/BE coupling formulation for (1)–(4) by performing (14) on finite dimensional subspaces of the used Sobolev spaces, $\mathbf{V}_h \subset \mathbf{H}^1(\Omega)$, $\mathbf{H}_h \subset \mathbf{H}^{-1/2}(\Gamma)$. The FE/BE coupling reads:
   Find $(\mathbf{u}_h, \boldsymbol{\phi}_h) \in \mathbf{V}_h \times \mathbf{H}_h =: \mathcal{H}_h$ s.t.

$$B((\mathbf{u}_h, \boldsymbol{\phi}_h), (\mathbf{v}_h, \boldsymbol{\psi}_h)) = L(\mathbf{v}_h, \boldsymbol{\psi}_h), \quad \forall(\mathbf{v}_h, \boldsymbol{\psi}_h) \in \mathcal{H}_h. \tag{15}$$

Next, we introduce an *error estimator of residual type* which we use as a device to steer an adaptive refinement algorithm. As usual we must assume that Ω has a regular triangulation $\mathcal{T}_h$ into tetrahedral or hexahedral elements. Further we assume that the mesh Γ_h on the interface boundary Γ is the trace mesh of $\mathcal{T}_h$. With $\mathcal{P}_k$ denoting the space of polynomials of degree k we take the discrete spaces

$$\mathbf{V}_h := \{p_h \in C(\Omega) \mid p_h|_T \in \mathcal{P}_1, \text{ for all } T \in \mathcal{T}_h\}$$

$$\mathbf{H}_h := \{p_h \in L^\infty(\Gamma) \mid p_h|_E \in \mathcal{P}_0, \text{ for all } E \in \Gamma_h\}.$$

Let $\mathbf{n}$ denote the exterior normal to Γ and on the element faces, let $[\sigma(\mathbf{u}_h) \cdot \mathbf{n}]$ denote the jump of the normal traction. With the surface gradient ∇_Γ we define the error estimator of residual type

$$\eta_h^2 := \sum_{T \in \mathcal{T}_h} \eta_h^2(T) = \sum_{T \in \mathcal{T}_h} h_T^2 \|\mathbf{F} + \operatorname{div}\sigma(\mathbf{u})\|^2_{\mathbf{L}^2(T)} + \sum_{F_i \in \mathcal{S}_h} h_{F_i} \|[\sigma(\mathbf{u}_h) \cdot \mathbf{n}]\|^2_{\mathbf{L}^2(F_i)}$$

$$+ \sum_{F_e \in \Gamma_h} h_{F_e} \|\mathbf{t}_0 - \sigma(\mathbf{u}_h) \cdot \mathbf{n} + \frac{1}{2}W(\mathbf{u}_0 - \mathbf{u}_h|_\Gamma) - \frac{1}{2}(K' - I)\boldsymbol{\phi}_h\|^2_{\mathbf{L}^2(F_e)}$$

$$+ \sum_{F_e \in \Gamma_h} h_{F_e} \|\nabla_\Gamma\{(I - K)(\mathbf{u}_0 - \mathbf{u}_h|_\Gamma) - V\boldsymbol{\phi}_h\}\|^2_{\mathbf{L}^2(F_e)},$$

where h_T, h_{F_i}, h_{F_e} are the diameters of the element T, of the face F_i and of the outer face F_e. With these terms we obtain an a posteriori error estimate for the Galerkin error:

There exists a constant $C > 0$ independent of the mesh size h, such that reliability holds, i.e.

$$\|\mathbf{u} - \mathbf{u}_h\|_{\mathbf{H}^1(\Omega)} + \|\boldsymbol{\phi} - \boldsymbol{\phi}_h\|_{\mathbf{H}^{-1/2}(\Gamma)} \leq C\eta_h \tag{16}$$

where $(\mathbf{u}, \boldsymbol{\phi})$ solves (14) and $(\mathbf{u}_h, \boldsymbol{\phi}_h)$ solves (15). Assuming quasi-uniform meshes there holds efficiency, i.e. with maximal (minimal) grid size on Γ $h_{\Gamma,\max}$ $(h_{\Gamma,\min})$ there holds with a constant c independent of h_T, h_{F_i}, h_{F_e}

$$\sum_{T \in \mathcal{T}_h} \eta_h(T)^2 \leq c\left[\|h_T(\mathbf{F} - \mathbf{F}_T)\|^2_{\mathbf{L}^2(\Omega)} + \|\sigma(\mathbf{u}) - \sigma(\mathbf{u}_h)\|^2_{\mathbf{L}^2(\Omega)}\right.$$

$$+ \frac{h_{\Gamma,\max}}{h_{\Gamma,\min}}(\|\mathbf{u} - \mathbf{u}_h\|^2_{\mathbf{H}^{1/2}(\Gamma)} + \|\boldsymbol{\phi} - \boldsymbol{\phi}_h\|^2_{\mathbf{H}^{-1/2}(\Gamma)})$$

$$\left. + \frac{h_{\Gamma,\max}^2}{h_{\Gamma,\min}}(\operatorname{dist}_{\mathbf{H}^1(\Gamma)}(\mathbf{u}, \mathbf{V}_h(\Gamma))^2 + \operatorname{dist}_{\mathbf{L}^2(\Gamma)}(S\mathbf{u}, \mathbf{H}_h(\Gamma))^2)\right] \tag{17}$$

The estimates (16), (17) extend the 2D results of Carstensen et al. [2] to 3D.

We have the *residual adaptive algorithm*: Let mesh $\mathcal{T}_{h,0}$ be uniform and compute $\mathbf{u}_h$, $\boldsymbol{\phi}_h$ from (15). Then compute $\eta_{h,k}(T)$ for all $T \in \mathcal{T}_{h,k}$, $k = 0, 1, 2, \ldots$ Next, mark $T \in \mathcal{T}_{h,k}$ for refinement if

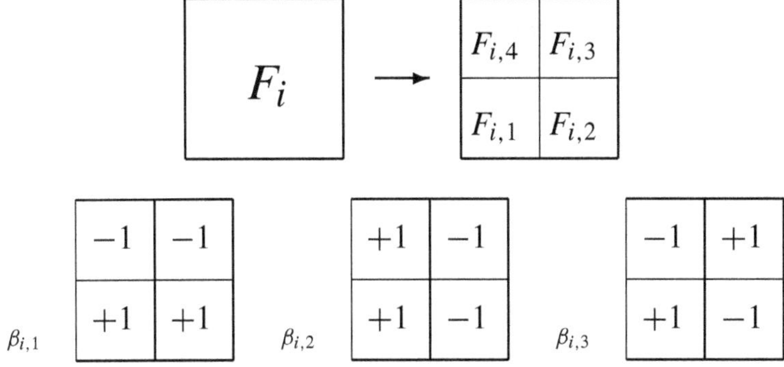

Fig. 1 New basis functions on boundary.

$$\eta_{h,k}(T) \geq \theta \max_{\widetilde{T} \in \mathcal{T}_{h,k}} \eta_{h,k}(\widetilde{T}), \quad \theta \in [0, 1]$$

Now, check whether all elements obey the "1-constraint rule". If necessary mark neighbouring elements for refinement. Create a new triangulation and compute a new Galerkin solution.

From a coarse mesh $\mathcal{T}_H$ we obtain a fine mesh $\mathcal{T}_h$ consisting of n elements T_i by subdividing all elements of $\mathcal{T}_H$ w.r.t. the xy-, xz- and yz-planes. Then we take the 2-level subspace decomposition for pw-linear functions on a hexahedral mesh

$$\widetilde{T}_h = \widetilde{T}_H \oplus D_h, \quad D_h := \widetilde{T}_1 \oplus \widetilde{T}_2 \oplus \cdots \oplus \widetilde{T}_n, \tag{18}$$

where the coarse (fine) mesh functions are given by $\widetilde{T}_H(\widetilde{T}_h)$ with

$$\widetilde{T}_H := \{\mathbf{u}_h \in C^0(\Omega) \ : \ \mathbf{u}_h \text{ pw. linear on } \mathcal{T}_H\},$$

$$\widetilde{T}_h := \{\mathbf{u}_h \in C^0(\Omega) \ : \ \mathbf{u}_h \text{ pw. linear on } \mathcal{T}_h\},$$

and $\widetilde{T}_i = \text{span}\{[b_i]^3\}$ denotes the 3D space of basis functions b_i on T_i where $b_i = 1$ at new node x_i, $b_i = 0$ at all other nodes of $\mathcal{T}_h$. Correspondingly for the boundary element functions on boundary mesh Γ_h

$$\tau_h = \tau_H \oplus \lambda_h, \quad \lambda_h := \bigoplus_{i=1}^{m} \tau_i^3, \tau_i = \bigoplus_{j=1}^{3} \tau_{i,j}, \tag{19}$$

where the coarse mesh functions in τ_H are pw. constant functions on $\mathcal{T}_H$ and the fine mesh functions in λ_h are defined by subdividing each square $\Gamma_i = $ face of T_i into 4 subsquares and using the new basis functions $\beta_{i,j}$ ($j = 1, 2, 3$), i.e. $\tau_{i,j} = \text{span}\{\beta_{i,1}, \beta_{i,2}, \beta_{i,3}\}$ is spanned by the new basis functions as displayed in Fig. 1.

Suppose a saturation assumption holds. Then there holds the *a posteriori error estimate with error estimator of hierarchical type, i.e. there exists constants* $c_1, c_2 >$

0 and $k_0 \in \mathbb{N}_0$, such that with the mesh size h_k of level k

$$c_1 h_k \left(\sum_{i=1}^n \Theta_{i,k}^2 + \sum_{j=1}^m \theta_{j,k}^2 \right)^{1/2}$$

$$\leq \|(\mathbf{u} - \mathbf{u}_k, \boldsymbol{\phi} - \boldsymbol{\phi}_k)\|_{\mathcal{H}} \leq c_2 h_k^{-1} \left(\sum_{i=1}^n \Theta_{i,k}^2 + \sum_{j=1}^m \theta_{j,k}^2 \right)^{1/2}, \qquad (20)$$

where

$$\Theta_{i,k} := \|\mathbf{e}_{k,i}\|_{\mathbf{H}^1(\Omega)}, \theta_{i,k} := \|\varepsilon_{j,k}\|_{\mathbf{V}},$$

and $\mathbf{e}_{k,i} \in T_i^3$ solves

$$(\mathbf{e}_{k,i}, \mathbf{v})_{\mathbf{H}^1(\Omega)} = L(\mathbf{v}, 0) - B((\mathbf{u}_H, \boldsymbol{\phi}_H), (v, 0)), \quad \forall \mathbf{v} \in T_i^3, \qquad (21)$$

and $\varepsilon_{k,i} \in \tau_i^3$ solves

$$V(\varepsilon_{k,i}, \boldsymbol{\psi}) = L(0, \boldsymbol{\psi}) - B((\mathbf{u}_H, \phi_H), (0, \boldsymbol{\psi})), \quad \forall \boldsymbol{\psi} \in \tau_i^3. \qquad (22)$$

We have the *hierarchical adaptive algorithm*:

First, compute $(\mathbf{u}_k, \boldsymbol{\phi}_k) \in T_k \times \tau_k$. Then for any $T_i \in \mathcal{T}_k$ ($F_i \in \Gamma_k$) compute local error indicator $\Theta_{i,k}$ ($\theta_{i,k}$). For given tolerance Tol stop if

$$\eta_k := \left(\sum_{i=1}^n \Theta_{i,k}^2 + \sum_{j=1}^m \theta_{j,k}^2 \right)^{1/2} < \text{Tol}.$$

Next, compute $\eta_{\max_1} := \max_{T_i \in \mathcal{T}_k} \{\Theta_{i,k}\}$, $\eta_{\max_2} := \max_{F_i \in \Gamma_k} \{\theta_{i,k}\}$ and bisect T_i and F_i if $\Theta_{i,k} \geq \vartheta \eta_{\max_1}$, $\theta_{i,k} \geq \vartheta \eta_{\max_2}$ with some fixed ϑ. Then refine neighbouring elements. Next, compute new $(\mathbf{u}_{k+1}, \phi_{k+1})$.

There are several ways to refine elements, namely a regular refinement, a directionwise one and one with layers, all leading to regular meshes without hanging nodes. The performance of the above adaptive algorithms is shown by some numerical experiments performed with the software package MAIPROGS developed at IfAM, Leibniz University Hannover.

Example 1. We consider the interface problem

$$\Delta \mathbf{u}_1 = 0 \ \text{ in } \ \Omega := [-1, 1]^3, \quad \Delta \mathbf{u}_2 = 0 \ \text{ in } \ \mathbb{R}^3 \backslash \overline{\Omega},$$

$$\mathbf{u}_0 = \mathbf{u}_1 - \mathbf{u}_2, \quad \frac{\partial \mathbf{u}_0}{\partial \mathbf{n}} = \frac{\partial \mathbf{u}_1}{\partial \mathbf{n}} - \frac{\partial \mathbf{u}_2}{\partial \mathbf{n}} \ \text{ on } \ \partial \Omega, \qquad (23)$$

with

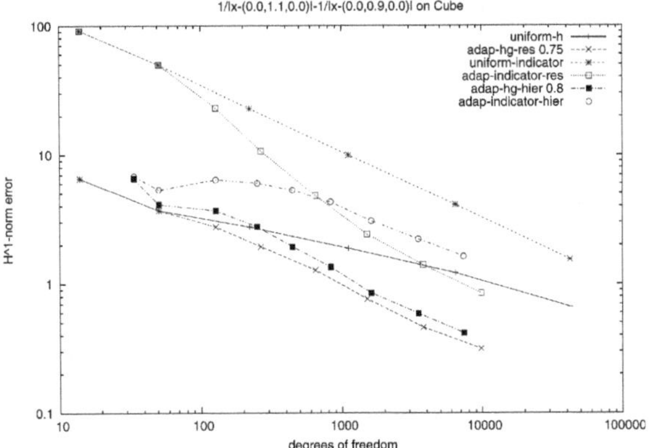

Fig. 2 H^1-error for uniform and adaptive (residual, hierarchical) refinement, error indicators for Example 1.

$$\mathbf{u}_1 = \left(x_1^2 + (x_2 - \frac{11}{10})^2 + x_3^2\right)^{-1/2}, \quad \mathbf{u}_2 = \left(x_1^2 + (x_2 - \frac{9}{10})^2 + x_3^2\right)^{-1/2}.$$

Uniform refinement is not optimal. The adaptive refinements (residual, hierarchical) yield the almost optimal convergence rate $1/3$ (cf. Fig. 2, Table 1).

Example 2. Same problem as above but $\Omega := [-1, 1]^3 \setminus [0, 1]^3$, the Fichera cube. We use

$$\mathbf{u}_1 = ((x_1 - 0.1)^2 + (x_2 - 0.1)^2 + (x_3 - 0.1)^2)^{-1/2},$$

$$\mathbf{u}_2 = (x_1 + 0.1)^2 + (x_2 + 0.1)^2 + (x_3 + 0.1)^2)^{-1/2}.$$

Again, the adaptive refinements approach the optimal rate $1/3$ and the effectivity indices $\eta_N / \mathbf{e}_{H^1}$ remain bounded (cf. Fig. 3, Table 2).

Next, we present a least squares FE/BE coupling method for linear elasticity. It is known that least squares methods need not to fulfill a discrete inf-sup condition and hence they are very interesting for mixed formulations. For nearly incompressible material the standard finite elements lead to locking effects. This is avoided in a least squares approach also for transmission problems when both finite elements and boundary elements are used (see [14, 18]).

Hooke's law gives a relation between stress and strain

$$\sigma_{ij} = 2\mu\varepsilon_{ij}(\mathbf{u}) + \lambda\delta_{ij}\operatorname{div}\mathbf{u}$$

with Kronecker delta δ_{ij} and Lamé parameter μ and λ for linear elasticity. If $\lambda \longrightarrow \infty$, i.e. Poisson ratio

$$\nu = \frac{\lambda}{2(\lambda + \mu)} \longrightarrow 1/2$$

Table 1 Degrees of freedom N, H^1-error, indicators, efficiency index, convergence rate α for the residual error estimator (above) and hierarchical error estimator (below) (Example 1).

N	e_{H^1}	η_N	η_N/e_{H^1}	α
14	6.5453999	91.325167		
51	3.6859984	49.700306	13.48	0.444
128	2.7564320	22.981763	8.338	0.316
267	1.9331828	10.681064	5.525	0.483
649	1.2754529	4.8104814	3.772	0.468
1513	0.7621123	2.4060009	3.157	0.608
3796	0.4610562	1.3986021	3.033	0.546
9840	0.3150961	0.8472700	2.689	0.400
26407	0.2356810	0.5381950	2.284	0.294

N	e_{H^1}	η_N	η_N/e_{H^1}	α
34	6.5453999	6.8461109		
51	4.1003257	5.3445779	1.303	1.153
128	3.6859984	6.4355118	1.746	0.116
251	2.7564320	6.0390225	2.191	0.432
448	1.9377480	5.3010328	2.736	0.608
833	1.3435567	4.2813467	3.187	0.590
1619	0.8470384	3.0541318	3.606	0.694
3504	0.5898188	2.2128713	3.752	0.469
7380	0.4177930	1.6309529	3.904	0.463

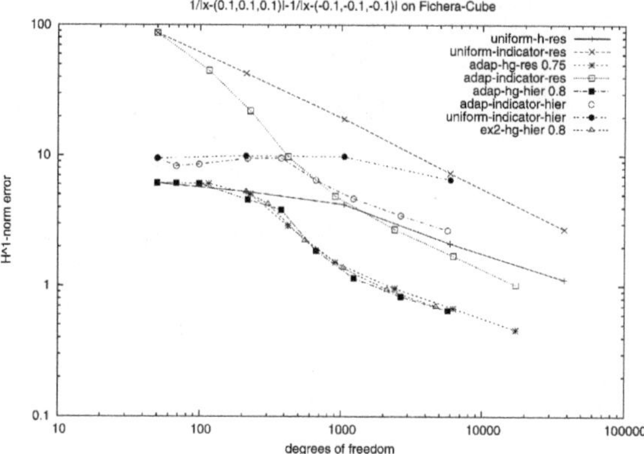

Fig. 3 H^1-error for uniform and adaptive (residual, hierarchical) refinement, error indicators for Example 2.

we have almost incompressible material. If $\nu \approx 1/2$ we have $\lambda \gg \mu$. This forces the ratio of the constants in the Céa-Lemma to become very large and we have locking when the exact solution is approximated, and the classical FE fails. Hence introduce a Lagrange multiplier $p = -\lambda \mathrm{div}\,\mathbf{u}$ and insert p in the definition of σ yielding

Table 2 Degrees of freedom N, H^1-error, indicators, efficiency index, convergence rate α for the residual error estimator (above) and hierarchical error estimator (below) (Example 2).

N	e_{H^1}	η_N	η_N/e_{H^1}	α
50	6.1469894	86.721906		
116	6.0632190	44.794101	7.388	0.016
228	5.0455398	22.026088	4.365	0.272
421	2.8911192	9.7537464	3.374	0.908
908	1.5201312	4.8657948	3.201	0.836
2410	0.9544718	2.7174849	2.847	0.477
6258	0.6724819	1.7001051	2.528	0.367
17281	0.4584471	1.0187227	2.222	0.377

N	e_{H^1}	η_N	η_N/e_{H^1}	α
213	5.2725031	9.8960902		
304	4.2420059	9.9955465	2.356	0.611
552	2.2397150	6.8901423	3.076	1.071
1036	1.3818762	4.9914741	3.612	0.767
2125	0.9363773	3.6505149	3.899	0.542
4708	0.6975674	2.7464362	3.937	0.370

$\sigma_{ij}(\mathbf{u}, p) = 2\mu\varepsilon_{ij}(\mathbf{u}) - \delta_{ij} p$. Now the equilibrium condition $-\operatorname{div}\sigma = \mathbf{F}$ in the elastic interior domain Ω becomes

$$-\operatorname{div}\sigma(\mathbf{u}, p) = \mathbf{F}, \quad \text{in} \quad \Omega, \tag{24}$$

with side condition

$$\frac{1}{\lambda}p + \operatorname{div}\mathbf{u} = 0. \tag{25}$$

In the transmission problem we give $\mathbf{F}$ and the functions $\mathbf{u}_0, \mathbf{t}_0$ on $\Gamma = \partial\Omega$ and look for $\mathbf{u}_1$ satisfying (24) and (25) in Ω and $\mathbf{u}_2$ satisfying (24) in Ω^c together with the interface conditions (3) and decay condition at infinity (4).

As above, we use the Somigliana representation formula for $\mathbf{u}_2$ in Ω^c and the jump relations and introduce a new unknown $\boldsymbol{\phi} = \sigma(\mathbf{u}, p) \cdot \mathbf{n}$ on Γ. The least squares formulation of the problem leads to minimize the functional

$$J(\mathbf{v}, q, \boldsymbol{\tau}) = \|\mathcal{L}(\mathbf{v}, q, \boldsymbol{\tau}) - \mathfrak{F}\|^2_{X'}$$

$$= \left\| \mathcal{L}(\mathbf{v}, q) - \mathbf{F} + \frac{1}{2}\delta_\Gamma \otimes [W(\mathbf{v} - \mathbf{u}_0) - 2\mathbf{t}_0 - (I - K')(\boldsymbol{\tau} - \mathbf{t}_0)] \right\|^2_{\widetilde{\mathbf{H}}^{-1}(\Omega)}$$

$$+ \left\| \frac{1}{\lambda}q + \operatorname{div}\mathbf{v} \right\|^2_{L^2(\Omega)} + \|(I - K)(\mathbf{v} - \mathbf{u}_0) + V(\boldsymbol{\tau} - \mathbf{t}_0)\|^2_{\mathbf{H}^{1/2}(\Gamma)},$$

where the $\mathcal{L}(\mathbf{u}, p) : \mathbf{H}^1(\Omega) \times L^2(\Omega) \longrightarrow \widetilde{\mathbf{H}}^{-1}(\Omega)$ is defined by

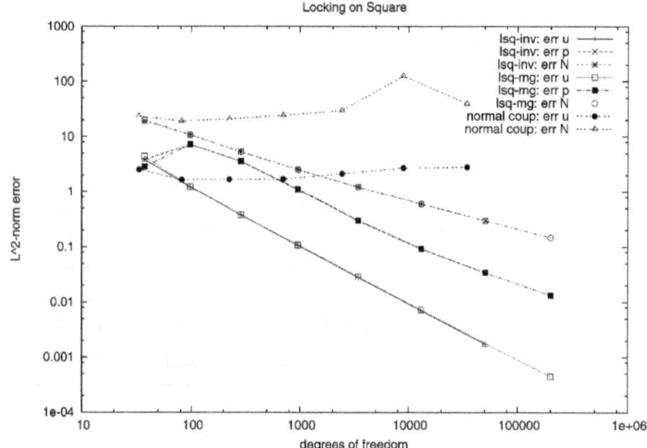

Fig. 4 No locking effect for least squares.

Table 3 L^2-errors and convergence rates for $\mathbf{u}$, p and $\boldsymbol{\phi}$.

DOF	L^2err u	α_u	L^2err p	α_p	L^2err $\boldsymbol{\phi}$	α_ϕ
38	4.3883		2.8534		19.868	
98	1.2329	1.3400	7.1823	−0.974	10.749	0.6484
290	0.3773	1.0914	3.5771	0.6425	5.2594	0.6589
962	0.1080	1.0429	1.1071	0.9781	2.5098	0.6170
3458	0.0286	1.0395	0.3020	1.0152	1.2184	0.5649
13058	0.0072	1.0336	0.0925	0.8908	0.5993	0.5340
50690	0.0018	1.0268	0.0341	0.7349	0.2971	0.5173
199682	0.0004	1.0191	0.0133	0.6871	0.1479	0.5087

$$[\mathcal{L}(\mathbf{u}, p); \mathbf{w}] = \int_\Omega \sigma(\mathbf{u}, p) : \varepsilon(\mathbf{w})dx, \quad \forall \mathbf{w} \in \mathbf{H}^1(\Omega).$$

In [14, 18] finite element subspaces $\mathbf{V}_h \subset \mathbf{H}^1(\Omega)$, $H_h \subset L^2(\Omega)$, $\mathbf{M}_h \subset \mathbf{H}^{-1/2}(\Gamma)$ are introduced together with preconditioners $\mathcal{B}_h$, $\mathcal{C}_h$ for realizing the $\mathbf{H}^{-1}(\Omega)$ -and $\mathbf{H}^{1/2}(\Gamma)$-scalar products. These preconditioners for the CG method can be the inverse of various subblocks of the FE/BE coupling Galerkin matrix or the use of multigrid (MG) and are analysed together with the solvability of the discrete version of the least squares functional. The numerical experiments show clearly no locking for the least squares solution (Fig. 4) and that the computing time can be reduced drastically with the MG preconditioner.

2 FE Analysis of a Process with Delamination

In industrial manufacturing processes, like sheet rolling, effects like friction or delamination can occur. The exact nature of these effects has a significant impact on the process. Thus, a careful treatment of the boundary interaction is in order. Some of these effects have been analyzed extensively. There exist efficient methods for the numerical treatment of contact problems with Tresca or Coulomb friction, where finite element or boundary element methods are chosen for the discretization [4,5]. Friction is typically included by regularization of the friction terms.

In this section, we reduce the model such that only one non-linear, non-smooth part needs to be considered. For this, we assume that the material exposes a linear-elastic behavior and undergoes only small deformations in a quasi-stationary process.

Let $\Omega \subset \mathbb{R}^d$ be an open, bounded domain ($d = 2, 3$). The boundary $\Gamma = \partial\Omega$ is decomposed into a Dirichlet boundary Γ_D, a Neumann boundary Γ_N and a potential contact boundary Γ_C. On Γ, we denote by $\mathbf{n}$ the exterior normal vector.

Let boundary forces $\mathbf{t}$ be given on Γ_N, assume zero displacement on Γ_D and impose a non-penetration condition on Γ_C, namely

$$\mathbf{u}(\mathbf{x}) \cdot \mathbf{n} \leq 0 \quad \text{a.e. } \mathbf{x} \in \Gamma_C \, .$$

We are looking for a solution $\mathbf{u}$ that satisfies

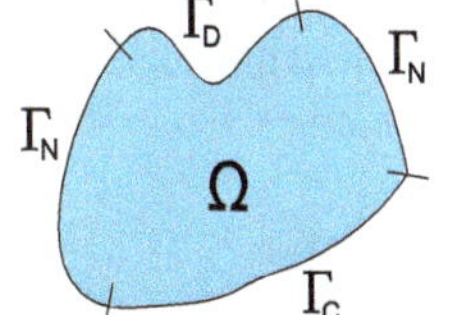

$$
\begin{aligned}
-\operatorname{div} \sigma(\mathbf{u}) &= 0 && \text{in } \Omega \subset \mathbb{R}^d && (26) \\
\sigma(\mathbf{u}) \cdot \mathbf{n} &= \mathbf{g} && \text{on } \Gamma_N && (27) \\
\mathbf{u} &= 0 && \text{on } \Gamma_D && (28) \\
\mathbf{u} \cdot \mathbf{n} &=: u_N \leq 0 && \text{on } \Gamma_C && (29)
\end{aligned}
$$

with $\sigma := \mathbb{C} : \varepsilon(\mathbf{u})$, where $\mathbb{C}$ is the Hooke tensor.

Additionally, we need to apply the interface law on Γ_C.

First, the surface law $\mathbf{b}$ can be decomposed into normal and tangential parts b_N and $b_{T,i}$. Assume now that $b : \Gamma_C \to \mathbb{R}$ is a piecewise Lipschitz function with finite jumps. Then we can define the "envelope" function $\hat{b}$ by

$$\hat{b}(t) := \left[\liminf_{\tau \to 0} b(t + \tau), \, \limsup_{\tau \to 0} b(t + \tau) \right] \, .$$

For example, a non-smooth delamination law can then be written as

$$\sigma_N(\mathbf{x}) \in \hat{b}(\mathbf{u}(\mathbf{x}) \cdot \mathbf{n}) \quad \text{a.e. } \mathbf{x} \in \Gamma_C$$

Define the space of energetically admissible displacements V and the convex cone $\mathcal{K}$ of geometrically admissible displacements:

$$V := \left\{ \mathbf{u} \in \left[H^1(\Omega) \right]^d : \mathbf{u}|_{\Gamma_D} = 0 \right\} \tag{30}$$

$$\mathcal{K} := \left\{ \mathbf{u} \in V : \mathbf{u}|_{\Gamma_V} \leq 0 \right\} \tag{31}$$

We introduce the following shorthand notations:

$$a(\mathbf{u}, \mathbf{v}) := \int_{\Omega} \varepsilon(\mathbf{u}) : \mathbb{C} : \varepsilon(\mathbf{v}) ; \tag{32}$$

$$g(\mathbf{u}) := \int_{\Gamma_N} \mathbf{t} \cdot \mathbf{u} \, \mathrm{d}s_x; \tag{33}$$

$$\langle \xi, \Pi\mathbf{u} \rangle_C := \int_{\Gamma_C} \xi(\mathbf{x}) \, (\Pi\mathbf{u})(\mathbf{x}) \mathrm{d}s_x \tag{34}$$

with Π linear, e.g. extracting the normal component from $\mathbf{u}$. Then we can derive a *hemivariational inequality*, introducing a new function ξ on Γ_C:

Find $\mathbf{u} \in \mathcal{K}$ s.t. for all $\mathbf{v} \in \mathcal{K}$,

$$a(\mathbf{u}, \mathbf{v} - \mathbf{u}) + \langle \xi, \Pi\mathbf{v} - \Pi\mathbf{u} \rangle_C \geq g(\mathbf{v} - \mathbf{u}) \quad \forall \mathbf{v} \in \mathcal{K} \tag{35}$$

$$\xi(\mathbf{x}) \in \hat{b}((\Pi\mathbf{u})(\mathbf{x})) \quad \text{a.e. } \mathbf{x} \in \Gamma_C \tag{36}$$

If some mild conditions are imposed on b, Haslinger et al. showed that $\xi \in L^2(\Gamma_C)$ [11, remark 3.8]. This is especially the case if we choose b piecewise Lipschitz.

With a triangulation $\mathcal{T}$ of Ω, we can construct the usual finite element space of piecewise linear functions, $V_h \subset V$. An approximation of $\mathcal{K}$ is done by imposing the non-penetration condition in the corner/face midpoints $\mathbf{m}_i$ on Γ_C, resulting in $\mathcal{K}_h \subset V_h$. The space $L^2(\Gamma_C)$ can be discretized by piecewise constant functions on the boundary corners/faces, $\mathbb{P}_h^0(\Gamma_C)$. This leads to a discrete system:

Find $(\mathbf{u}_h, \xi_h) \in \mathcal{K}_h \times \mathbb{P}_h^0(\Gamma_C)$ s.t.

$$a(\mathbf{u}_h, \mathbf{v}_h - \mathbf{u}_h) + \langle \xi_h, \Pi\mathbf{v}_h - \Pi\mathbf{u}_h \rangle_C \geq g(\mathbf{v}_h - \mathbf{u}_h) \quad \forall \mathbf{v}_h \in \mathcal{K}_h \tag{37}$$

$$\xi_h(\mathbf{m}_i) \in \hat{b}((\Pi\mathbf{u}_h)(\mathbf{m}_i)) \quad \forall \mathbf{m}_i \tag{38}$$

The hemivariational inequality (35), (36) can be re-cast as a minimization problem. For this, we seek a minimizer of an augmented potential function on $\mathcal{K}$,

$$\Phi(\mathbf{v}) := \frac{1}{2}a(\mathbf{v}, \mathbf{v}) - g(\mathbf{v}) + \int_{\Gamma_C} \int_0^{(\Pi\mathbf{u})(\mathbf{x})} b(t) \, \mathrm{d}t \, \mathrm{d}s_x . \tag{39}$$

Likewise, we can set up a potential function of the discretized problem:

Associate the coefficient vector $\vec{v_h} \in \mathbb{R}^N$ with the function $\mathbf{v}_h \in \mathcal{K}_h$, so we get a convex cone $K_h \subset \mathbb{R}^N$. Let A denote the standard stiffness matrix and $\vec{g}$ the load vector. Γ_C can be decomposed into the edges e_i, $i = 1, \ldots, M$. The matrix $\Lambda \in \mathbb{R}^{M \times N}$ performs a similar operation as Π in the continuous formulation. We get that

$$\left(\Lambda \vec{v_h} \right)_i = \left(\Pi\mathbf{v}_h \right)(\mathbf{m}_i) ;$$

for delamination, Λ extracts normal displacements in the midpoints $\mathbf{m}_i$.

We seek a minimizer over K_h of

$$\Phi_h(\vec{v_h}) := \frac{1}{2}\vec{v_h} \cdot A \cdot \vec{v_h} - \vec{g} \cdot \vec{v_h} + \sum_{e_i \subset \Gamma_C} |e_i| \int_0^{(\Lambda \vec{v_h})_i} b(t) \, \mathrm{d}t . \tag{40}$$

Minimizers of the potential Φ_h can be found using the Bundle–Newton algorithm [13].

Note that Φ and Φ_h are, in general, neither convex nor differentiable, so there can be multiple solutions.

If b is Lipschitz and the Lipschitz constant is bounded by a certain value, we can however prove that Φ is strictly convex:

Theorem 1 (Conditions on b for strict convexity, [16]). *Let $b \in Lip(\mathbb{R})$ with Lipschitz constant c_L. If this constant satisfies for $a(.,.)$ in (32)*

$$(c_L + \varepsilon) \leq \min_{\substack{\mathbf{d} \in V : \\ \|\Pi\mathbf{d}\|_{0,\Gamma_C} = 1}} a(\mathbf{d}, \mathbf{d})$$

for an arbitrarily small $\varepsilon > 0$, the function Φ is convex.

The fact that $\Phi(\mathbf{v}) \to \infty$ as $\|\mathbf{v}\| \to \infty$ then yields the existence of a unique solution, as proven in [17, theorem C.1.1].

If b is only piecewise Lipschitz continuous with bounded constant c_L as above, for positive jumps $c_J > 0$, a solution is still unique (i). For negative jumps $c_J < 0$, the additional conditions in (ii) ensure uniqueness:

Theorem 2 (Uniqueness of a minimizer, [16]). *Let b be the sum of a Lipschitz function s with constant c_L and a scaled, shifted Heaviside function:*

$$b(t) := s(t) + c_J \Theta(t - t^*)$$

Let $\mathbf{u}$ be a minimizer of Φ.
If one of the following conditions is met, $\mathbf{u}$ is unique:

(i) *The jump c_J is non-negative, and c_L satisfies*

$$c_L < \min_{\mathbf{d} \in V \setminus \ker \Pi} \frac{a(\mathbf{d}, \mathbf{d})}{\|\Pi\mathbf{d}\|_{0,\Gamma_C}^2} .$$

(ii) *The jump c_J is negative,*

$$(\Pi\mathbf{u})(\mathbf{x}) \leq t^* - \frac{1}{4} \quad \forall \mathbf{x} \in \Gamma_C, \quad and \quad c_L - 2c_J < \min_{\mathbf{d} \in V \setminus \ker \Pi} \frac{a(\mathbf{d}, \mathbf{d})}{\|\Pi\mathbf{d}\|_{0,\Gamma_C}^2} .$$

This theorem can be extended to a finite number of jumps.

Numerical experiments were performed with an own finite element code in 2D and 3D.

The 2D benchmark was a rectangular block with height 1, width 10 and the material parameters $E = 210 \times 10^9$ and $\nu = 0.3$. The block was fixed on the left (Dirichlet) boundary, and various constant forces were applied on the rightmost 1 units of the upper boundary. A simple delamination law b was used, being linear in the first two intervals (with jumps down) and zero outside these intervals.

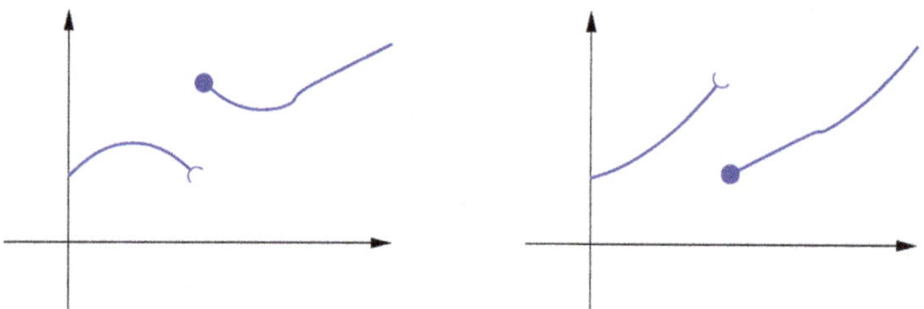

Fig. 5 (i) Piecewise Lipschitz continuous function with positive jump. (ii) Piecewise Lipschitz continuous function with negative jump.

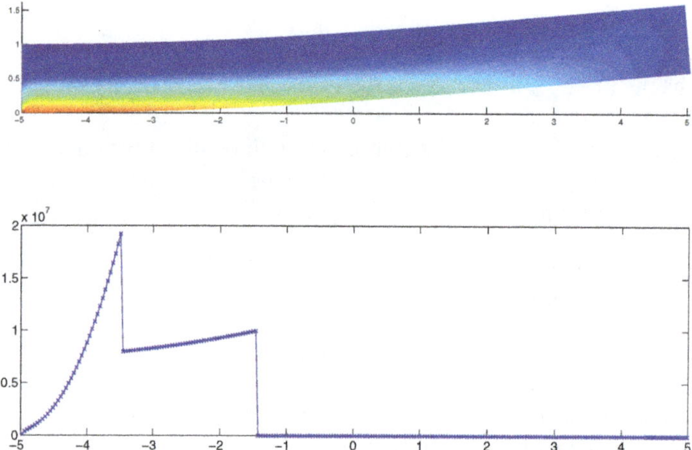

Fig. 6 Deformed mesh for the 2D benchmark; reaction forces.

The 3D benchmark was a $1 \times 5 \times 10$ block with the same material parameters. The block was fixed on one side, and various constant forces were applied on a 1×1 square on the upper surface. The same delamination law b as in the 2D case was used. A (so far) experimental residual error estimator was used here, introducing an additional term with the reaction force $-b(\Pi \mathbf{u}_h)\mathbf{n}$:

$$\eta_T^2 := \sum_{e \in \mathcal{E}_{int}} h_e \| [\sigma(\mathbf{u}_h) \cdot \mathbf{n}] \|^2 + \sum_{e \in \mathcal{E}_N} h_e \| \sigma(\mathbf{u}_h) \cdot \mathbf{n} - \mathbf{g} \|^2$$

$$+ \sum_{e \in \mathcal{E}_C} h_e \| \sigma(\mathbf{u}_h) \cdot \mathbf{n} + b(\Pi \mathbf{u}_h)\mathbf{n} \|^2$$

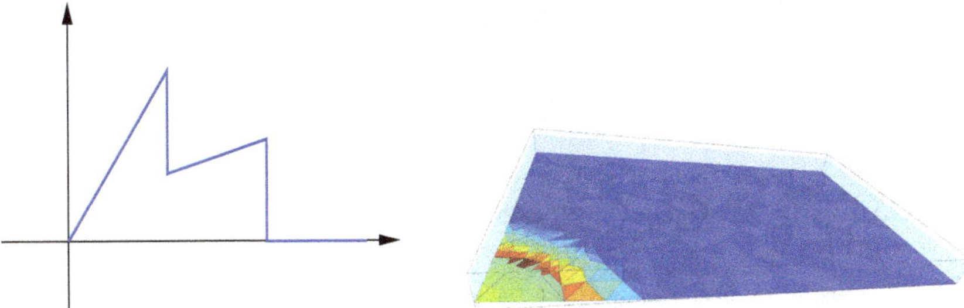

Fig. 7 Delamination law; reaction forces on the 3D contact interface (adaptive refinement).

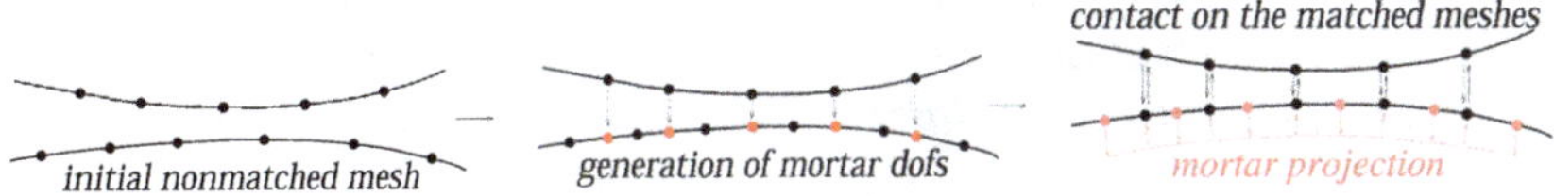

Fig. 8 Basic scheme of information transfer with the mortar projection.

3 Contact Problems with Friction

The accurate and efficient numerical simulation of frictional contact problems has wide industrial applications, e.g. in automobile industry, in metal forging, etc. The numerical experiments are commonly performed with the low order finite element methods. We introduce a new high order FE/BE technique for multibody contact problems with friction.

Classical formulation: Find $\mathbf{u} \in C^2(\Omega)$ such that

$$\operatorname{div} \sigma(\mathbf{u}) = 0 \quad \text{equilibrium in } \Omega,$$
$$\mathbf{u} = 0 \quad \text{on } \Gamma_D(\text{fixed}),$$
$$\sigma(\mathbf{u}) \cdot \mathbf{n} = \hat{\mathbf{t}} \quad \text{on } \Gamma_N(\text{surf. force}),$$
$$\left.\begin{array}{l} \sigma_n \le 0, \ [u_n] \le 0, \ \sigma_n[u_n] = 0, \\ |\sigma_t| \le \mathcal{F}, \ \sigma_t[u_t] + \mathcal{F}|[u_t]| = 0, \ \mathcal{F} = \mu_f|\sigma_n| \end{array}\right\} \quad \text{on } \Gamma_C(\text{contact}).$$

Often, it is very convenient to use independent discretizations of the bodies, subjected to their particularities. Furthermore, in case of *non-matched* meshes on the contact interfaces, independent mesh optimization (refinement) of the bodies can be performed. Therefore, specific methods, realizing communication between independently discretized interfaces, must be developed. A high order mortar projection operator and its adjoint are employed for transferring the information across the contact boundary.

Let $\mathcal{W}_{hp}^1$ be the piecewise polynomial space on mesh 1 on Γ_C. The mortar projection operator π_{hp}^1 onto mesh 1 is defined as follows:

For $\varphi \in H^{1/2}(\Gamma_C)$ the function $\Phi^1 = \pi_{hp}^1 \varphi \in \mathcal{W}_{hp}^1$ is its projection, iff

$$\Phi^1 = \varphi \text{ in end points of } \Gamma_C,$$

$$\int_{\Gamma_C} \Phi^1 \Psi^1 \, ds = \int_{\Gamma_C} \varphi \Psi^1 \, ds, \quad \forall \Psi^1 \in \mathcal{M}_{hp}^1$$

Here $\mathcal{M}_{hp}^1 \subset \mathcal{W}_{hp}^1$ has reduced polynomial degree in the end intervals.

A special emphasis is made on the high order boundary element method (BEM). This method has some significant advantages over the finite elements (FEM). In BEM, only the boundary of the body is needed to be discretized, which *reduces essentially the number of unknowns*. The internal elastic behaviour inside the body is provided by the boundary integral operators, which are based on the fundamental solution of the elasticity operator. Furthermore, the task of mesh generation is much simpler for BE, since there is *no need to build a mesh inside the body*. Unlike in the FEM, the *resulting matrix in the BEM is dense*, due to non-locality of the boundary integral operators, which increases the computational effort. This drawback can be overcome with so-called fast-BE methods (panel clustering, H-matrices, etc.)

Discrete weak formulation (pure BE): Find $\mathbf{U} \in K_{hp} := \{U_n^1 \leq \pi_{hp}^1 U_n^2\}$:

$$\int_{\partial\Omega} (\hat{S}\mathbf{U}) \cdot (\mathbf{\Phi} - \mathbf{U}) \, ds + \int_{\Gamma_C} \mathcal{F}(|\mathbf{\Phi}| - |\mathbf{U}|) \, ds \geq \int_{\Gamma_N} \hat{\mathbf{t}} \cdot (\mathbf{\Phi} - \mathbf{U}) \, ds$$

$$\forall \mathbf{\Phi} \in \mathcal{K}_{hp} \qquad (41)$$

with the non-local Steklov–Poincaré operator $\hat{S}$.

For a constant C independent of h and p, we have the estimate [5]

$$\|\mathbf{u} - \mathbf{U}\|_{\mathbf{H}^{1/2}(\partial\Omega)} \leq C(h^{1/4} + p^{-1/4})\|\mathbf{u}\|_{\mathbf{H}^{3/2}(\partial\Omega)} \, . \qquad (42)$$

Note that the inequality constraints in (41) are often inconvenient for implementation. The penalty formulation allows to avoid such inequality constraints in the set of admissible solutions and to obtain a variational equation. The penalty form of frictionless contact is [4]:

Find $\mathbf{u}^\varepsilon \in \tilde{H}^{1/2}(\Gamma_N \cup \Gamma_C)$ such that

$$\langle S\mathbf{u}^\varepsilon, \mathbf{v} \rangle - \langle p^\varepsilon, v_n \rangle = L(\mathbf{v}) \quad \forall \mathbf{v} \in \tilde{H}^{1/2}(\Gamma_N \cup \Gamma_C) \, , \qquad (43)$$

where $p^\varepsilon := -\frac{1}{\varepsilon}(u_n^\varepsilon - g)^+$ with the preset penalty parameter $\varepsilon > 0$. It holds that (see [4])

$$\|\mathbf{u}^\varepsilon - \mathbf{U}^\varepsilon\|_{\tilde{H}^{1/2}(\Gamma_N \cup \Gamma_C)} = \mathcal{O}\left(\left(\frac{h}{p}\right)^{1-\eta}\right), \quad \eta > 0$$

where $\mathbf{U}^\varepsilon$ is the boundary element Galerkin solution of (43). Adaptive simulations for the hp-versions of (41) and (43) are presented in [6].

4 Fluid-Structure Interaction

Here we consider the FE/BE coupling for an interface scattering problem.

Let $\Omega \subset \mathbb{R}^3$ be a bounded region with boundary Γ and exterior $\Omega^+ = \mathbb{R}^3 \setminus \bar{\Omega}$. Ω represents an inhomogeneous elastic obstacle and Ω^+ is a compressible, non-viscous, homogeneous fluid. $\mathbf{u}$ represents the displacement in Ω and $P = p + p^0$ the fluid pressure in Ω^+ with incident field p^0. Let $\sigma(\mathbf{u})$ denote the stress tensor for linear elasticity, ρ a positive function in Ω, and ρ_0, c_0, ω are positive constants. p^0 is a solution of $\Delta p^0 + \omega^2/c_0^2 p^0 = 0$ in $\mathbb{R}^3$. Then we seek $\mathbf{u}$ and p satisfying:

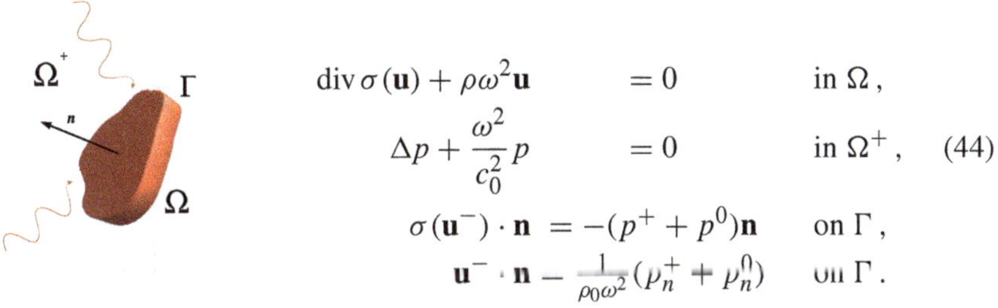

$$
\begin{aligned}
\operatorname{div}\sigma(\mathbf{u}) + \rho\omega^2\mathbf{u} &= 0 && \text{in } \Omega, \\
\Delta p + \frac{\omega^2}{c_0^2}p &= 0 && \text{in } \Omega^+, \quad (44)\\
\sigma(\mathbf{u}^-)\cdot\mathbf{n} &= -(p^+ + p^0)\mathbf{n} && \text{on } \Gamma, \\
\mathbf{u}^-\cdot\mathbf{n} - \frac{1}{\rho_0\omega^2}(p_n^+ + p_n^0) && && \text{on } \Gamma.
\end{aligned}
$$

$\mathbf{n}$ denotes the exterior normal vector on Γ, p_n is the normal derivative of p and p^+, $\mathbf{u}^-$ indicates the limits from Ω^+ and Ω.

Similarly to (14) one obtains the variational form for (44), where now the integral operators V, K, K' and W are given via the Green's function

$$
\Gamma(\mathbf{x} - \mathbf{y}) = \frac{1}{4\pi}\frac{e^{ik|x-y|}}{|x-y|}
$$

for the Helmholtz equation (see [1]):

For $\operatorname{Im}\alpha \neq 0$, find $\mathbf{u} \in H^1(\Omega)^3$, $\phi \in H^{1/2}(\Gamma)$ such that $\forall \mathbf{v} \in H^1(\Omega)^3$, $\psi \in H^{1/2}(\Gamma)$:

Table 4 Residual error estimator η_R and effectivity index q calculated for $k = 3.5$ and $k = 5$ with $\alpha = \imath/k$.

h	N	$k = 3.5$		$k = 5.0$	
		η_r	$q = \eta_r/e$	η_r	$q = \eta_r/e$
1	26	6.3622	1.2257	29.9753	3.6887
1/2	66	5.5580	1.4366	31.5406	3.4438
1/4	194	4.7024	1.9920	20.8456	2.2508
1/8	642	3.2281	2.3674	13.0691	2.1793
1/16	2306	1.9494	2.5368	7.1497	2.2010
1/32	8706	1.1588	2.6715	3.7212	2.1949
1/64	33794	0.7152	2.9234	1.9314	2.1949

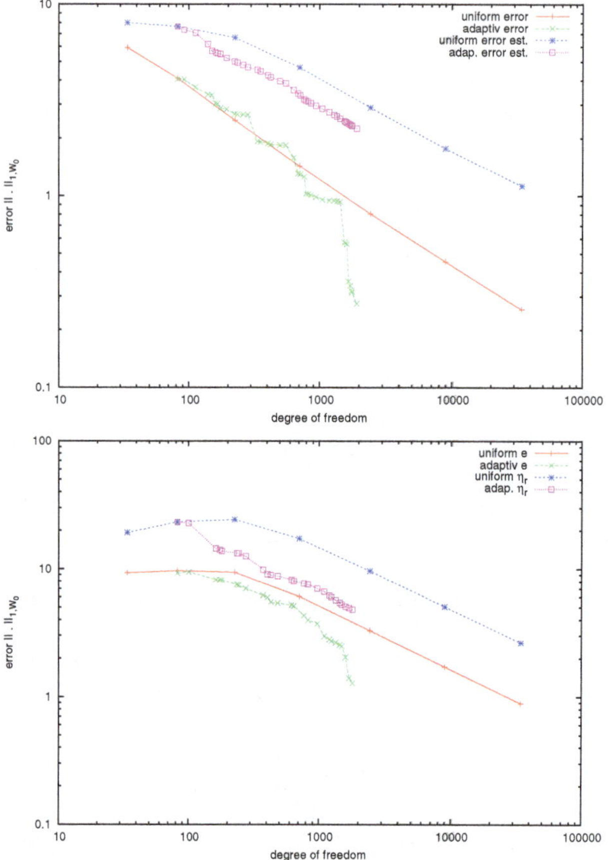

Fig. 9 Errors and residual error estimator, $\alpha = i/k$: $-+-$ e using uniform refinement, $-*-$ residual error estimator using uniform refinement, $-\times-$ e using residual adaptive refinement, $-\Box-$ residual error estimator using adaptive refinement.

$$\left(\sigma(\mathbf{u}), \varepsilon(\bar{\mathbf{v}})\right) - \rho\omega^2(\mathbf{u}, \bar{\mathbf{v}}) + \langle V\phi\mathbf{n}, \bar{\mathbf{v}}\rangle + \alpha\left\langle\left(K + \frac{1}{2}\right)\phi\mathbf{n}, \bar{\mathbf{v}}\right\rangle = -\langle p^0\mathbf{n}, \bar{\mathbf{v}}\rangle \quad (45)$$

$$-\langle\phi\mathbf{n}, \bar{\mathbf{v}}\rangle + \frac{1}{\rho_0\omega^2}\left\langle\left(K' - \frac{1}{2}\right)\phi, \bar{\psi}\right\rangle - \alpha\langle W\phi, \bar{\psi}\rangle = -\frac{1}{\rho_0\omega^2}\left\langle\frac{\partial p^0}{\partial\mathbf{n}}, \bar{\psi}\right\rangle \quad (46)$$

In [8–10], the variational form (45), (46) is solved with finite elements in Ω and boundary elements on Γ and *a posteriori* error estimators of residual and hierarchical type are derived for the Galerkin error, analogously to those in Section 1. Error controlled adaptive numerical simulations are given there for 2D and 3D problems. Below a 2D numerical example is given which shows the applicabiltity of the coupling approach (Table 4, Fig. 9).

Numerical example in 2D [8]: *Consider a square-shaped, homogeneous, isotropic, elastic scatterer made of steel with $\bar{\Omega} = [-1, 1]^2$. The scatterer possesses the following material parameters: Poisson's ratio $v = 0.28$, Young's module $E = 200$ GPa and $\rho = 7800$ kg/m^3. The scatterer is submerged in sea water and is subject to a plane incident wave $p^0(x, y) = e^{ikx}$. Furthermore, we assume for sea water a density $\rho_0 = 1020$ kg/m^3 and a sound velocity $c_0 = 1500$ m/s.*

References

1. J. Bielak and R.C. McCamy. Symmetric finite element and boundary element coupling methods for fluid-solid interaction. *Quart. Appl. Math.*, 107–119, 1991.
2. Carsten Carstensen, Stefan A. Funken, and Ernst P. Stephan. A posteriori error estimates for *hp*-boundary element methods. *Appl. Anal.*, 61(3-4):233–253, 1996.
3. Carsten Carstensen and Ernst P. Stephan. Adaptive coupling of boundary elements and finite elements. *RAIRO Modél. Math. Anal. Numér.*, 29(7):779–817, 1995.
4. A. Chernov, M. Maischak, and E.P. Stephan. A priori error estimates for hp penalty bem for contact problems in elasticity. *Comp. Methods Appl. Mech. Eng.*, 196:3871–3880, 2007.
5. A. Chernov, M. Maischak, and E.P. Stephan. *hp*-mortar boundary element method for two-body contact problems with friction. *Math. Meth. Appl. Sci.*, 31:2029–2054, 2008.
6. A. Chernov and E.P. Stephan. Adaptive BEM for contact problems with friction. In P. Wriggers and U. Nackenhorst (Eds.), *IUTAM Symposium on Computational Contact Mechanics*, pages 113–122, Springer, Dordrecht, 2007.
7. M. Costabel and E.P. Stephan. Coupling of finite and boundary element methods for an elastoplastic interface problem. *SIAM J. Numer. Anal.*, 27:1212–1226, 1990.
8. C. Domínguez. Finite element and boundary element coupling for fluid-structure interaction. PhD Thesis, Hannover, 2010.
9. C. Domínguez and E.P. Stephan. A posteriori hierarchical error estimates for a FE/BE coupling of a fluid-structure interaction problem, to appear.
10. C. Domínguez and E.P. Stephan. A posteriori residual error estimates for a FE/BE coupling of a fluid-structure interaction problem, to appear.
11. J. Haslinger, M. Miettinen, and P.D. Panagiotopoulos. *Finite Element Method for Hemivariational Inequalities*. Nonconvex Optimization and Its Applications. Kluwer Academic Publishers, Dordrecht, 1999.
12. G.C. Hsiao and W.L. Wendland. *Boundary Integral Equations*. Springer, Heidelberg, 2008.
13. L. Lukšan and J. Vlček. A Bundle–Newton method for nonsmooth unconstrained minimization. *Math. Progr.*, 83:373–391, 1998.
14. M. Maischak, S. Oestmann, and E.P. Stephan. A least squares FE-BE coupling method for linear elasticity, 2010, submitted.
15. M. Maischak and E.P. Stephan. A least squares coupling method with finite elements and boundary elements for transmission problems. *Comput. Math. Appl.*, 48(7-8):995–1016, 2004.
16. L. Nesemann. Finite element and boundary element methods for contact with adhesion. PhD Thesis, Hannover, 2010.
17. C. Niculescu and L.-E. Persson. *Convex Functions and Their Applications*. CMS Books in Mathematics. Canadian Mathematical Society, Halifax, 2006.
18. S. Oestmann. Error controled adaptive FE-BE couply methods and applications. PhD Thesis, Hannover, 2006.
19. E.P. Stephan. Coupling of finite elements and boundary elements for some nonlinear interface problems. *Comp. Meth. Appl. Mech. Engrg*, 101:61–72, 1992.
20. E.P. Stephan. Coupling of boundary element methods and finite element methods. In: E. Stein, R. de Borst, and T.J.R. Hughes (Eds.), *Encyclopedia of Computational Mechanics*. John Wiley & Sons, 2004.

Efficient Solvers for Mixed Finite Element Discretizations of Nonlinear Problems in Solid Mechanics

Gerhard Starke

Abstract A common goal of our projects in the three phases of GRK 615 was, among other issues, the development of efficient solvers for different mixed finite element approaches to nonlinear problems in solid mechanics. In the first phase, the PEERS ('plane elasticity element with reduced symmetry') was studied for elastoplastic deformation models. The nonlinear algebraic systems were solved with a fixed point iteration leading to a linear elasticity problem in each step which was treated by suitable constraint preconditioners. The treatment of elastoplastic deformations by least squares mixed finite element methods was the subject of the project in the second phase. In particular, appropriate regularizations for the non-smoothness of the nonlinear problems were investigated. In the third phase, the least squares finite element formulation of contact problems was studied. For the Signorini problem, the quadratic minimization problems under affine constraints were treated by an active set strategy. Preconditioned conjugate gradient iterations for a null space formulation were used for the systems arising in each step.

1 Efficient Solvers and A-posteriori Error Estimators for Mixed Problems in Elastoplasticity

For the numerical treatment of elastoplastic deformation models, mixed finite element methods are particularly valuable since they provide accurate stress approximations directly. The PEERS ('plane elasticity element with reduced symmetry') approach constitutes one of the earliest and one of the simplest stable element combinations. For the linear elasticity case, the PEERS approach is based on the saddle point problem of finding $(\sigma, \mathbf{u}, \gamma) \in H_{\Gamma_N}(\mathrm{div}, \Omega)^2 \times L^2(\Omega)^2 \times L^2(\Omega)$ such that

Gerhard Starke
Institut für Angewandte Mathematik, Leibniz Universität Hannover, Welfengarten 1, 30167 Hannover, Germany; e-mail: gcs@ifam.uni-hannover.de

E. Stephan & P. Wriggers (Eds.): Modelling, Simulation and Software Concepts, LNACM 57, pp. 201–209.
springerlink.com © Springer-Verlag Berlin Heidelberg 2011

$$(\mathcal{C}^{-1}(\boldsymbol{\sigma}^N + \boldsymbol{\sigma}), \boldsymbol{\tau})_{0,\Omega} + (\mathbf{u}, \operatorname{div} \boldsymbol{\tau})_{0,\Omega} + (\gamma, \operatorname{as} \boldsymbol{\tau})_{0,\Omega} = 0 \tag{1}$$

$$(\operatorname{div}(\boldsymbol{\sigma}^N + \boldsymbol{\sigma}) + \mathbf{f}, \mathbf{v})_{0,\Omega} = 0 \tag{2}$$

$$(\operatorname{as}(\boldsymbol{\sigma}^N + \boldsymbol{\sigma}), \eta)_{0,\Omega} = 0 \tag{3}$$

holds for all $(\boldsymbol{\tau}, \mathbf{v}, \eta) \in H_{\Gamma_N}(\operatorname{div}, \Omega)^2 \times L^2(\Omega)^2 \times L^2(\Omega)$. Here $\boldsymbol{\sigma}^N \in H(\operatorname{div}, \Omega)^2$ is a suitable extension of the boundary values, i.e. satisfying $\boldsymbol{\sigma}^N \cdot \mathbf{n} = \mathbf{g}$ on Γ_N. This mixed formulation is based on the Hellinger–Reissner principle where the symmetry of the stress tensor is only weakly enforced. Linear systems of the form (1), (2), (3) arise in each step of a fixed point iteration in each time step of an incremental elastoplastic modelling.

For the efficient solution of these linear systems, constraint preconditioning strategies are appropriate (cf. [10]). These preconditioners also possess a saddle point structure of the form

$$\mathcal{G} = \begin{pmatrix} G & B^T & C^T \\ B & 0 & 0 \\ C & 0 & 0 \end{pmatrix}.$$

The upper left block G should consist of some reasonable approximation of the corresponding block in the PEERS system, e.g. its diagonal part. In that case, we may write

$$\mathcal{G} = \begin{pmatrix} D & \bar{B}^T \\ \bar{B} & 0 \end{pmatrix} = \begin{pmatrix} I & 0 \\ \bar{B}D^{-1} & I \end{pmatrix} \cdot \begin{pmatrix} D & 0 \\ 0 & \bar{B}D^{-1}\bar{B}^T \end{pmatrix} \cdot \begin{pmatrix} I & D^{-1}\bar{B}^T \\ 0 & I \end{pmatrix}.$$

In an implementation, the inverse of the Schur complement matrix $\bar{B}D^{-1}\bar{B}^T$ is required which may be based on the Cholesky decomposition $\bar{B}D^{-1}\bar{B}^T = R^T R$. In this step, it is important to use an appropriate reordering algorithm like minimum degree ordering in order to significantly lower the computational cost. The numerical results presented in [5] show the good performance of this preconditioning approach in terms of GMRES iterations and overall computational cost.

Alternatively, positive definite block diagonal preconditioners could be used which have the advantage that a Krylov subspace method with short recurrences like MINRES can be employed [12]. On the other hand, the number of iterations is often quite a bit higher than for constraint preconditioning.

Further details of these investigations are contained in the dissertation by Geilenkothen [6] which was finished just a couple of months after the completion of the GRK project he was funded from.

2 Least Squares Mixed Finite Element Methods for Elastoplastic Problems

As a continuation of the project from the first phase, the least squares mixed finite element method was used for the computation of elastoplastic deformation models. The least squares formulation constitutes a generalization from the linear elasticity case investigated in [2–4]. A detailed analysis of the least squares method for elastoplastic problems was presented in [17].

The least squares formulation, associated with the elastoplastic deformation model consists in minimizing the functional

$$\mathcal{F}(\mathbf{u}, \boldsymbol{\sigma}; \boldsymbol{\sigma}^{\mathrm{old}}, \alpha^{\mathrm{old}}) = \|\mathrm{div}(\boldsymbol{\sigma}^{\mathrm{old}} + \boldsymbol{\sigma})\|_{0,\Omega}^2 + \|\mathcal{C}^{-1/2}(\boldsymbol{\sigma} - \mathcal{R}(\boldsymbol{\varepsilon}(\mathbf{u})))\|_{0,\Omega}^2 \quad (4)$$

among all suitable $(\mathbf{u}, \boldsymbol{\sigma}) \in H^1(\Omega)^3 \times H(\mathrm{div}, \Omega)^3$. For the case of von Mises plasticity with isotropic hardening, the stress response is given by

$$\mathcal{R}(\boldsymbol{\varepsilon}) = \mathcal{C} \left(\boldsymbol{\varepsilon} - \frac{1}{2\mu} \gamma_R (\mathrm{dev}(\boldsymbol{\sigma}^{\mathrm{old}} + \mathcal{C}\,\boldsymbol{\varepsilon})) \frac{\mathrm{dev}(\boldsymbol{\sigma}^{\mathrm{old}} + \mathcal{C}\,\boldsymbol{\varepsilon})}{|\mathrm{dev}(\boldsymbol{\sigma}^{\mathrm{old}} + \mathcal{C}\,\boldsymbol{\varepsilon})|} \right), \quad (5)$$

where the return parameter $\gamma_R(\mathrm{dev}(\boldsymbol{\sigma}^{\mathrm{old}} + \mathcal{C}\,\boldsymbol{\varepsilon}))$ is implicitly defined as the solution of the nonlinear equation

$$\gamma_R = |\mathrm{dev}(\boldsymbol{\sigma}^{\mathrm{old}} + \mathcal{C}\,\boldsymbol{\varepsilon})| - \sqrt{\frac{2}{3}} K \left(\alpha^{\mathrm{old}} + \sqrt{\frac{2}{3}} \frac{\gamma_R}{2\mu} \right), \quad (6)$$

if $|\mathrm{dev}(\boldsymbol{\sigma}^{\mathrm{old}} + \mathcal{C}\,\boldsymbol{\varepsilon})| > \sqrt{2/3} K(\alpha^{\mathrm{old}})$ and $\gamma_R(\mathrm{dev}(\boldsymbol{\sigma}^{\mathrm{old}} + \mathcal{C}\,\boldsymbol{\varepsilon})) = 0$ otherwise. The hardening parameter is updated by

$$\alpha = \alpha^{\mathrm{old}} + \sqrt{\frac{2}{3}} \frac{\gamma_R(\mathrm{dev}(\boldsymbol{\sigma}^{\mathrm{old}} + \mathcal{C}\,\boldsymbol{\varepsilon}))}{2\mu}. \quad (7)$$

Here $\mathrm{dev}(\,\cdot\,)$ denotes the deviatoric (trace-free) part of a matrix. Due to the fact that $\mathcal{R}(\boldsymbol{\varepsilon}(\mathbf{u}))$ is not differentiable everywhere, the subdifferential $\partial\mathcal{R}(\boldsymbol{\varepsilon}(\mathbf{u}))$ needs to be employed in the formulation of the variational problem. The minimum of (4) is given by the solution of the variational problem

$$(\mathrm{div}(\boldsymbol{\sigma}^{\mathrm{old}} + \boldsymbol{\sigma}), \mathrm{div}\,\boldsymbol{\tau}) + (\mathcal{C}^{-1}(\boldsymbol{\sigma} - \mathcal{R}(\boldsymbol{\varepsilon}(\mathbf{u})), \boldsymbol{\tau}) = 0 \quad (8)$$

$$0 \in (\mathcal{C}^{-1}(\boldsymbol{\sigma} - \mathcal{R}(\boldsymbol{\varepsilon}(\mathbf{u})), \partial\mathcal{R}(\boldsymbol{\varepsilon}(\mathbf{u}))[\boldsymbol{\varepsilon}(\mathbf{v})]) \quad (9)$$

for all $\boldsymbol{\tau} \in H_{\Gamma_N}(\mathrm{div}, \Omega)^3$ and $\mathbf{v} \in H_{\Gamma_D}^1(\Omega)^3$ (see e.g. [8, Section 4] for the use of the subdifferential in the context of plasticity models). If $K(\alpha)$ satisfies

$$K(\alpha) \geq K_0 > 0, \; K'(\alpha) \geq K_1 > 0 \quad (10)$$

204   G. Starke

for all $\alpha > 0$, then it is known that (4) possesses a unique minimizer (cf. [8, section 8]).

For the minimization of the nonlinear least squares functional (4), a Gauss–Newton type iteration is the natural approach. This consists in minimizing a quadratic least squares functional of the form

$$\mathcal{F}^{(k)}(\delta\mathbf{u}, \delta\boldsymbol{\sigma}) = \|\mathrm{div}(\boldsymbol{\sigma}^{\mathrm{old}} + \boldsymbol{\sigma}^{(k)} + \delta\boldsymbol{\sigma})\|^2$$
$$+ \|\mathcal{C}^{-1/2}(\boldsymbol{\sigma}^{(k)} + \delta\boldsymbol{\sigma} - \mathcal{R}(\boldsymbol{\varepsilon}(\mathbf{u}^{(k)})) - \mathcal{S}^{(k)}[\boldsymbol{\varepsilon}(\delta\mathbf{u})]\|^2$$

in each step and setting $(\boldsymbol{\sigma}^{(k+1)}, \mathbf{u}^{(k+1)}) = (\boldsymbol{\sigma}^{(k)}, \mathbf{u}^{(k)}) + (\delta\boldsymbol{\sigma}, \delta\mathbf{u})$. Since $\mathcal{R}(\boldsymbol{\varepsilon}(\mathbf{u}^{(k)}))$ is not necessarily differentiable, it is not clear which element of the subdifferential should be chosen as $\mathcal{S}^{(k)}[\boldsymbol{\varepsilon}(\delta\mathbf{u})]$. For $|\mathrm{dev}(\boldsymbol{\sigma}^{\mathrm{old}} + \mathcal{C}\,\boldsymbol{\varepsilon}(\mathbf{u}^{(k)}))| > \sqrt{2/3}K(\alpha^{\mathrm{old}})$, differentiating (5) yields

$$\mathcal{S}^{(k)}[\boldsymbol{\varepsilon}(\delta\mathbf{u})] = \mathcal{C}\boldsymbol{\varepsilon}(\delta\mathbf{u})$$
$$- \frac{1}{2\mu}\left(\gamma_R'(\boldsymbol{\xi}^{(k)})[\boldsymbol{\varepsilon}(\delta\mathbf{u})] + \gamma_R(\boldsymbol{\xi}^{(k)})\left(\frac{\boldsymbol{\varepsilon}(\delta\mathbf{u})}{|\boldsymbol{\xi}^{(k)}|} - \frac{(\boldsymbol{\xi}^{(k)}:\boldsymbol{\varepsilon}(\delta\mathbf{u}))\boldsymbol{\xi}^{(k)}}{|\boldsymbol{\xi}^{(k)}|^3}\right)\right)$$

where $\boldsymbol{\xi}^{(k)} = \mathrm{dev}(\boldsymbol{\sigma}^{\mathrm{old}} + \mathcal{C}\,\boldsymbol{\varepsilon}(\mathbf{u}^{(k)}))$. Differentiating (6) leads to

$$\gamma_R'(\boldsymbol{\xi}^{(k)})[\boldsymbol{\varepsilon}(\delta u)] = \frac{\boldsymbol{\xi}^{(k)} : \boldsymbol{\varepsilon}(\delta\mathbf{u})}{|\boldsymbol{\xi}^{(k)}|} - \frac{1}{3\mu}\gamma_R'(\boldsymbol{\xi}^{(k)})[\boldsymbol{\varepsilon}(\delta\mathbf{u})]\, K'\left(\alpha^{\mathrm{old}} + \sqrt{\frac{2}{3}}\frac{\gamma_R(\boldsymbol{\xi}^{(k)})}{2\mu}\right)$$

which implies the explicit formula

$$\gamma_R'(\boldsymbol{\xi}^{(k)})[\boldsymbol{\varepsilon}(\delta\mathbf{u})] = \left(1 + \frac{1}{3\mu}K'\left(\alpha^{\mathrm{old}} + \sqrt{\frac{2}{3}}\frac{\gamma_R(\boldsymbol{\xi}^{(k)})}{2\mu}\right)\right)^{-1}\frac{\boldsymbol{\xi}^{(k)} : \boldsymbol{\varepsilon}(\delta\mathbf{u})}{|\boldsymbol{\xi}^{(k)}|}$$

to be used for the computation of $\mathcal{S}^{(k)}[\boldsymbol{\varepsilon}(\delta\mathbf{u})]$. If $|\mathrm{dev}(\boldsymbol{\sigma}^{\mathrm{old}} + \mathcal{C}\,\boldsymbol{\varepsilon}(\mathbf{u}^{(k)}))| < \sqrt{2/3}K(\alpha^{\mathrm{old}})$ holds, then we are inside the elastic domain and

$$\mathcal{S}^{(k)}[\boldsymbol{\varepsilon}(\delta\mathbf{u})] = \mathcal{C}\boldsymbol{\varepsilon}(\delta\mathbf{u})\,. \tag{11}$$

The problem was regularized by a smooth least squares formulation in [15]. However, it was later observed (see [18]) that one can simply use formula (11) also in the case $|\mathrm{dev}(\boldsymbol{\sigma}^{\mathrm{old}} + \mathcal{C}\,\boldsymbol{\varepsilon}(\mathbf{u}^{(k)}))| = \sqrt{2/3}K(\alpha^{\mathrm{old}})$ since this certainly constitutes an element of the subdifferential. The resulting non-smooth Newton iteration converges in at most three iterations throughout a loading cycle for an elastoplastic benchmark problem (cf. [19]).

The least squares formulation is extended to poro-plasticity in [14] based on the work in [13]. Further details of these studies are contained in the dissertation by Kubitz [15] which was finished almost timely at the completion of the GRK project he was funded from.

3 Least Squares Mixed Finite Element Solution of Contact Problems

For the treatment of the Signorini contact problem, the following modified bilinear form is used:

$$\mathcal{A}(\mathbf{u}, \boldsymbol{\sigma}; \mathbf{v}, \boldsymbol{\tau}) = (\operatorname{div} \boldsymbol{\sigma}, \operatorname{div} \boldsymbol{\tau})_{0,\Omega} + (\mathcal{C}^{-1/2}\boldsymbol{\sigma} - \mathcal{C}^{1/2}\boldsymbol{\varepsilon}(\mathbf{u}), \mathcal{C}^{-1/2}\boldsymbol{\tau} - \mathcal{C}^{1/2}\boldsymbol{\varepsilon}(\mathbf{v}))_{0,\Omega}$$
$$+ \frac{1}{2}\langle \mathbf{n} \cdot \mathbf{u}, \mathbf{n} \cdot (\boldsymbol{\tau} \cdot \mathbf{n})\rangle_{\Gamma_C} + \frac{1}{2}\langle \mathbf{n} \cdot (\boldsymbol{\sigma} \cdot \mathbf{n}), \mathbf{n} \cdot \mathbf{v}\rangle_{\Gamma_C},$$

where Γ_C is the contact part of the boundary $\partial\Omega = \Gamma_D \cup \Gamma_N \Gamma_C$. This bilinear form may be shown to be coercive with respect to $H^1_{\Gamma_D}(\Omega)^d \times H(\operatorname{div}, \Omega)^d$ (cf. [1]) and its minimization subject to the contact constraints

$$\mathbf{n} \cdot (\mathbf{u}^D + \mathbf{u}_h) - g \le 0, \tag{12}$$
$$\mathbf{n} \cdot ((\boldsymbol{\sigma}^N + \boldsymbol{\sigma}_h) \cdot \mathbf{n}) \le 0, \tag{13}$$
$$\mathbf{t} \cdot ((\boldsymbol{\sigma}^N + \boldsymbol{\sigma}_h) \cdot \mathbf{n}) = 0 \tag{14}$$

on Γ_C provides the unique solution of the Signorini contact problem (cf. [11, sections 5.3, 5.5]). The numerical analysis of this least squares finite element approach for the Signorini contact problem using H^1 conforming and $H(\operatorname{div})$ conforming finite elements for the displacement and stress components, respectively, is presented in [1]. Here, some of the practical aspects of this approach, in particular, concerning the efficient solution of the associated systems of algebraic equations, are described.

The contact conditions (14) need to be implemented appropriately (see [20]). In the lowest-order case, this is simply done by treating the constraints for the displacements at all vertices and the constraints for the stresses at all edge midpoints of the contact boundary. Since the normal stress is piecewise constant on the boundary (assumed to be polygonal), this is an exact implementation of the contact conditions (14). For higher order elements, the contact conditions can only be fulfilled approximately, in general.

Let us denote the set of displacements and stresses which satisfy the contact constraints as admissible set

$$\mathcal{K} = \{(\mathbf{v}_h, \boldsymbol{\sigma}_h) \in \mathbf{V}_h \times \boldsymbol{\Sigma}_h : (14) \text{ satisfied}\}.$$

$\mathcal{K}$ is a convex subset of the finite element space $\mathbf{V}_h \times \boldsymbol{\Sigma}_h$. The least squares finite element formulation consists in finding $(\mathbf{u}_h, \boldsymbol{\sigma}_h) \in \mathcal{K}$ in such a way that

$$\mathcal{A}(\mathbf{u}^D + \mathbf{u}_h, \boldsymbol{\sigma}^N + \boldsymbol{\sigma}_h; \mathbf{u}^D + \mathbf{u}_h, \boldsymbol{\sigma}^N + \boldsymbol{\sigma}_h) - \langle g, \mathbf{n} \cdot ((\boldsymbol{\sigma}^N + \boldsymbol{\sigma}_h) \cdot \mathbf{n})\rangle_{\Gamma_C} \tag{15}$$

is minimized. Introducing a basis for $\mathbf{V}_h \times \boldsymbol{\Sigma}_h$, this minimization may be written with respect to $\mathbf{R}^n$. Let us denote the bases for the spaces $\mathbf{V}_h$ and $\boldsymbol{\Sigma}_h$ as follows:

$$\mathbf{V}_h = \operatorname{span}\{\boldsymbol{\phi}_1, \ldots, \boldsymbol{\phi}_{n_u}\}, \tag{16}$$

$$\Sigma_h = \mathrm{span}\{\boldsymbol{\psi}_1, \ldots, \boldsymbol{\psi}_{n_s}\} \tag{17}$$

$(n_u + n_s = n)$. With the matrix

$$G = \begin{pmatrix} G_{uu} & G_{su}^T \\ G_{su} & G_{ss} \end{pmatrix} = \begin{pmatrix} \left[\mathcal{A}(\boldsymbol{\phi}_j, 0; \boldsymbol{\phi}_i, 0)\right]_{i,j} & \left[\mathcal{A}(0, \boldsymbol{\psi}_j; \boldsymbol{\phi}_i, 0)\right]_{i,j} \\ \left[\mathcal{A}(\boldsymbol{\phi}_j, 0; 0, \boldsymbol{\psi}_i)\right]_{i,j} & \left[\mathcal{A}(0, \boldsymbol{\psi}_j; 0, \boldsymbol{\psi}_i)\right]_{i,j} \end{pmatrix}$$

and the vectors

$$\mathbf{c}^D = \begin{pmatrix} \left[\mathcal{A}(\mathbf{u}^D, 0; \boldsymbol{\phi}_i, 0)\right]_{1 \le i \le n_u} \\ \left[\mathcal{A}(\mathbf{u}^D, 0; 0, \boldsymbol{\psi}_i)\right]_{1 \le i \le n_s} \end{pmatrix}, \tag{18}$$

$$\mathbf{c}^N = \begin{pmatrix} \left[\mathcal{A}(0, \boldsymbol{\sigma}^N; \boldsymbol{\phi}_i, 0)\right]_{1 \le i \le n_u} \\ \left[\mathcal{A}(0, \boldsymbol{\sigma}^N; 0, \boldsymbol{\psi}_i)\right]_{1 \le i \le n_s} \end{pmatrix}, \tag{19}$$

$$\mathbf{c}^C = \begin{pmatrix} [0]_{1 \le i \le n_u} \\ \left[-\langle g, \mathbf{n} \cdot ((\boldsymbol{\sigma}^N + \boldsymbol{\psi}_i) \cdot \mathbf{n})\rangle_{\Gamma_C}\right]_{1 \le i \le n_s} \end{pmatrix} \tag{20}$$

this becomes the minimization of

$$\frac{1}{2}\mathbf{x}^T G \mathbf{x} + \mathbf{x}^T \mathbf{c} \quad \text{with } \mathbf{c} = \mathbf{c}^D + \mathbf{c}^N + \mathbf{c}^C$$

for $\mathbf{x} = (\mathbf{x}_u, \mathbf{x}_s)^T \in \mathbf{R}^n$ under the constraints (14). These constraints may also be written in matrix notation with respect to $\mathbf{x}$:

$$A_u \mathbf{x} \le \mathbf{b}_u , \tag{21}$$

$$A_{sn} \mathbf{x} \le \mathbf{b}_{sn} , \tag{22}$$

$$A_{st} \mathbf{x} = \mathbf{b}_{st} \tag{23}$$

with

$$A_u = \begin{pmatrix} \left[\mathbf{n} \cdot \boldsymbol{\phi}_j|_i\right]_{1 \le i \le m_u, 1 \le j \le n_u} & 0 \\ 0 & 0 \end{pmatrix}, \quad \mathbf{b}_u = \begin{pmatrix} \left[g - \mathbf{n} \cdot \mathbf{u}^D|_i\right]_{1 \le i \le m_u} \\ \mathbf{0} \end{pmatrix},$$

$$A_{sn} = \begin{pmatrix} 0 & 0 \\ 0 & \left[\mathbf{n} \cdot (\boldsymbol{\psi}_j \cdot \mathbf{n})|_i\right]_{1 \le i \le m_{sn}, 1 \le j \le n_s} \end{pmatrix}, \quad \mathbf{b}_{sn} = \begin{pmatrix} \mathbf{0} \\ \left[-\mathbf{n} \cdot (\boldsymbol{\sigma}^N \cdot \mathbf{n})|_i\right]_{1 \le i \le m_{sn}} \end{pmatrix},$$

$$A_{st} = \begin{pmatrix} 0 & 0 \\ 0 & \left[\mathbf{t} \cdot (\boldsymbol{\psi}_j \cdot \mathbf{n})|_i\right]_{1 \le i \le m_{st}, 1 \le j \le n_s} \end{pmatrix}, \quad \mathbf{b}_{st} = \begin{pmatrix} \mathbf{0} \\ \left[-\mathbf{t} \cdot (\boldsymbol{\sigma}^N \cdot \mathbf{n})|_i\right]_{1 \le i \le m_{st}} \end{pmatrix}.$$

Obviously, this constitutes a quadratic minimization problem under affine constraints which may be handled by active set strategies (cf. [16, chapter 16]).

Iteration k of the active set strategy consists of the solution of a quadratic minimization problem under affine equality constraints: Minimize

$$\frac{1}{2}(\mathbf{x}^{(k)} + \mathbf{p})^T G(\mathbf{x}^{(k)} + \mathbf{p}) + (\mathbf{x}^{(k)} + \mathbf{p})^T \mathbf{c}$$

among all $\mathbf{p} \in \mathbf{R}^n$ satisfying the conditions

$$A_u(I_u^{(k)}, :)\mathbf{p} = \mathbf{b}_u(I_u^{(k)}) - A_u(I_u^{(k)}, :)\mathbf{x}^{(k)} , \tag{24}$$

$$A_{sn}(I_{sn}^{(k)}, :)\mathbf{p} = \mathbf{b}_{sn}(I_{sn}^{(k)}) - A_{sn}(I_{sn}^{(k)}, :)\mathbf{x}^{(k)} , \tag{25}$$

$$A_{st}\mathbf{p} = \mathbf{b}_{st} - A_{st}\mathbf{x}^{(k)} . \tag{26}$$

Here, $I_u^{(k)}$ and $I_{sn}^{(k)}$ denote approximations to the corresponding sets of indices. Since the active set strategy only considers indices in $I_u^{(k)}$ and $I_{sn}^{(k)}$, respectively, for which the corresponding constraint for $\mathbf{x}^{(k)}$ is active, the corresponding conditions are also homogeneous and the constraints simplify to

$$A_u(I_u^{(k)}, :)\mathbf{p} = \mathbf{0} , \tag{27}$$

$$A_{sn}(I_{sn}^{(k)}, :)\mathbf{p} = \mathbf{0} , \tag{28}$$

$$A_{st}\mathbf{p} = \mathbf{0} \tag{29}$$

The KKT conditions for this minimization problem leads to the linear system of equations

$$\begin{pmatrix} G & A_u(I_u^{(k)}, :)^T & A_{sn}(I_{sn}^{(k)}, :)^T & A_{st}^T \\ A_u(I_u^{(k)}, :) & 0 & 0 & 0 \\ A_{sn}(I_{sn}^{(k)}, :) & 0 & 0 & 0 \\ A_{st} & 0 & 0 & 0 \end{pmatrix} \begin{pmatrix} \mathbf{p} \\ \lambda_u \\ \lambda_{sn} \\ \lambda_{st} \end{pmatrix} = \begin{pmatrix} -\mathbf{c} - G\mathbf{x}^{(k)} \\ \mathbf{0} \\ \mathbf{0} \\ \mathbf{0} \end{pmatrix} ,$$

or, abbreviated,

$$\begin{pmatrix} G & A^T \\ A & 0 \end{pmatrix} \begin{pmatrix} \mathbf{p} \\ \lambda \end{pmatrix} = \begin{pmatrix} -\mathbf{c} - G\mathbf{x}^{(k)} \\ \mathbf{0} \end{pmatrix} .$$

This constitutes a linear system with saddle point structure, with symmetric but indefinite coefficient matrix. On the other hand, the matrix $A \in \mathbf{R}^{m \times n}$ ($m = m_u + m_{sn} + m_{st}$) possesses only a rather small number of rows (one for each constraint) compared to the overall dimension of the system. This structure can be utilized if an effective preconditioner M for G is available. Then,

$$\begin{pmatrix} M & A^T \\ A & 0 \end{pmatrix}$$

provides an effective preconditioner for the above saddle point problem. This saddle point preconditioner can be implemented in an efficient way using the factorization

$$\begin{pmatrix} M & A^T \\ A & 0 \end{pmatrix} = \begin{pmatrix} I & 0 \\ AM^{-1} & I \end{pmatrix} \begin{pmatrix} M & 0 \\ 0 & -AM^{-1}A^T \end{pmatrix} \begin{pmatrix} I & M^{-1}A^T \\ 0 & I \end{pmatrix} .$$

This preconditioner involves the solution of linear systems of equations with M as well as with the matrix $AM^{-1}A^T$, the latter one being of rather small dimension.

An additional problem associated with our least squares finite element formulation for contact problems is due to the fact that the matrix G is not positive definite, in general. This needs to be considered in the construction of suitable preconditioners for G. An alternative is therefore the restriction to approximations which satisfy the constraint $A\mathbf{p} = \mathbf{0}$. This leads to the so-called null space methods (see [16, chapter 16]). These methods are based on a matrix $Z \in \mathbf{R}^{n \times (n-m)}$ whose columns form a basis of the null space of A. For the Signorini contact problem considered here, the construction of this matrix Z is relatively cheap. For each degree of freedom which is not associated with the contact boundary, the corresponding column is simply a unit vector. For the degrees of freedom on the contact boundary one needs to distinguish, if the corresponding constraint is contained in the current index sets $I_u^{(k)}$ or $I_{sn}^{(k)}$, respectively, or not. In the former case, the associated column of Z consists of two entries which may be determined from the corresponding values in A. In the latter case we again have unit vectors.

The saddle point problem turns into $Z^T G Z \mathbf{q} = -Z^T(\mathbf{c} + G\mathbf{x}^{(k)})$. The matrix $Z^T G Z$ is positive definite, if the index sets $I_u^{(k)}$ and $I_{sn}^{(k)}$ "sufficiently rich" which may be assumed for a reasonable approximation. Afterwards, the Lagrange parameters can be computed from

$$AA^T\lambda = -A(c + G(x^{(k)} + p)) \, .$$

In our case, AA^T is non-singular, since the affine constraints associated with the rows of A are linearly independent. For the treatment of the positive definite system with $Z^T G Z$ different preconditioners are appropriate. In order to get good results also for larger systems one needs to treat particularly the divergence-free components of the Raviart–Thomas spaces used for the stress approximations. Since the corresponding subspace is known explicitly (see e.g. [7]), standard preconditioners may be extended with appropriate basis functions. Details of these studies are contained in the dissertation by Astrid Intas [9] which was finished a couple of months after the completion of the GRK project she was funded from.

References

1. F. S. Attia, Z. Cai, and G. Starke. First-order system least squares for the Signorini contact problem in linear elasticity. *SIAM J. Numer. Anal.*, 47:3027–3043, 2009.
2. Z. Cai, J. Korsawe, and G. Starke. An adaptive least squares mixed finite element method for the stress-displacement formulation of linear elasticity. *Numer. Methods Partial Differential Equations*, 21:132–148, 2005.
3. Z. Cai and G. Starke. First-order system least squares for the stress-displacement formulation: Linear elasticity. *SIAM J. Numer. Anal.*, 41:715–730, 2003.
4. Z. Cai and G. Starke. Least squares methods for linear elasticity. *SIAM J. Numer. Anal.*, 42:826–842, 2004.

5. A. Geilenkothen. Constraint preconditioners for linear systems in elasticity. *Proc. Appl. Math. Mech.*, 2:481–482, 2003.
6. A. Geilenkothen. Efficient solvers and error estimators for a mixed method in elastoplasticity. PhD Thesis, Leibniz Universität Hannover, 2004.
7. R. Hiptmair. Multigrid method for $H(\mathrm{div})$ in three dimensions. *Electr. Trans. Numer. Anal.*, 6:133–152, 1997.
8. W. Han and B. D. Reddy. *Plasticity: Mathematical Theory and Numerical Analysis*. Springer, New York, 1999.
9. A. Intas. Finite-Element-Ausgleichsprobleme für Kontaktprobleme: Effiziente Löser. PhD Thesis, Leibniz Universität Hannover, 2010.
10. C. Keller, N.I.M. Gould, and A.J. Wathen. Constraint preconditioning for indefinite linear systems. *SIAM J. Matrix Anal. Appl.*, 21:1300–1317, 2000.
11. N. Kikuchi and J.T. Oden. *Contact Problems in Elasticity: A Study of Variational Inequalities and Finite Element Methods*. SIAM, Philadelphia, 1988.
12. A. Klawonn and G. Starke. A preconditioner for the equations of linear elasticity discretized by the PEERS element. *Numer. Linear Algebra Appl.*, 11:493–510, 2004.
13. J. Korsawe and G. Starke. A least squares mixed finite element method for Biot's consolidation problem in porous media. *SIAM J. Numer. Anal.*, 43:318–339, 2005.
14. J. Kubitz and G. Starke. An adaptive mixed finite element method for elasto-plastic consolidation. In: O. Kolditz et al. (Eds.), *Proceedings 5th Workshop on Porous Media*. ZAG Publisher, 2005.
15. J. Kubitz. Gemischte Least-Squares-FEM für Elastoplastizität. PhD Thesis, Leibniz Universität Hannover, 2007.
16. J. Nocedal and S.J. Wright. *Numerical Optimization*, 2nd edition. Springer, New York, 2006.
17. G. Starke. An adaptive least-squares mixed finite element method for elasto-plasticity. *SIAM J. Numer. Anal.*, 45:371–388, 2007.
18. G. Starke. Adaptive least squares finite element methods in elasto-plasticity. In: *LSSC 2009*, Lecture Notes in Computer Science, Vol. 5910, pp. 671–678. Springer, Heidelberg, 2010.
19. E. Stein, P. Wriggers, A. Rieger, and M. Schmidt. Benchmarks. In: E. Stein (Ed.), *Error-controlled Adaptive Finite Elements in Solid Mechanics*, chapter 11, pages 385–404. John Wiley and Sons, 2002.
20. P. Wriggers. *Computational Contact Mechanics*. Wiley, Chichester, 2002.

Computational Differential Geometry Contributions of the Welfenlab to GRK 615

Franz-Erich Wolter, Philipp Blanke, Hannes Thielhelm and Alexander Vais

Abstract This chapter presents an overview on contributions of the Welfenlab to GRK 615. Those contributions partial to computational differential geometry include computations of geodesic medial axis, cut locus, geodesic Voronoi diagrams, ("shortest") geodesics joining two given points, "focal sets and conjugate loci" in Riemannian manifolds and the application of the medial axis on metal forming simulation. The chapter includes also the computation of Laplace spectra of surfaces, solids and images and the application of those Laplace spectra to recognize the respective objects in large collections of surfaces, solids and images. Beyond that this article touches also on the origin of the afore-mentioned works including research done at the Welfenlab as well as works that can be traced back to the graduate studies of the first author.

1 Introduction

The occasion of writing a report on the contributions of the Welfenlab to GRK 615 gives the first author of this article an opportunity to look back to those years when many of his research projects related to the above-mentioned contributions had its origin. Those were the years of F.-E. Wolter's own graduate studies in Berlin in the late seventies and early eighties of the last century. There was a time in the 1970s when research on the Riemannian Laplacian operator and its eigenvalues was extremely popular in global differential geometry even more than today. In those days, many members in the community of differential geometry still had in their ears Lipman Bers' tersely formulated question "Can one hear the shape of a drum?"

In other words, is the shape of a two-dimensional bounded region determined by the eigenvalues of its corresponding Laplacian operator?

Franz-Erich Wolter · Philipp Blanke · Hannes Thielhelm · Alexander Vais
Division of Computer Graphics, Leibniz Universität Hannover, Welfengarten 1, 30165 Hannover, Germany; e-mail: {few, blanke, thielhel, vais}@welfenlab.de

E. Stephan & P. Wriggers (Eds.): Modelling, Simulation and Software Concepts, LNACM 57, pp. 211–235.
springerlink.com   © Springer-Verlag Berlin Heidelberg 2011

Indeed it was to a significant extent this question and the partly available theoretical knowledge in those days that motivated Wolter early on in the eighties to consider it being an exciting project to investigate via numerical experiments *if Laplace spectra could be employed to define feature vectors that could be used as fingerprints to practically distinguish different objects in large collections of similar surfaces and solids.* That exciting project had to wait until around 1997 when there was a chance to pursue it at the Welfenlab that had been built up by the first author in 1995 when he came to the University of Hannover. It was in 1997 when he asked two students, Herbst and Sust, to present two seminars at the Welfenlab dealing with Gordon's examples of 1992 [17] and (the respective theoretical background) showing the existence of planar polygonal regions that are isospectral but not congruent. This was the starting point for a series of diploma theses investigating Laplace spectra of planar regions; see [3, 20] for curved surfaces aiming at distinguishing those geometric entities by their Laplace spectra. This work was taken up again in 2000 by Peinecke and Reuter studying in their respective diploma theses how Laplace spectra could be used to classify geometric entities like planar domains and surfaces while Peinecke studied the respective problem for images. Peinecke and Reuter became "Kollegiaten" in the GRK 615. Their diploma and especially their PhD research extending and deepening their diploma research built up one important line of research partial to the area of computational differential geometry contributed by the Welfenlab to the GRK 615. We will give more details on this work in Section 4.

In the years of Wolter's graduate research the aforementioned popularity of research on Laplacian spectra was not shared by research related to the Riemannian cut locus – a subject that finally became the center of Wolter's PhD research [45]. The latter research works probably had its origin in Wolter's unsupervised studies on geodesics as presented in the book *Variational Theory of Geodesics* by Postnikov [33]. This book contains an error on page 101, stating there that the squared Riemannian distance function with respect to any given reference point p is differentiable everywhere on a complete Riemannian manifold M. In fact the latter squared distance function $d^2(p,x)$ is not differentiable in a dense subset $Se(p)$ of the cut locus $C(p)$ of p, implying that a complete Riemannian manifold must be diffeomorphic to $\mathbb{R}^n$ in case there exists one point p on M such that the squared distance function $d^2(p,x)$ is differentiable on all M. Here $Se(p)$ contains those points in $C(p)$ having at least two distinct shortest geodesic joins to p. Those results firstly observed by Wolter [44] subsequently lead to new characterizations of the cut locus in terms of differentiability properties of the distance function. In his diploma thesis [43], Wolter looked into the problem of generalizing classical geodesics in Riemannian manifolds to geodesics in bordered Riemannian manifolds. Here the geodesics being locally shortest paths joining any two points in the manifold may have contact with its boundary but must stay inside the manifold. Like in the classical case, the intrinsic distance of any two given points in the bordered manifold may be defined as infimum of lengths of continuously differentiable paths joining the two given points. It turns out that various basic concepts like distance functions and cut loci may be transferred into the situation of bordered manifolds as well. However, there occur new phenomena and new complications as shortest paths in bordered manifolds may bifurcate at boundary points. This implies that initial direction and length will not determine uniquely the end point of the geodesic related to the respective start point. Clearly, this causes difficulties for efforts aiming at defining generalizations of classical exponential maps used to control the paths of geodesics.

All those works may be seen as partial to a new approach studying global and local relations between the shape of bordered and unbordered Riemannian geometric objects and its respective intrinsic distance geometry. Here, shape would include topological properties as well as properties determining the isometry type of a geometric object or even more specifically the congruence type, i.e., the geometry type up to a Euclidean motion. In that sense the theory built via the (Riemannian) medial axis and cut locus serves the purpose of global and local shape cognition, reconstruction and classification. The latter property may be seen as an aspect of shape cognition. Note that research on the shape cognition problem could be viewed as a central goal of the computational efforts pertaining to the studies of spectral geometry done at the Welfenlab that were mentioned in the beginning of this introduction.

All the afore-mentioned works of Wolter prior to 1987 were essentially theoretical considerations and it was in 1988 at Purdue University where Wolter firstly created a software system to be used for computational differential geometry [48]. This research was partial to the ARO (US Army Office of Research) funded "Project Riemann" yielding a software system implemented in C, essentially capable of real time computing and visualizing geodesics and curvature lines on parametrically and implicitly defined surfaces being described by the user in a very flexible way via symbolically defined elementary functions. Most of the software development in Project Riemann was done by undergraduate students supervised by Wolter who explained in a summer course to those students the theoretical background and the algorithms being implemented in the system. It is remarkable that Project Riemann ended up as a state-of-the-art system for the respective computational differential geometry tasks as in those days apparently no competing system existed that could perform those computations in a similar generality.

Later on at MIT starting in early 1989, Wolter pursued research in the area of computational differential geometry with an emphasis on applications related to geometric modeling. In fact Wolter's contributions in those days may be viewed as efforts to appropriately transfer concepts from local and global classical differential geometry to computational geometry to be used in geometric modeling. *All those works were an effort to establish the new area of computational differential geometry. Such an enterprise was still in the very beginning in the years prior to 1990.*

Although the focus of Wolter's research on computational differential geometry during his years at MIT (1989–1994) were not those topics that he had pursued in his theoretical thesis works, some basic steps were taken preparing for later works that dealt with computational efforts in the area of the medial axis and its geodesic counterpart. In his diploma and doctoral theses [43,45], Wolter had developed the foundations for the theory of geodesics and cut loci in the general setting of bordered and unbordered Riemannian manifolds with mathematical rigor. Since this technically involved presentation was difficult to digest for computer scientists and engineers, a condensed version focusing on solids in $\mathbb{R}^3$ was written as MIT report stemming from December 1991 being later presented as a Sea Grant report in the national Sea Grant Library [46]. This report contains the mathematical foundation for various fundamental results on the medial axis and it explains also the relation between the medial axis and the much older concept of the cut locus.

Two results from [46], stated further down, were the basis for later works at the Welfenlab, as well as in the community pursuing research on the medial axis. Those two results involve the medial axis of a solid.

The first result to be mentioned here is called *topological shape theorem* of the medial axis. It states that the medial axis of a solid with twice continuously differentiable boundary in Euclidean space may be viewed as deformation retract of the solid. This result even holds under weaker regularity assumptions for the solid's boundary. The second basic result is the *shape reconstruction theorem*. It states that any solid can be reconstructed from its medial axis transform. The latter result later on lead to the so-called "medial modeller" useful to efficiently design 3D solids in real time via modifying their medial axis and respective radius function. An early simple prototype of this 3D modeller was presented at the Welfenlab in Howind's diploma thesis in 1998 [19]. A more advanced medial modeller was developed in the diploma thesis of Böttcher in 2004 [11, 50].

The variety of cut locus applications arises from the topological flexibility of the reference set. In case of a solid S, the cut locus or the medial axis transformation provides a compressed representation of S, that allows for intuitive shape modeling [49, 50]. According to the topological shape theorem mentioned above the medial axis itself preserves topological properties of the reference solid. The cut locus of a single point p on a complete Riemannian manifold can be interpreted as the natural glueing seam of charts of geodesic polar coordinates with respect to p and is therefore of natural interest when it comes to distance computations.

More precisely, any compact or complete Riemannian manifold may be obtained by (glueing together) identifying points on the boundary of a disc in the cut locus points C_p of a point p on the manifold. This glueing seam concept holds also for the construction of solids with smooth boundary. In the latter case the interior normal collar whose border is given by the solids boundary and by an offset surface (curve respectively) of the solids boundary is topologically glued as to become the solid. Here the respective glueing seam is defined by the solids medial axis, containing intersections points of segments created by (interior) normals to the solid's boundary.

At the Welfenlab since its foundation in 1994, a whole line of computational differential geometry research was involved with the medial axis, the cut locus and closely related concepts in Riemannian and Euclidean settings. Here, the geodesic medial axis is defined with regard to a bordered n-dimensional Riemannian submanifold S of a complete n-dimensional Riemannian manifold and contains all centers of maximal geodesic balls contained in S.

Medial sets consist of points being equidistantial with respect to two or more reference sets. First computations of medial curves on regions in the Euclidean plane

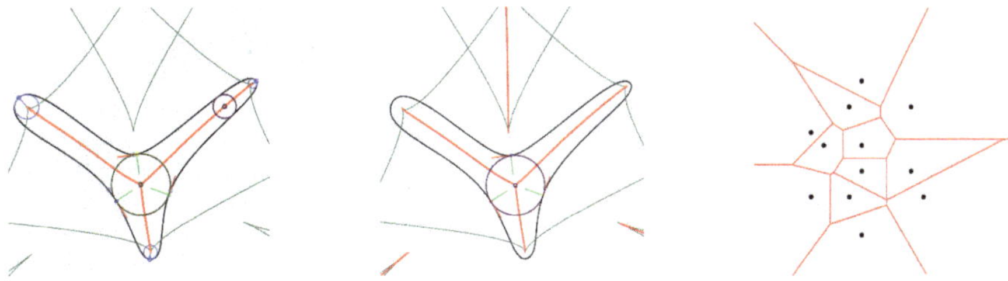

Fig. 1 Medial axis, cut locus and voronoi diagram.

were done by Wolter in 1990 and later on in 1995 on surfaces at the Welfenlab, see [35]. This lead to the computation of Voronoi diagrams for points on parametric surfaces by Kunze [22] and the computation of the geodesic (Riemannian) medial axis of bordered subsets of parametric surfaces in Euclidean 3-space [5]. Many results from these work are part of the PhD thesis of Rausch [34]. Funded by the GRK 615, this line of research was later on pursued in Naß' PhD thesis [18] and finally in the still ongoing thesis works of Thielhelm. The ongoing thesis works of Blanke that had been funded by GRK 615 is dealing with applications of the medial axis in two and three dimensions to be used for rapid modeling of metal forming processes. We will present an outline of the latter line of research in Section 3.

2 Medial Sets in Euclidean and Riemannian Spaces

For the purpose of clarity we shall start with a short explanation of the aforementioned geometric concepts.

The *medial axis* $M(S)$ of a reference solid $S \in \mathbb{R}^d$ is defined by the set of all points being centers of maximal balls contained in S. The function $r : M(S) \to \mathbb{R}_{\geq 0}$ assigning to any medial axis point p the radius of the maximal ball with center p and radius $r(p)$ is called *radius function* of the medial axis. The pair $(M(S), r)$ of medial axis and respective radius function constitutes the *medial axis transform* of a solid.

The *medial set* $MS(A, B)$ of two closed reference sets A, B is the set of all points with equal distance to A and B.

To investigate more general situations it is convenient to introduce the *cut locus* $C(A)$ of a given reference set $A \subseteq \mathbb{R}^d$ as the closure of the set of all points, that have at least two shortest paths to the reference set. In fact, the medial axis can be understood as a special case of the cut locus, since we have $M(S) = C(\partial S) \cap S$, where ∂S denotes the topological border of the solid S [46].

For a discrete and finite set of points $A = \{p_1, \ldots, p_n\}$ the cut locus of A is usually referred to as the *Voronoi diagram* of A, which has found numerous applications reaching from geophysics to physiology, that usually base on a distance related partition of $\mathbb{R}^d$ with respect to A.

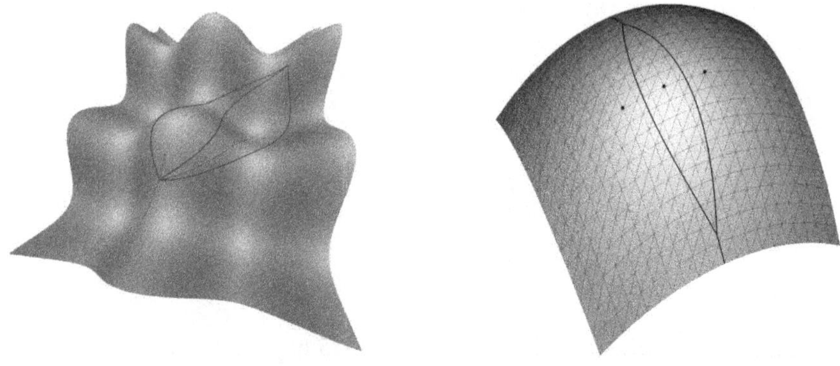

Fig. 2 Complications arising from non-Euclidean situations.

A generalization of the cut locus concept to non-Euclidean spaces M with metric d_M requires the existence of so-called distance realizing paths, which are paths that connect two given points $p, q \in M$ with length $d_M(p, q)$. A special class of metric spaces of natural interest are Riemannian manifolds. A Riemannian manifold is defined as a differentiable manifold M together with a family of metric tensors g_p. Among Riemannian manifolds are those ones of particular importance that are complete as metric spaces. Those spaces are called complete Riemannian manifolds. According to a theorem of Hopf and Rinow a complete Riemannian manifold may also be characterized by the property that every geodesic ray may be extended up to infinity. The above-mentioned theorem of Hopf and Rinow also says that in a complete Riemannian manifold any two given points can be joined by a distance realizing path. Such a path is often called distance minimizer. Uniqueness of minimal geodesic joins however, which holds in the Euclidean case, can not be guaranteed and this leads to significant difficulties in the context of geodesic coordinates and distance computation. We would like to illustrate this with an example shown in Fig. 2. Here the left part shows that on a surface the cut locus of a single point indicated by the red colored set can have a complicated structure while the cut locus of a point in the plane is empty. The right part of Fig. 2 shows that on surfaces a Voronoi diagram (being the cut locus of a finite point set) may contain compact proximity regions bounded by two edges only while in the Euclidean plane a compact proximity region of some point has at least three edges.

To be more precise we shall give some basics on the connection of metric, distance function, metric tensor and the length of curves in n-dimensional Riemannian manifolds M with *metric tensor* $g_p : T_p M \times T_p M \to \mathbb{R}_{\geq 0}$. Here $T_p M$ denotes the tangent space of a Riemannian manifold M in the point p. In local coordinates the metric tensor g_p can be described by a matrix g_{ij} depending on the (foot) point p of the respective tangent space $T_p M$. For the sake of simplicity we focus on differentiable curves and refer to [15] or especially [43, 45] for a more detailed introduction and discussion. The length L of a curve $c : [0, 1] \to M$ is given by

$$L(c) := \int_0^1 \sqrt{g(c'(t), c'(t))}\,dt,$$

and minimizing over all curves that connect two arbitrary points $p, q \in M$ we obtain a metric space in the sense of point set topology by defining

$$d_M(p,q) := \inf\{L(c) | c(0) = p, c(1) = q\}$$

on M, the so called *Riemannian distance*. The respective minima are called *distance minimizers* or *shortest paths*. To determine the distance for an arbitrary pair of points one obviously has to compute the corresponding distance minimizing path joining the points. Unfortunately it is usually very difficult to compute globally shortest paths.

However, their local counterparts, so-called geodesics, can be computed using the *geodesic differential equations*:

$$\gamma_k''(s) + \sum_{i,j} \Gamma_{ij}^k \, \gamma_i'(s)\gamma_j'(s) = 0, \tag{1}$$

where the *Christoffel symbols* are the local coefficients of the *Levi-Civita connection*:

$$\Gamma_{ij}^k = \frac{1}{2}\sum_m g^{mk}\left(\frac{\partial}{\partial x_j}g_{im} + \frac{\partial}{\partial x_i}g_{jm} - \frac{\partial}{\partial x_m}g_{ij}\right).$$

Here g^{ij} denotes the inverse of the metric tensor matrix g_{ij}. *Note that a (globally) shortest path joining two points is always a geodesic but not vice versa.* The descriptions above make use of a local parametrization $X : \mathbb{R}^n \to M$ of M, that maps the coordinates $x_1, \ldots, x_n$ diffeomorphic to M. Here we assume for simplicity that M is a submanifold of $\mathbb{R}^k$, but the concept holds also within a more general setting. A *geodesic* starting at $p = X(p_1, \ldots, p_n) \in M$ with the initial direction $v = DX \cdot (v_1, \ldots, v_n) \in T_pM$ is given by $\gamma(s) := X(x(s)) := X(\gamma_1(s), \ldots, \gamma_n(s))$, when choosing the initial values of (1) according to $\gamma_k(0) = p_k$, $\gamma_k'(0) = v_k$.

To simplify the notation we introduce the *exponential map* $\exp_p : T_pM \to M$ by $\exp_p(v) := \gamma(1)$, i.e. mapping an initial starting direction to a corresponding point $q = \exp_p(v) \in M$. The exponential map enables us to use coordinates of the tangent space T_pM to parametrize M, via so-called *geodesic coordinates*. Using for example polar coordinates (s, φ), $\varphi = (\varphi_1, \ldots, \varphi_{n-1})$ to parametrize T_pM leads to *geodesic polar coordinates* denoted by

$$O_p(s, \varphi) := \exp_p(v(s, \varphi)).$$

O_p is usually referred to as *offset function* of p. Figure 3 shows an example of isolines of geodesic polar coordinates.

For a more complicated reference set, represented by a d-dimensional submanifold $N \subset M$ with local parametrization $\xi : \mathbb{R}^d \to N$ and coordinates $\xi = (\xi_1, \ldots, \xi_d)$, the corresponding offset is given by

$$O_N(\xi, s, \varphi) := \exp_\xi(v(s, \varphi)), \tag{2}$$

with $v \in T_p N^\perp$ where the tangent space $T_p M$ splits according to $T_p M = T_p N \oplus T_p N^\perp$ and $\varphi = (\varphi_1, \ldots, \varphi_{n-d-1})$.

The offset function O_p and its partial derivative $\partial O_p / \partial s$ is computed by numerically tracing the ODE (1). In the context of distance or medial computations it is also necessary to obtain the partial derivatives with respect to the parameters ξ or φ, i.e. to compute the variation of geodesic coordinates. More generally, consider a one parameter family of geodesics $c : I \times [0, a] \to M$ where each curve $s \mapsto c_\eta(s) := c(\eta, s)$ is a geodesic. The derivative

$$w(s) := \frac{\partial}{\partial \eta} c(\eta_0, s)$$

defines a vector field along $\gamma := c_{\eta_0}$, which is a so-called *Jacobi field* that satisfies the *Jacobi equation*

$$\frac{D^2}{ds^2} w + R(w, \gamma') \gamma' = 0, \tag{3}$$

where R is the *Riemannian curvature tensor* and D/ds is the *covariant derivative* along γ. For a detailed definition and description of these two central concepts of differential geometry we refer to [15]. The vector field w can be easily decomposed into two components one parallel to the geodesic and the other orthogonal to the geodesic. In the two-dimensional (surface) case the orthogonal component being contained in a one-dimensional subspace of the tangent plane can be described by a real number $y(s)$ at the point $\gamma(s)$ of the geodesic γ. Hence here $y(s)$ describes the oriented length of the Jacobi field, characterizing it completely. The function y satisfies the simplified equation

$$y''(s) + K(\gamma(s)) y(s) = 0, \tag{4}$$

with K being the Gaussian curvature along γ. For general n-dimensional Riemannian manifolds solving the Jacobi equation (3) boils down to solving an n-dimensional second order linear system of differential equations along a geodesic γ. This is equivalent to solving a $2n$-dimensional first order system of differential equations along $\gamma(s)$. Here in addition to the initial values of geodesics (cf. (1)) we need to provide also the initial values of the vector field w. These, however arise from the special form of variation.

Of particular interest are the points where the differential of O_p becomes singular. These points make up the so-called *focal set* of the reference object here being a point p. In this case the focal set of p is also called (first) conjugate locus of p. Within our setting, points located on the focal set (or conjugate locus) of p can be characterized by the condition

$$\det DO_p(s, \varphi) = 0 \text{ with } DO_p = (\partial_s O_p, \partial_\phi O_p) = (\partial_s O_p, y(s)(\partial_s O_p)^\perp),$$

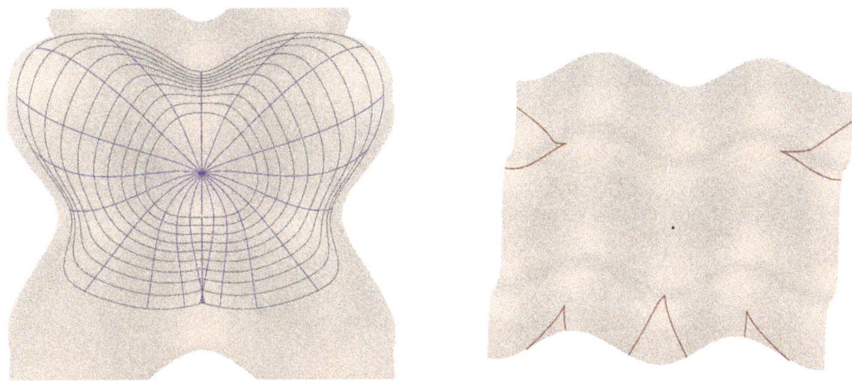

Fig. 3 Geodesics and focal curves.

implying that the focal set of p in two dimensions can described by the implicit equation $y(s, \varphi) = 0$. We use the latter equation for implicitly describing a focal curve being a (connected) component of the focal set of p and get

$$y(s(t), \varphi(t)) = 0. \tag{5}$$

We want to trace the above-mentioned focal curve by integrating its tangent vector $(s'(t), \varphi'(t))$. The latter may be obtained from equation (5) by differentiating with respect to t and applying the chain rule. Therefore in two dimensions the focal set can be computed by numerically tracing the zero set of y using the implicit differential equation

$$\frac{\partial y}{\partial s} s'(t) + \frac{\partial y}{\partial \varphi} \varphi'(t) = 0$$

That leads to a solution $s(t), \varphi(t)$ which describes the focal set in polar coordinates with respect to p [35]. (The respective detailed computations needed to compute tangent vectors of the focal curve are quite involved. They also employ derivatives of the Gaussian curvature.) An example of resulting focal curves is shown in Fig. 3.

The tools of differential geometry presented above are used to state and solve problems such as the shortest-distance problem or the computation of medial sets in higher dimensional Riemannian manifolds. (For the sake of simplicity we keep the same symbolic notation with $\varphi = (\varphi_1, \ldots, \varphi_{n-1})$ now referring to a vectorial parameter.)

For example, to determine the distance of two arbitrary points $p, q \in M$ we can reduce the challenge of finding the shortest path from p to q to computing all geodesics that connect p and q, since every shortest path has to be a geodesic. This translates to finding tuples of geodesic parameters (s^j, φ^j) that satisfy

$$O_p(s^j, \varphi^j) = q. \tag{6}$$

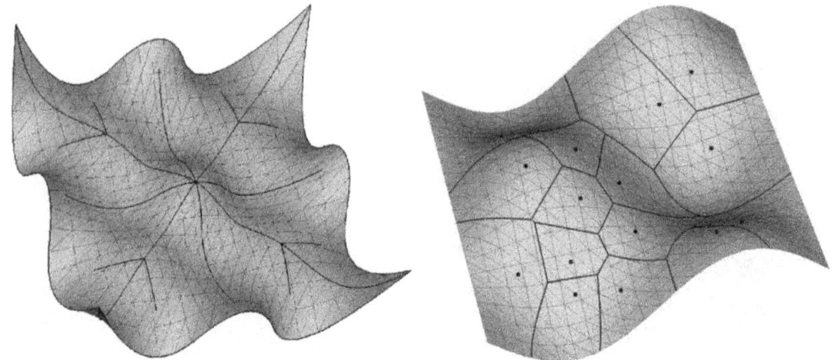

Fig. 4 Medial axis and Voronoi diagram on 2D manifolds ([5,22,49]).

This boundary value problem can be looked upon as the problem to solve a nonlinear system of n equations with the n unknowns $s^j, \varphi_1^j, \ldots, \varphi_{n-1}^j$.

The computation of the medial set of two reference sets A and B in M translates to finding tuples of geodesic parameters $(\xi, \eta, \varphi, \psi, s)$ that satisfy

$$F(\xi, \eta, \varphi, \psi, s) := O_A(\xi, s, \varphi) - O_B(\eta, s, \psi) = 0,$$

where O_A, O_B are the generalized offset functions defined in (2). By differentiation we obtain a differential equation, called *medial differential equation*, that can be used to trace (isolines) of the medial set. For example, in case A and B are two points and using t as the parameter of a component in the one-dimensional medial set we obtain

$$\left(\left. \frac{\partial O_A}{\partial \varphi} \right| - \frac{\partial O_B}{\partial \psi} \left| \frac{\partial O_A}{\partial s} - \frac{\partial O_B}{\partial s} \right) \cdot \frac{d}{dt} \begin{pmatrix} \varphi(t) \\ \psi(t) \\ s(t) \end{pmatrix} = 0 \right.$$

In the years between 1996 and 1998/99 the research on computing geodesic medial curves, geodesic medial axes on bordered subsurfaces of spline patches as well as computing geodesic Voronoi diagrams on parametric surfaces had reached some maturity. Among the tools employed for the computation, three basic ingredients stand out. The first one is the so-called geodesic medial differential equation already present in a basic form in [45, pp. 171–174], later on used within a computational setting in [35]. The second one is the computational description of focal curves [35] and the third ingredient, is the observation that on a bordered C^2-smooth surface assembled from finitely many real analytic surface patches with real analytic boundary arcs, the medial axis would be topologically a graph [49]. The end points of the latter graph would be focal points with respect to the surface's boundary curve [34,49]. Since the distance between the surface boundary curve and the focal curve has local minima at the end points of the medial axis graph, a tracing method could be implemented starting at those points.

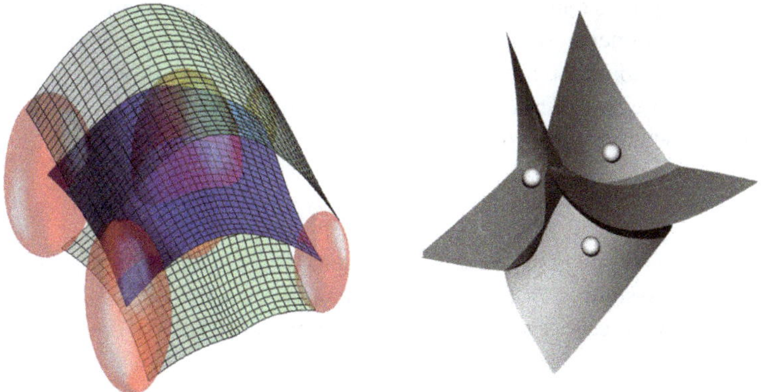

Fig. 5 Medial axis and Voronoi diagram on 3D Riemannian manifolds.

Combining the computational instruments and observations presented above, it was possible to develop prototype software that could compute medial sets in fairly challenging cases of bordered subsurfaces, e.g. spline surfaces or surfaces being C^2-smoothly assembled of real analytic patches [5,22,34,49]. The tools presented above allow also computing of a Voronoi diagram on a parametric surface [22] provided that the bisectors involved to describe the Voronoi diagram do not meet the focal set of one of the two points defining the bisector (see Fig. 4 for an example). For a more comprehensive survey of the presented research prior to the year 2000, see also [49].

2.1 Medial Computations Since 2000 during the Years of GRK 615

Since the year 2000, the contributions of of the Welfenlab with respect to medial axis computations brought significant extensions beyond that what had been achieved before. This was essentially possible through research works supported by GRK 615.

First, the restriction to two dimensions for the respective computations on geodesic medial sets and geodesic Voronoi diagrams could be removed. Thus, it was possible to develop methods that would work in Riemannian worlds of dimension three and higher and it was possible to present prototypical implementations for the computation of sample cases for medial sets in dimension three. Furthermore, several examples of geodesic Voronoi diagrams of point sets P in Riemannian manifolds of dimension three could be obtained [18,27,41] as shown in Fig. 5.

All those computations needed some substantial extensions of the methods that had been developed by the year 2000. For instance, the Jacobi equation had to be solved in its general form (3) instead of its simpler two dimensional special case (4). Another significant extension was caused by the problem that finding a geodesic joining any two points could not be done any more by a simple shooting method that one might use in dimension two. The additional degrees of freedom in the dimension of the space describing the initial directions made it necessary to employ homotopy methods. For a general introduction to those methods, we refer to [2, 16]. In our case, the nonlinear equation (6) is embedded into a homotopy

$$H(s, \varphi, \lambda) = O_p(s, \varphi) - c(\lambda)$$

where $c : [0, 1] \to M$ is a curve connecting an arbitrary starting point $c(0)$ with the point $c(1) = q$. Assuming (s, φ, λ) to be a function of an additional parameter t and differentiating with respect to t we obtain the implicit differential equation

$$\left(\frac{\partial O_p}{\partial s} \left| \frac{\partial O_p}{\partial \varphi} \right| - c'(\lambda) \right) \cdot \frac{d}{dt} \begin{pmatrix} s(t) \\ \varphi(t) \\ \lambda(t) \end{pmatrix} = 0$$

that can be used to trace the zero set of H whose intersection with the plane $\lambda = 1$ contains the sought solutions. For more details, we refer to [18, 26, 27].

However, all considerations in the years from 1996 to 2007 were focussing on the simplified case, where shortest paths are unique. In elaborated experiments it was discovered that the traced solution paths $x(t) = (s(t), \varphi(t), \lambda(t))$ satisfying $H(x(t)) = 0$ can turn around with respect to λ in points where $\partial O_p / \partial \varphi$ vanishes, i.e. in points where the curve c meets the focal curve of p transversally by construction, see Fig. 6. Therefore if we introduce a generalized homotopy curve c which contains the point q in its interior, the approach can yield multiple solutions. The curve $c(\lambda(\varphi))$ describes end points of a (continuous) family of geodesics starting in p whose initial direction continuously depends on an angle φ. In case there are multiple solutions we obtain for different (initial) angles φ^k the same parameter $\lambda = \lambda(\varphi^k)$ related to geodesics ending up in the same end point $c(\lambda(\varphi^k)) = c(\lambda)$.

In typical situations as depicted in Fig. 6, the focal curve (red-coloured) separates regions of the surface where the number of solutions changes. More concretely we have a unique (geodesic) connection outside of the region bordered by the focal curve (cyan-coloured), two connections on the border (green-coloured) and three connections inside (blue-coloured). The right part of Fig. 6 indicates how to collect different geodesics corresponding to different angular parameters φ and intersecting in the point $c(\lambda)$. Since the structure of focal curves shows some variety, a classification of relevant situations where the number of near by geodesics can be found precisely is subject of ongoing research. In this context the just described method appears to be a promising approach for the computation of "near by" (and under additional assumptions of all) geodesics joining two points $p, q \in M$. Thus the computation of $d_M(p, q)$ is feasible in a direct manner with respect to the definition of d_M. Apparently the latter approach has not been described in the respective literature.

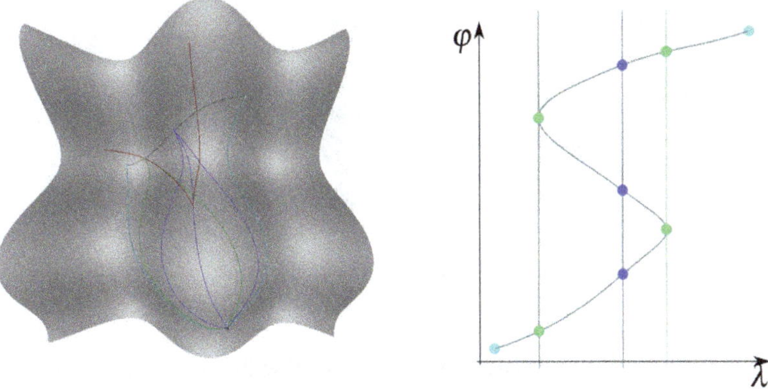

Fig. 6 Several geodesics.

2.2 Remarks

The discussions and methods presented in the preceding section (and also all the underlying respective research until today) make some simplifying assumptions implying the omission of crucial difficulties. Those simplifying assumptions were taken for granted for computations in the two-dimensional surface case or even in the planar case. Therefore in the latter two-dimensional cases important and difficult computational problems related to computations regarding medial axis and Voronoi diagrams in the Riemannian or even in the Euclidean case are still subject of our ongoing research activities. We illustrate this statement with a few examples:

1. For the computation of Voronoi diagrams, the generating point set was assumed to be "benign", meaning that the related medial sets and bisectors would stay away from the (first) focal sets of their generators.
2. Furthermore, an n-dimensional Riemannian version of the respective Euclidean "general position" assumption was made for the point set generating the Voronoi diagram. This means vertices of the respective geodesic Voronoi diagram were assumed to be centers of uniquely defined geodesic distance spheres containing exactly $n+1$ generator points partial to the point set generating the Voronoi diagram.
3. The computation of the medial axis close to an end point was usually done by a fairly crude approximation whenever the angle between intersecting geodesic normals became very small close to the respective end point of the medial axis.
4. The analysis of the situation where one wants to compute all geodesics joining a reference point p with points close to the cut locus of p where those geodesics have already or will soon reach the (first) focal locus of p appears to have never been done appropriately in a systematical way. The latter situation is crucial for computing minimal geodesics joining two given points in a complete Riemannian manifold.

Fig. 7 Sequence of forging steps from billet to final product.

The last two afore-mentioned issues 3 and 4 also fall within the scope of our current research projects.

3 Application of 3D-Medial Axis on Metal Forming Simulation

One application of the medial axis was researched in the graduate college 615, using the medial axis transform as description for the geometry of forging dies in hot drop forging. This serves as basis for rapid backwards simulation of material flow.

In hot metal drop forging, a heated semi-finished part is formed by pressing two forging dies which contain the negative final shape. If the design of the forging dies or the layout of the process is incorrect, the quality of the final product will be severely reduced.

Since the design of the tools is a very cost-intensive part of forging, computer aided techniques are used to reduce design time and to decrease the number of iterations until the final layout is reached. Usually, a number of pre-forms are needed in order to achieve the final complex shape from the initial simple shape with optimal properties and within a geometrical tolerance.

The prediction of preforms from the final product is what we call inverse or "backward" simulation. There exist several approaches, based on the Finite-Element-Method and backward tracking of solutions [9, 21] or upper boundary methods [12]. These algorithms have quite severe drawbacks, since they have to be fitted closely to the problem at hand and have to our knowledge not been utilized in practical applications.

The medial axis approach is based on experimental observations and elementary plasticity theory, see Mathieu et al. [23]. In drop forging experiments, Mathieu noticed that the material flow followed specific paths, which can be described as medial axis of the die gap. Based on Mathieu's observations, algorithms were developed which simulate material transport along these displacement paths [6, 25]. It is important to note that these simulations only provide an approximation to the velocity field of the material and the filling of the forging form. They will not yield local stresses, strains or temperature, and do not allow the compuation of hardening phenomena. Thus, they can provide only limited assistance for the simulation of the forging process. Nevertheless, this approach is a good basis for backwards simulation where a prediction of the preform is needed as implemented by Wienstroer for the 2D case [42]. The velocity field of the material is not computed explicitly, rather an iterative method based on flow resistance along the displacement paths is used to determine the distribution of material between the cells.

Computation of the Medial Axis for CAD Objects

Forging dies are usually described by surface models constructed in CAD programs, but these programs do not offer a medial axis (MA) representation. Since there exist no production ready programs that offer the construction of the medial axis from (boundary) surfaces, it was necessary to develop a stable tool to this end. Our demands on the algorithm were that it should be fast and return a surface (mesh) representation of the MA. Speed is important because the backward simulation should give the user a first preform which is afterwards analysed and corrected. The simulation will iterate several time steps and in each the MA has to be computed again. Therefore the computation of the MA shoul be fast. The material transport in the simulation will take place on the MA, so it is crucial that its connection information is obtained. We chose to look for algorithms that take triangle mesh surfaces as input, since the formats of surface representations in CAD tools differ very much, but every system can export triangle meshes, giving us a wide range of possible applications.

Research on existing algorithms showed, that these can be classified into three categories: discrete, direct and indirect methods. Discrete methods discretize the surrounding space of the reference surface using, e.g., octrees or voxels, and then implement a discrete grassfire algorithm (i.e. a thinning operation) to determine a discrete representation of the MA. These methods do not provide the connectivity information and were therefore discarded.

In three dimensions, the MA is composed of medial faces (i.e. bordered surface patches) connected by medial seams (i.e. curve elements). Direct methods setup generalized Voronoi diagrams between elements of the reference surface (triangles) and intersect these to get the medial seams [13,40]. Their runtime is $O(n^2)$ which is bad for large input sets.

The indirect approach approximates the Medial Axis by filtering or pruning the Voronoi diagram of sample points on the reference surface [4, 14]. We have implemented such a strategy, robustly computing a 3D Delaunay tesselation of a sampled point set. Then, we filter the dual Voronoi diagram using a heuristic based on the assumption that there should exist a homeomorphism between the reference surface and the surface Delaunay triangles. Several steps have to be performed prior to the tesselation used to transform the given tool geometry to a boundary representation of the die gap and to analyze its features, e.g., detecting sharp edges. Finally a meshed medial surface of the die gap is computed as shown in Fig. 8.

Fig. 8 Forging dies and approximated medial axis of the die gap.

To this end, a fast data structure for the Delaunay triangulation has been implemented by Obydenna [29] and used by Algaier [1] to rapidly recompute the Voronoi diagram after each time step in which the dies move.

Connection between Medial Set and Material Flow

We could show that for viscous Bingham fluids flowing in a completely filled pipe, the maxima of material flow speed will lie on the Medial Axis of the pipe boundary [10]. Hot metal in forging can be modeled as a very viscous Bingham fluid.

Simulation Scheme

With the development of the Medial Axis computation software, the fundament of the simulation has been laid. Important parts of the framework, such as the partition of the die gap and graph-representation of the Medial Axis are already in place. The next steps will be the implementation of the geometric resistance model and the material transport algorithm.

4 Spectrum of Eigenvalues of the Laplace–Beltrami Operator

As pointed out in the introduction, it is known from theoretical research, that a substantial amount of geometric and topological information on a Riemannian Manifold M is contained in the spectrum of eigenvalues of its associated Laplacian. To become more concrete in the following discussion, let M denote a Riemannian manifold and let Δ denote its associated Laplacian. In case M is a subdomain of euclidean space equipped with cartesian coordinates, Δ is the well-known Laplace-Operator given by $\Delta = \sum_i \partial_{ii}$, assigning to a function the trace of the Hessian of that function. In the more general Riemannian setting, this operator becomes the *Laplace–Beltrami-Operator*, whose action on a function f can be defined using the metric tensor g of

M with respect to a local chart via

$$\Delta f := \operatorname{div} \operatorname{grad} f := \frac{1}{\sqrt{\det g}} \sum_{i,j} \partial_i(g^{ij}\sqrt{\det g}\,\partial_j f)$$

The *spectrum* of M consists of all scalars λ that satisfy the eigenvalue equation

$$-\Delta f = \lambda f \tag{7}$$

for a non-zero function f defined on the manifold, subject to appropriate boundary conditions of the Dirichlet or Neumann type. From the theory of compact elliptic operators it is known, that there is a countably infinite number of non-negative eigenvalues $\lambda_1 \le \lambda_2 \le \lambda_3 \le \dots$ accumulating at infinity. Each eigenvalue corresponds to a finite dimensional space of eigenfunctions.

Among the geometrical information determined by the spectrum we have the dimension and the volume of M, the volume of its boundary, the scalar curvature integral over M, the mean curvature integral over its boundary and the Euler characteristic of M in case of a two-dimensional surface or a planar domain with smooth boundary. This information can be extracted from the asymptotic expansion of the so-called *heat-trace* function

$$Z(t) = \sum_{i=1}^{\infty} \exp(-\lambda_i t) = (4\pi t)^{-\dim M/2}\left(\sum_{i=0}^{n} c_i t^{i/2} + o(t^{(n+1)/2})\right) \quad \text{for } t \to O^+,$$

where, according to a theorem by McKean and Singer [24], the first few coefficients are given by

$$c_0 = \operatorname{vol} M, \; c_1 = -\frac{\sqrt{\pi}}{2}\operatorname{vol}(\partial M) \text{ and } c_2 = \frac{1}{3}\int_M K - \frac{1}{6}\int_B J \tag{8}$$

where K is the scalar curvature of M and J is the mean curvature of the boundary of M. In case the dimension of M is two, then K coincides with the Gaussian curvature of M.

The spectrum is invariant under isometric transformations of M and changes continuously as the manifold is continuously deformed in a non-isometric way. Therefore the spectrum can be considered to be characteristic for the intrinsic shape of the underlying manifold. However it is also well-known that the spectrum does not completely determine the underlying manifold, as exemplified by the existence of pairs of isospectral but non-isometric manifolds. One such a pair in the case of planar domains was given by Gordon in [17] and is depicted in Fig. 9. An interesting property of these examples is, that isospectrality still holds with respect to the eigenvalues of the three dimensional Laplace operator in case the domains are extruded to three dimensional prisms. However, the boundaries of the prisms have different spectra with respect to their respective two-dimensional Laplace–Beltrami operators.

Leaving aside the rare phenomenon of isospectrality, the above-mentioned theoretical properties of the Laplace spectrum make it suitable for the construction of

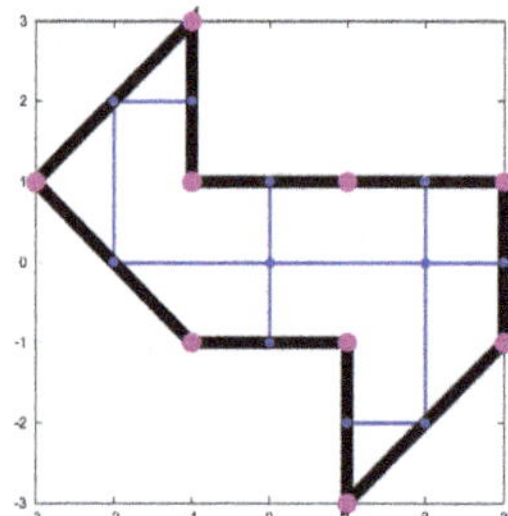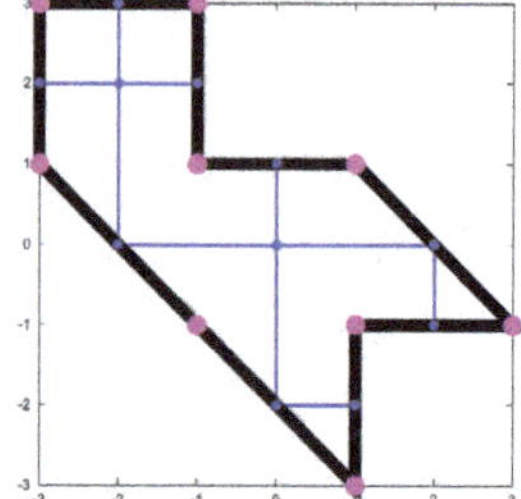

Fig. 9 Isospectral domains.

feature vectors that can be used as *fingerprints* of objects, as long as the objects under consideration can be represented or at least modeled as Riemannian manifolds. The above-mentioned fingerprint can be constructed from a (finite) initial part of the spectrum of the respective manifold and finds a natural application in the context of efficient retrieval of similar objects in large databases.

4.1 Laplace Spectra as Shape DNA for Surfaces and Solids

The two-dimensional surfaces and three-dimensional solids commonly encountered in CAD/CAE applications are instances of Riemannian manifolds with with an intrinsic metric that is naturally induced by their embedding in $\mathbb{R}^3$. Extending the work on mesh-based discrete Laplacians, the Laplace–Beltrami Operator can also be applied to free-form surfaces and solids. A collection of its first few smallest eigenvalues can be used as feature vectors that are invariant with respect to rotation and translation and any reparametrization parametrization of the object. Furthermore, it is known that a scaling transformation by the factor a results in scaled eigenvalues by the factor $1/a^2$. Therefore, by normalizing the eigenvalues, shape can be compared regardless of the objects scale.

By transforming the eigenvalue problem (7) into a variational formulation, the Finite Element Method can be employed in order to obtain on modern hardware within seconds an accurate set of eigenvalues for a fairly large variety of reasonably detailed objects. If the given surface or solid has a boundary, generally the Dirichlet boundary condition is applied. If objects with small holes or missing triangles are to be compared, the Neumann boundary condition can be used instead, because the unwanted holes appear to change the Neumann spectrum not as much as in the Dirichlet case.

In order to "show" how the Shape-DNA can help to distinguish many different surfaces the latter technique was applied to a database of 1000 randomly generated B-Spline surface patches [36]. For these patches the first 11 eigenvalues were calculated and stored with the shapes. By using the Euclidean distance of the normalized 11-dimensional vectors of eigenvalues, each patch could be uniquely identified even

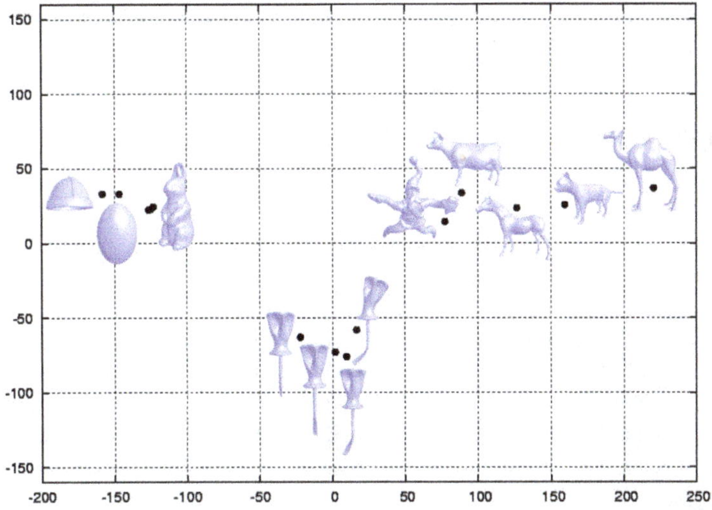

Fig. 10 Clustering of eigenvalues.

with deliberately different (not optimal) meshes introducing distinct calculation errors. Still, these inaccurate eigenvalues yielded distances of less than 0.02 between the original and the modified patch. Furthermore, from all the 500,000 possible pairs of different patches only 300 had a distance of less than 0.3 to each other, none was closer than 0.15. This confirmed the conjecture that the Laplace–Beltrami method is sensitive enough to be used for identifying patches even with reduced capacities for calculation (since only the first 11 eigenvalues were used).

In another experiment, the spectra of different objects were computed and multi-dimensional scaling was employed to obtain the two-dimensional projection in Fig. 10 which shows how similar objects cluster according to their eigenvalues. Of course projecting the high dimensional feature vectors to a very low dimensional space means a massive loss of information, resulting in the formation of additional clusters, and thus cannot serve more than purposes of illustration. For practical applications one should work with more than two dimensions.

Furthermore [36] contains some results with respect to the mutual independence of the eigenvalues and on the rapid convergence of the heat trace series. Especially the latter property made it possible to extract the volume, the boundary length and the Euler characteristic of a shape from its computed eigenvalues with high numerical accuracy. This numerical approach was novel and confirmed the theoretical results stated in (8).

As a biomedical application it was shown later in [28, 38, 39] that Laplace–Beltrami spectra posses the discriminatory power to distinguish two populations of female persons via the shapes of their respective caudate nuclei. In this biomedical application one population would contain normal control subjects while the other population would consist of subjects with with schizotypal personality disorder. The caudate nucleus of a person is a subcortical gray matter structure of the brain, involved in memory function, emotion processing, and learning.

4.2 Laplace Spectra as Image DNA

In order to use Laplace–Beltrami eigenvalues as fingerprints for images, these images have to be modeled as Riemannian manifolds. For example, a gray scale image can be represented as a surface defined by the graph of a height function being the gray scale intensity function of the image while a color image can be understood as a two-dimensional surface in a five-dimensional Euclidean space whose coordinates include the intensity parameters of the red, green, blue values assigned to any (x, y) pixel of the image. It is possible as well to understand other even higher dimensional signals as height functions and therefore as manifolds, whose Laplace–Beltrami spectra can be computed. These topics were studied by Peinecke during his PhD research.

Another approach pursued by Peinecke was an extension of the classical Laplacian eigenvalue problem (7) to the form

$$-\Delta f = \lambda \rho f$$

in which the gray-value information of an image is encoded in a *mass-density-function* ρ instead of the structure of the representing manifold [31].

Although using the discrete Laplace operator or more generally using eigenvalues of different operators and matrices derived from this operator is a well known and established technique in the community of shape and image recognition, typical applications employ discrete forms of the Laplacian directly instead of making use of the underlying continuous operator. An advantage of the continuum point of view is the independence of the particular discretization employed in the computation, as the results are stable under mesh refinement or change of image resolution.

In a series of example calculations, Peinecke observed that the discrete graph-based Laplace–Kirchhoff and the Laplace–Beltrami operator perform similarly in terms of run time. However, while the Laplace–Kirchhoff operator is more easily implemented, the Laplace–Beltrami variants open up the possibility to use a coarser mesh and thus save computation time. This observation fits into the general finding that for surfaces, images and solids Laplace spectra derived from a discrete model (instead of using the underlying continuous operator) typically have the disadvantage that they make it difficult to decide if two objects are similar when using spectra obtained from different discretizations and different resolutions. In the continuous differential geometric (parametrization invariant) setting the afore-mentioned decision is possible under resonable assumptions for images, surfaces, solids with or without boundary. Indeed for the latter objects Laplace spectra derived within a continuous (parametrization invariant) setting provide the gold standard and spectra obtained within a merely discrete combinatorial point set setting must be shown to

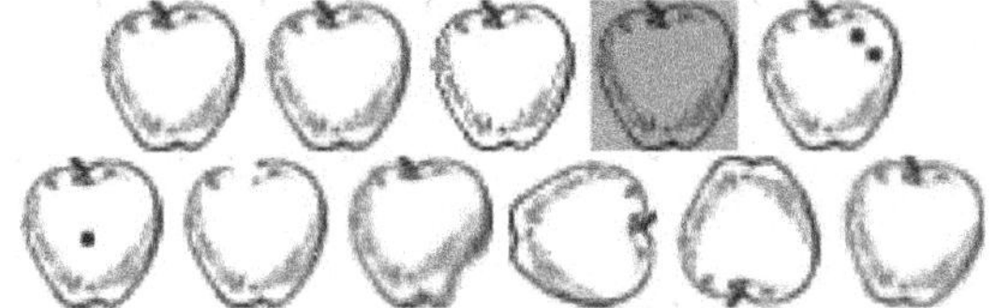

Fig. 11 Transformed versions of a test image.

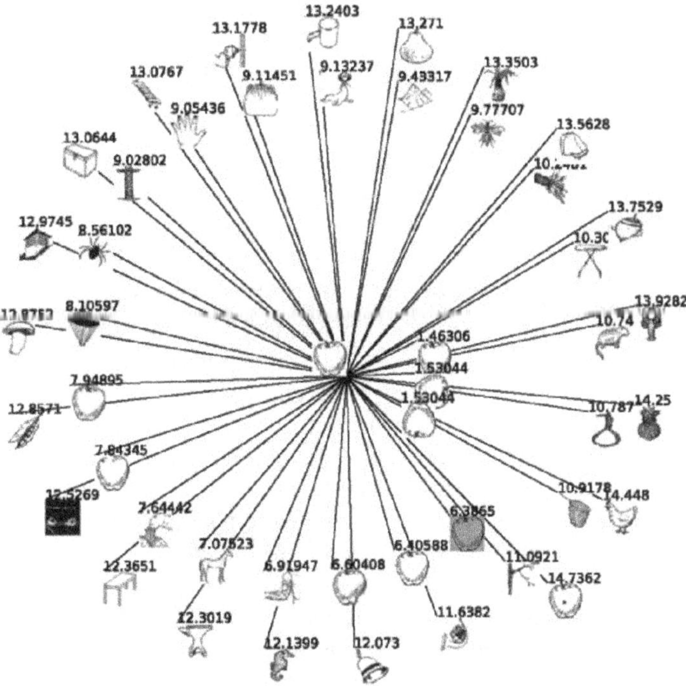

Fig. 12 Image classification results according to fingerprint distance.

converge to the spectrum of the respective continuous operator [37, 51]. The latter requirement typically causes problems for operators (and their spectra) in case the operators are obtained within a merely discrete combinatorial setting [37,51]. In the discussion of computational models for the computation of the Laplace operator and its eigenvalues one should bear in mind that there is only one Laplace operator of a function and this Laplace operator is the divergence of the gradient of the respective function defining the gold standard.

To test the robustness of the computed eigenvalues, a collection of images was considered and augmented with images differing in scale and contrast as shown exemplary in Fig. 11. For each image in the collection fingerprints were calculated using the Laplace–Beltrami operator obtained from the surface defined by the graph of the grey value function using finite elements. Afterwards the fingerprints were compared using Euclidean distances. A reliability of about 96% was obtained, meaning that in 96% of the cases it was possible to match an image with transformed copies of itself. For a more detailed discussion of the implementation and the results, we refer the reader to [30, 32] With respect to color images, the proposed methods were shown to be especially useful in the presence of rotations or color rotations, changes of contrast and scale, and combinations of all these operations, since the underlying calculations based on the continuous Laplace–Beltrami operator are invariant against such transformations. It was shown, that the proposed method uses substantially less information than established techniques for discriminating collections of images while maintaining a high reliability. This is especially useful for data bases of images where high dimensional searches are very cost intensive, see for example [7, 8].

5 Conclusions and Prospects

This survey chapter reviewed the contributions of the Welfenlab to GRK 615. All those contributions could essentially be viewed as being partial to a field that one might call "Computational Differential Geometry". This description would be justified because the respective research essentially presents analysis, discussion and applications of methods that would result in numerical computations of entities that mostly were originally introduced within the classical framework of differential geometry avoiding numerical computations.

The contributions in this chapter are limited to a specific selection of subjects including *Cut Locus, medial axis, geodesics, focal sets, conjugate loci, geodesic Voronoi diagrams, Laplace Spectra of surfaces, solids and images*. Despite this limitation the contributions presented in this chapter cover important highlights of research the first author has been involved in since more than 30 years. Many of the mentioned geometric entities that 30 years ago would only exist as mental objects (resulting from mathematical definitions) can nowadays be efficiently numerically computed via works outlined in this chapter. Although those numerical computations are now to some extent possible in a number of relevant situations one should mention that many if not most difficult questions still remain open, see e.g. the remark at the end of Section 2. Certainly the latter point is one of the reasons why various research topics outlined in this chapter (decribed by the words presented in italics above) being – 20 years ago – initially pursued by a small number of computational researchers only (including computational geometers, computer scientists and engineers) nowadays constitute a substantial part of main stream research in the respective areas. Overall it has turned out that areas that say 25 years ago were viewed at as being sort of exotic in the respective communities – including researchers from computational geometry and computer graphics – mean while are moving into the center of attention in the respective communities. There are several reasons for this development. One is that the area of computational geometry is becoming more and more sophisticated. This holds because the respective researchers are realizing that elementary methods are tentatively exhausted and they are discovering the power of advanced mathematical tools contained in the theoretical achievements of local and global differential geometry. The other equally or perhaps even more important reason for this development is that sophisticated tools from differential geometry can help to make important progress for the central questions of geometric modeling, computer graphics and image processing. Those central poblems are *Shape and Image Cognition* and *(Re)-Construction and Compression* [47]. The application of tools of differential geometry to the afore-mentioned subjects will be the topic of an upcoming paper expanding the referenced keynote lecture [47].

References

1. R. Algaier. Schnelle dynamische Voronoi-Diagramme mit History-DAG. Diplomarbeit, Leibniz Universität Hannover, December 2009.
2. E.L. Allgower and K. Georg. *Introduction to Numerical Continuation Methods*, Classics in Applied Mathematics, Vol. 45, SIAM, 2003.
3. T. Altschaffel. Untersuchungen zur Spektralanalyse von Freiformflächen. Diplomarbeit, Leibniz Universität Hannover, May 1999.
4. N. Amenta and R.K. Kolluri. The medial axis of a union of balls. *J. Computational Geometry*, 20:25–37, 2001.
5. M. Baer. Berechnung Medialer Achsen von einfach berandeten zusammenhängenden Teilstücken parametrisierter Flächen. Diplomarbeit, Leibniz Universität Hannover, December 1998. Available as Welfenlab Report 5.
6. W. Beneker. Rechnergestützte Simulation des Füllverhaltens beim Gesenkschmieden. *Fortschritt-Berichte VDI Reihe*, Vol. 20(187), VDI, 1995.
7. S. Berchtold, C. Böhm, B. Braunmüller, D.A. Keim, and H.-P. Kriegel. Fast parallel similarity search in multimedia databases. In *SIGMOD'97: Proceedings of the 1997 ACM SIGMOD International Conference on Management of Data*, pages 1–12, ACM Press, 1997.
8. S. Berchtold, D.A. Keim, and H.-P. Kriegel. The X-tree: An index structure for high-dimensional data. In T.M. Vijayaraman, A.P. Buchmann, C. Mohan, and N.L. Sarda (Eds.), *Proceedings of the Twenty-Second International Conference on Very Large Data Bases*, Mumbai (Bombay), India, pages 28–39, Morgan Kaufmann Publishers, Los Altos, CA, September 1996.
9. F.R. Biglari, N.P. O'Dowd, and R.T. Fenner. Optimum design of forging dies using fuzzy logic in conjunction with the backward deformation method. *International Journal of Machine Tools and Manufacture*, 38(8):981–1000, August 1998.
10. P. Blanke and F.-E. Wolter. Fast inverse forging simulation via medial axis transform. In *Proceedings of the 2007 International Conference on Cyberworlds, NASAGEM Workshop*, pages 396–403, IEEE Computer Society, Washington, DC, 2007.
11. G. Böttcher. Medial axis and haptics. Diplomarbeit, Leibniz Universität Hannover, October 2004.
12. A.N. Bramley. UBET and TEUBA: Fast methods for forging simulation and preform design. *Journal of Materials Processing*, 116, 2001.
13. T. Culver. Computing the medial axis of a polyhedron reliably and efficiently. PhD Thesis, Department of Computer Science, University of North Carolina, Chapel Hill, 2000.
14. T.K. Dey and W. Zhao. Approximating the medial axis from the Voronoi diagram with a convergence guarantee. *Algorithmica*, 38:179–200, 2004.
15. M.P. do Carmo. *Riemannian Geometry*. Birkhäuser, Boston, 1992.
16. C.B. Garcia and W.J. Zangwill. *Pathways to Solutions, Fixed Points, and Equilibria*. Prentice-Hall, 1984.

17. C. Gordon, D.L. Webb, and S. Wolpert. One cannot hear the shape of a drum. *Bull. Amer. Math. Soc.*, 26:134–138, 1992.
18. H. Naß. Computation of medial sets in Riemannian manifolds. PhD Thesis, Leibniz Universität Hannover, 2007.
19. A. Howind. Untersuchungen und Berechnungen zur Medialen Achse im Raum. Diplomarbeit, Leibniz Universität Hannover, October 1998.
20. T. Howind. Spektralanalyse von berandeten Gebieten. Diplomarbeit, Leibniz Universität Hannover, September 1998.
21. S.M. Hwang and S. Kobayashi. Preform design in disk forging. *International Journal of Machine Tool Design & Research* , 26(3):231–243, 1986.
22. R. Kunze, F.-E. Wolter, and T. Rausch. Geodesic Voronoi diagrams on parametric surfaces. In *Proceedings of CGI'97*, Vol. 6, IEEE Computer Society, 1997. Available as Welfenlab Report 2.
23. H. Mathieu. Ein Beitrag zur Auslegung der Stadienfolge beim Gesenkschmieden mit Grat. *Fortschritt-Berichte VDI-Reihe 2*, 213, VDI, 1991.
24. H.P. McKean and I.M. Singer. Curvature and the eigenvalues of the Laplacian. *J. Diff. Geom*, 1:43–69, 1967.
25. M. Michael. Konstruktionsbegleitende Modellierung von Schmiedeprozessen. PhD Thesis, Leibniz Universität Hannover, 1999.
26. H. Naß F.-E. Wolter, C. Doğan, and H. Thielhelm. Medial axis (inverse) transform in complete 3-dimensional Riemannian manifolds. In *Proceedings of the 2007 International Conference on Cyberworlds, NASAGEM Workshop*, pages 386–395, IEEE Computer Society, Washington, DC, 2007.
27. H. Naß, F.-E. Wolter, H. Thielhelm, and C. Doğan. Computation of geodesic voronoi diagrams in 3-space using medial equations. In *Proceedings of the 2007 International Conference on Cyberworlds, NASAGEM Workshop*, pages 376–385, IEEE Computer Society, Washington, DC, 2007.
28. M. Niethammer, M. Reuter, F.-E. Wolter, S. Bouix, N. Peinecke, M.-S. Ko, and M.E. Shenton. Global medical shape analysis using the Laplace–Beltrami-spectrum. In *Proceedings of MICCAI07, 10th International Conference on Medical Image Computing and Computer Assisted Intervention*, 2007.
29. N. Obydenna. Implementierung einer Datenstruktur für simpliziale 3D-Netze. Studienarbeit, Leibniz Universität Hannover, April 2009.
30. N. Peinecke. Eigenwertspektren des Laplaceoperators in der Bilderkennung. PhD Thesis, Leibniz Universität Hannover, 2006.
31. N. Peinecke and F.-E. Wolter. Mass density Laplace-spectra for image recognition. In *Proceedings of the 2007 International Conference on Cyberworlds, NASAGEM Workshop*, pages 409–416, IEEE Computer Society, Washington, DC, 2007.
32. N. Peinecke, F.-E. Wolter, and M. Reuter. Laplace-spectra as fingerprints for image recognition. *Computer-Aided Design*, 6(39):460–476, 0 2007.
33. M.M. Postnikov. *The Variational Theory of Geodesics*. Saunders, Philadelphia, PA, 1967.
34. T. Rausch. Analysis and computation of the geodetic medial axis of bordered surface patches (Untersuchungen und Berechnungen zur geodätischen Medialen Achse bei Berandeten Flächenstücken). PhD Thesis, Leibniz Universität Hannover, 1999.
35. T. Rausch, F.-E. Wolter, and O. Sniehotta. Computation of medial curves in surfaces. In *Proceedings of Conference on the Mathematics of Surfaces VII*, pages 43–68, Institute of Mathematics and Applications, 1996. Also available as Welfenlab Report 1.
36. M. Reuter, F.-E. Wolter, and N. Peinecke. Laplace–Beltrami spectra as shape DNA of surfaces and solids. *Computer-Aided Design*, 4(38):342–366, 2006.
37. M. Reuter, S. Biasotti, D. Giorgi, G. Patanè, and M. Spagnuolo. Discrete Laplace–Beltrami operators for shape analysis and segmentation. *Computers & Graphics*, 33(3):381–390, 2009.
38. M. Reuter, M. Niethammer, F.-E. Wolter, S. Bouix, N. Peinecke, M.-S. Ko, and M.E. Shenton. Global medical shape analysis using the volumetric laplace spectrum. In *Proceedings of the 2007 International Conference on Cyberworlds, NASAGEM Workshop*, pages 417–426, IEEE Computer Society, Washington, DC, 2007.

39. M. Reuter, F.-E. Wolter, M. Shenton, and M. Niethammer. Laplace–Beltrami eigenvalues and topological features of eigenfunctions for statistical shape analysis. *Computer-Aided Design*, 41(10):739 – 755, 2009.
40. E.C. Sherbrooke. 3D shape interrogation by medial axis transform. PhD Thesis, Department of Ocean Engineering, MIT, Boston, MA, 1995.
41. H. Thielhelm. Geodätische Voronoi Diagramme. Diplomarbeit, Leibniz Universität Hannover, May 2007.
42. M. Wienströer. Konstruktionsintegrierte Prozessimulation der Stadienfolge beim Schmieden mittels Rückwärtssimulation. PhD Thesis, Leibniz Universität Hannover, 2004.
43. F.-E. Wolter. Interior metric, shortest paths and loops in Riemannian manifolds with not necessarily smooth boundary. Diplomarbeit, Freie Universität Berlin, 1979.
44. F.-E. Wolter. Distance function and cut loci on a complete Riemannian manifold. *Arch. Math.*, 1(32):92–96, 1979.
45. F.-E. Wolter. Cut loci in bordered and unbordered Riemannian manifolds. PhD Thesis, Technical University of Berlin, 1985.
46. F.-E. Wolter. Cut locus & medial axis in global shape interrogation & representation. MIT Design Laboratory Memorandum 92-2 and MIT National Sea Grant Library Report, 1992. MIT, December 1993 (revised version).
47. F.-E. Wolter. Shape and image cognition, (Re)-construction and compression via tools from differential geometry, Keynote Lecture presented at CGI 2010, Singapore. Available at http://cgi2010.miralab.unige.ch/CGI_InvitedSpeakers.html, 2010.
48. F.-E. Wolter, S. Cutchin, T. Hausman, B. Johnson, S. Goehring, and L. Lambers. Project Riemann source code. Available at http://www.welfenlab.de/en/research/projects/riemann, 1988.
49. F.-E. Wolter and K.-I. Friese. Local and global geometric methods for analysis interrogation, reconstruction, modification and design of shape. In *Proceedings of Computer Graphics International 2000*, Geneva, Switzerland, pages 137–151, IEEE Computer Society, 2000. Also available as Welfenlab Report 3.
50. F.-E. Wolter, M. Reuter, and N. Peinecke. Geometric modeling for engineering applications. In E. Stein, R. de Borst, and T.J.R. Hughes (Eds.), *Encyclopedia of Computational Mechanics. Part 1: Fundamentals*, chapter 16, John Wiley & Sons, 2007.
51. G. Xu. Discrete Laplace–Beltrami operators and their convergence. *Computer Aided Geometric Design*, 21:767–784, 2004.

Analysis of a Mathematical Model Describing Necrotic Tumor Growth

Joachim Escher, Anca-Voichita Matioc and Bogdan-Vasile Matioc

Abstract A model describing the growth of necrotic tumors in different regimes of vascularisation is studied. The tumor consists of a necrotic core of death cells and a surrounding shell which contains life-proliferating cells. The blood supply provides the nonnecrotic region with nutrients and no inhibitor chemical species are present. The corresponding mathematical formulation is a moving boundary problem since both boundaries delimiting the nonnecrotic shell are allowed to evolve in time. We determine all radially symmetric stationary solutions and reduce the moving boundary problem into a nonlinear evolution equation for the functions parameterising the boundaries of the shell. Parabolic theory provides a suitable context for proving local well-posedness of the problem for small initial data.

1 The Mathematical Model

In this chapter we study a moving boundary problem describing the growth of a necrotic tumor in the absence of inhibitors. The model purposed initially in [4, 15, 17] was reformulated by using algebraic manipulations [5, 12] to describe evolution of tumors in all regimes of vascularisation. Nevertheless, the analysis in [5, 12] is simplified by the assumption that the tumor core is nonnecrotic. Our aim is to abandon this simplification. Following [10, 17, 18], we assume that the tumor consists of a core of death cells (necrotic core) and a shell of life-proliferating cells surrounding the core (nonnecrotic shell). The blood supply provides the nonnecrotic region with nutrients, while the necrotic region is not vascularised and the concentration of nutrients is at a constant level which cannot sustain cell proliferation. However, the model presented here includes two moving boundaries, one parametrising the boundary of the necrotic core and one for the outer boundary of the tumor, both

Joachim Escher · Anca-Voichita Matioc · Bogdan-Vasile Matioc
Institut für Angewandte Mathematik, Leibniz Universität Hannover, Welfengarten 1,
30167 Hannover, Germany; e-mail: {escher, matioca, matioc}@ifam.uni-hannover.de

E. Stephan & P. Wriggers (Eds.): Modelling, Simulation and Software Concepts, LNACM 57, pp. 237–250.
springerlink.com   © Springer-Verlag Berlin Heidelberg 2011

of them having infinitely many degrees of freedom. This fact makes the problem more involved in comparison to other models which either neglect the necrotic core [3, 6–10], or consider only the radially symmetric problem when the tumors are annular domains [10, 17, 18].

The mathematical model is given by the following system of equations

$$
\begin{cases}
\Delta\psi = \psi & \text{in } \Omega(t), \ t \geq 0, \\[4pt]
\Delta p = 0 & \text{in } \Omega(t), \ t \geq 0, \\[4pt]
\psi = G & \text{on } \Gamma_1(t), t \geq 0, \\[4pt]
\psi = G - \psi_0 & \text{on } \Gamma_2(t), t \geq 0, \\[10pt]
p = \kappa_{\Gamma_1(t)} - AG\dfrac{|x|^2}{4} & \text{on } \Gamma_1(t), t \geq 0, \\[14pt]
p = \kappa_{\Gamma_2(t)} - AG\dfrac{|x|^2}{4} - \psi_0 & \text{on } \Gamma_2(t), t \geq 0, \\[14pt]
V_i(t) = \partial_{v_i}\psi - \partial_{v_i}p - AG\dfrac{v_i \cdot x}{2} & \text{on } \Gamma_i(t), \ t > 0, i = 1,2, \\[10pt]
\Omega(0) = \Omega_0,
\end{cases}
\tag{1}
$$

where $\Omega(t) \subset \mathbb{R}$ is the domain occupied by the nonnecrotic shell, ψ is the rate at which nutrient is added to $\Omega(t)$ over the outer boundary $\Gamma_1(t)$, by the vascularisation, p is the pressure, $\Gamma_2(t)$ is the interior boundary enclosing the necrotic core, v_i is the restriction of the outward orientated normal at $\partial\Omega(t)$ to $\Gamma_i(t)$, and $\kappa_{\Gamma_i(t)}$ the curvature of $\Gamma_i(t)$, $i = 1,2$. By convention, $\kappa_{\Gamma_1(t)}$ is positive and $\kappa_{\Gamma_2(t)}$ negative if $\Gamma_i(t)$ are close to circles. Moreover, $V_i(t)$ stands for the normal velocity of $\Gamma_i(t)$, while the constants $A, G \in \mathbb{R}$ have biological relevancy being related to cell proliferation, cell apoptosis, and vascularisation. The scalar $\psi_0 > 0$ corresponds to the nutrient concentration assumed constant within the necrotic region. The initial tumor domain is given by Ω_0 and x is the position vector in $\mathbb{R}^2$. For a precise deduction of the system (1) and its biological meaning, we refer to [5, 12], the only difference to the model presented there being the consideration of the interior necrotic region bounded by $\Gamma_2(t)$. The first main result of this chapter is the following theorem:

Theorem 1 (Radially symmetric stationary solutions). *Given $(R_1, R_2) \in (0, \infty)^2$ with $R_2 < R_1$, let ψ_0^c be the constant defined by (10). There exists $A \in \mathbb{R}$ and $G \in \mathbb{R} \setminus \{0\}$, such that the annulus*

$$
A(R_1, R_2) := \{x \in \mathbb{R}^2 \ : \ R_2 < |x| < R_1\},
$$

is a stationary solution of problem (1) provided $\psi_0 \neq \psi_0^c$. Moreover, A and G are uniquelly determined by R_1, R_2, and ψ_0.

If $G = 0$, then problem (1) has no radially symmetric stationary solutions.

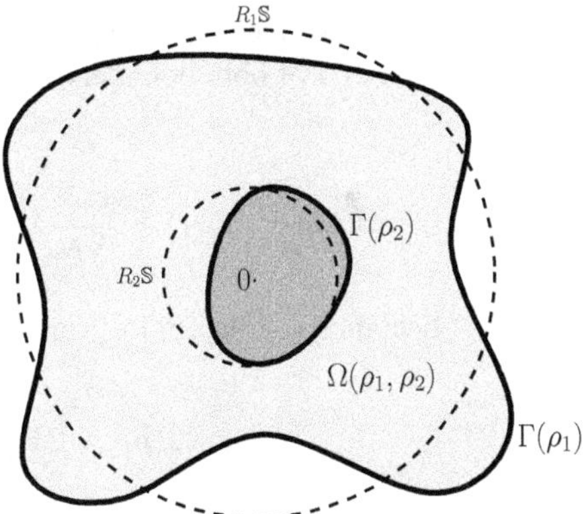

Fig. 1 Parametrisation of the tumor domain.

In contrast to [5, 12], where the radially symmetric stationary tumors are circles with radius which depends only on the constant Λ, the radii of the stationary annular tumors found in Theorem 1 depend on both constants A and G, cf. (11).

In order to prove local well-posedness of the moving boundary problem (1) (see Theorem 2 below) we introduce first a parametrisation for the interfaces $\Gamma_1(t)$ and $\Gamma_2(t)$, which are the main unknowns of system (1). Let $0 < R_2 < R_1$ be given and fix $\alpha \in (0,1)$. We set

$$\mathscr{V} := \{\rho \in h^{4+\alpha}(\mathbb{S}) : \|\rho\|_{C(\mathbb{S})} < a\},$$

where

$$a < \frac{R_1 - R_2}{R_1 + R_2}.$$

The small Hölder space $h^{m+\beta}(\mathbb{S})$, $\beta \in (0,1)$ and $m \in \mathbb{N}$, is defined as the completion of the smooth functions in $C^{m+\beta}(\mathbb{S})$. Each pair $(\rho_1,\rho_2) \in \mathscr{V}^2$ parametrises a $C^{4+\alpha}$-domain

$$\Omega(\rho_1,\rho_2) := \left\{y \in \mathbb{R}^2 : R_2(1 + \rho_2(y/|y|)) < |y| < R_1(1 + \rho_1(y/|y|))\right\}.$$

The condition on a ensures that the boundary portions of $\Omega(\rho_1,\rho_2)$

$$\Gamma(\rho_i) := \{x : |x| = R_i(1 + \rho_i(x/|x|))\},$$

$i = 1,2$, are disjoint (see Fig. 1) for any choice of $(\rho_1,\rho_2) \in \mathscr{V}^2$. Moreover, they can be seen to be zero level sets, $\Gamma(\rho_i) = N_{\rho_i}^{-1}(0)$, where $N_{\rho_i} : \mathbb{R}^2 \setminus \{0\} \to \mathbb{R}, i = 1,2$, are defined by

$$N_{\rho_i}(x) = |x| - R_i - R_i\rho_i(x/|x|), \qquad x \neq 0.$$

Hence, the outward unit normal at $\partial\Omega(\rho_1,\rho_2)$ is given by

$$v_{\rho_1} = \frac{\nabla N_{\rho_1}}{|\nabla N_{\rho_1}|} \text{ on } \Gamma(\rho_1) \quad \text{and} \quad v_{\rho_2} = -\frac{\nabla N_{\rho_2}}{|\nabla N_{\rho_2}|} \text{ on } \Gamma(\rho_2).$$

If the function $(\rho_1, \rho_2) : [0,T] \to \mathcal{V}^2$ describes the motion of the tumor boundaries, then we can express the normal velocity of both boundary components in terms of ρ_i by

$$V_1(t) = -\frac{\partial_t N_{\rho_1}}{|\nabla N_{\rho_1}|} \text{ on } \Gamma(\rho_1(t)) \quad \text{and} \quad V_2(t) = \frac{\partial_t N_{\rho_2}}{|\nabla N_{\rho_2}|} \text{ on } \Gamma(\rho_2(t)).$$

With this notation, system (1) becomes a problem having also ρ_1 and ρ_2 as unknowns:

$$\begin{cases}
\Delta \psi = \psi & \text{in } \Omega(\rho_1, \rho_2),\, t \geq 0, \\
\Delta p = 0 & \text{in } \Omega(\rho_1, \rho_2),\, t \geq 0, \\
\psi = G & \text{on } \Gamma(\rho_1), \quad t \geq 0, \\
\psi = G - \psi_0 & \text{on } \Gamma(\rho_2), \quad t \geq 0, \\
p = \kappa_{\Gamma(\rho_1)} - AG\dfrac{|x|^2}{4} & \text{on } \Gamma(\rho_1), \quad t \geq 0, \\
p = \kappa_{\Gamma(\rho_2)} - AG\dfrac{|x|^2}{4} - \psi_0 & \text{on } \Gamma(\rho_2), \quad t \geq 0, \\
\partial_t N_{\rho_i} = -\langle \nabla \psi - \nabla p - AG\dfrac{x}{2} | \nabla N_{\rho_i} \rangle & \text{on } \Gamma(\rho_i), \quad t > 0,\, i = 1,2, \\
\rho_1(0) = \rho_{01}, \\
\rho_2(0) = \rho_{02},
\end{cases} \tag{2}$$

with $(\rho_1(0), \rho_2(0))$ describing the initail shape of the tumor. A pair $(\rho_1, \rho_2, \psi, p)$ is called *classical solution* of (1) on $[0,T], T > 0$, if

$$\rho_i \in C([0,T], \mathcal{V}) \cap C^1([0,T], h^{1+\alpha}(\mathbb{S})),\, i = 1,2,$$

$$\psi(t, \cdot), p(t, \cdot) \in buc^{2+\alpha}(\Omega(\rho_1(t), \rho_2(t))),\, t \in [0,T],$$

and if $(\rho_1, \rho_2, \psi, p)$ solves (2) pointwise. Given $U \subset \mathbb{R}^2$ open, we set $buc^{2+\alpha}(U)$ to be the closure of the smooth functions with bounded and uniformly continuous derivatives $BUC^\infty(U)$ within $BUC^{2+\alpha}(U)$ (if U is also bounded then $BUC^{2+\alpha}(U) = C^{2+\alpha}(\overline{U})$).

Concerning well-posedness of system (1), our second main result states that problem (1) possesses a unique solution provided that initially the tumor is close to an annulus (which must not be necessarily a stationary solution).

Theorem 2 (Local well-posedness). *Let $0 < R_2 < R_1$ and $(A, G, \psi_0) \in \mathbb{R}^3$ be given.*
There exists an open neighbourhood $\mathcal{O} \subset \mathcal{V}$ such that for all $(\rho_1, \rho_2) \in \mathcal{O}^2$, problem (2) possesses a unique classical solution defined on a maximal time interval $[0, T(\rho_{01}, \rho_{02}))$ and which satisfies $(\rho_1, \rho_2)(t) \in \mathcal{O}^2$ for all $t \in [0, T(\rho_{01}, \rho_{02}))$.

The outline of this chapter is as follows: we study in Section 2 the radially symmetric free boundary problem which describes the stationary solutions of system (1) and prove Theorem 1. In the last section we prove the local well-posedness result, Theorem 2.

2 Radially Symmetric Stationary Solutions

We determine in this section the radially symmetric steady-state solutions of (1), situation when the nonnecrotic shell is a steady annulus.

The most simple situation is the case $G = 0$, when the problem is invariant under translations and rotations. Then, the annulus $A(R_1, R_2)$ centred in zero with radii $R_1 > R_2$, is a stationary solution of system (1) if and only if

$$p'(R_i) = \psi'(R_i), \qquad i = 1, 2,$$

where p is the solution of the problem

$$\begin{cases} p'' + \dfrac{1}{r}p' = 0, & R_2 < r < R_1, \\ p(R_1) = R_1^{-1} - AGR_1^2/4, \\ p(R_2) = -R_2^{-1} - AGR_2^2/4 - \psi_0, \end{cases} \tag{3}$$

when $G = 0$. System (3) corresponds to the Dirichlet problem for the pressure p in (1) (the second, fifth, and sixth equations of (1)), where we used polar coordinates when expressing the Laplacian. Notice that the boundary data are constants, thus p depends only on r, the distance to the origin.

Given $G \in \mathbb{R}$, the solution of (3) is given by the relation $p(r) = a_{R_1 R_2} \ln(r) + b_{R_1 R_2}$, $R_2 \le |r| \le R_1$, with

$$a_{R_1 R_2} = \frac{R_1^{-1} + R_2^{-1} + AG\left(R_2^2 - R_1^2\right)/4 + \psi_0}{\ln(R_1/R_2)},$$

$$b_{R_1 R_2} = R_1^{-1} - AGR_1^2/4 - a_{R_1 R_2}\ln(R_1).$$

Furthermore, ψ is the solution of the problem

$$\begin{cases} \psi'' + \dfrac{1}{r}\psi' - \psi = 0, & R_2 < r < R_1, \\ \psi(R_1) = G, \\ \psi(R_2) = G - \psi_0, \end{cases} \tag{4}$$

when $G = 0$. Also, for fixed $G \in \mathbb{R}$, the solution of (4) can be written as linear combination of modified Bessel functions of first and second kind $\psi = c_{R_1 R_2}^1 I_0 + c_{R_1 R_2}^2 K_0$, with scalars

$$c^1_{R_1 R_2} = \frac{GK_0(R_2) + (\psi_0 - G)K_0(R_1)}{I_0(R_1)K_0(R_2) - I_0(R_2)K_0(R_1)}, \qquad c^2_{R_1 R_2} = \frac{-GI_0(R_2) - (\psi_0 - G)I_0(R_1)}{I_0(R_1)K_0(R_2) - I_0(R_2)K_0(R_1)}.$$

Consequently, $A(R_1, R_2)$ is a steady-state solution of (1) when $G = 0$ if and only if

$$\frac{\frac{1}{R_1} + \frac{1}{R_2} + \psi_0}{\ln(R_1/R_2)} \frac{1}{R_i} = \psi_0 \frac{K_0(R_1)I_1(R_i) + I_0(R_1)K_1(R_i)}{I_0(R_1)K_0(R_2) - I_0(R_2)K_0(R_1)}, \qquad i = 1, 2, \qquad (5)$$

where we used the relations $I'_0 = I_1$ and $K'_0 = -K_1$. It follows then easily that the system consisting of equations (5) has solutions (R_1, R_2) with $R_1 > R_2$ exactly when

$$\frac{R_2}{R_1} = \frac{K_0(R_1)I_1(R_1) + I_0(R_1)K_1(R_1)}{K_0(R_1)I_1(R_2) + I_0(R_1)K_1(R_2)}. \qquad (6)$$

Equation (6) is obtained by expressing ψ_0 in both relations (5) and setting them to be equal. We show now that equality holds in the relation above only when $R_1 = R_2$. Indeed, fix $R_1 > 0$ and consider the auxiliary function $g : (0, R_1] \to \mathbb{R}$ with

$$g(r) = K_0(R_1)rI_1(r) + I_0(R_1)rK_1(r) - R_1(K_0(R_1)I_1(R_1) + I_0(R_1)K_1(R_1))$$

for $0 < r \leq R_1$. Obviously $g(R_1) = 0$. If we show that the derivative g' has constant sign on $(0, R_1]$ then we are done, that is there is no positive $R_2 < R_1$ such that (R_1, R_2) solves (6). Well-known properties of the modified Bessel functions (see [2]) lead to

$$g'(r) = K_0(R_1)I_1(r) + I_0(R_1)K_1(r)$$
$$+ K_0(R_1)r(I_0(r) - (1/r)I_1(r)) + I_0(R_1)r(-K_0(r) - (1/r)K_1(r))$$
$$= r(I_0(r)K_0(R_1) - I_0(R_1)K_0(r)) < 0$$

for all $r \in (0, R_1)$. That the last expression is negative is a consequence of the following facts: I_0 and K_0 are both positive functions, I_0 is strictly increasing, and K_0 is strictly decreasing. Hence, problem (1) has no radially symmetric stationary solutions when $G = 0$.

Let now $G \neq 0$. In this case $A(R_1, R_2)$ is a steady-state solution of (1) exactly when

$$\psi'(R_i) - p'(R_i) - AG\frac{R_i}{2} = 0, \qquad i = 1, 2. \qquad (7)$$

Using again the relations $I'_0 = I_1$ and $K'_0 = -K_1$, the identities (7) re-write

$$c^1_{R_1 R_2} I_1(R_i) - c^2_{R_1 R_2} K_1(R_i) - a_{R_1 R_2} \frac{1}{R_i} - AG\frac{R_i}{2} = 0, \qquad i = 1, 2,$$

which seem to be very involved as expressions of variables R_1 and R_2 when trying to solve the system consisting of both of them. However, they can be viewed as equations for A and G

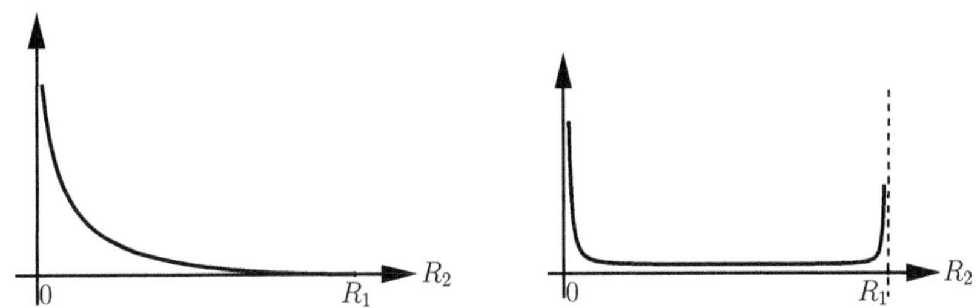

Fig. 2 The expression $a_1 b_2 - a_2 b_1$ and ψ_0^c, for fixed R_1, as a function of the variable $R_2 \in (0, R_1)$.

$$a_i G + b_i A G = c_i, \qquad i = 1, 2, \tag{8}$$

with coefficients a_i, b_i, and c_i given by

$$a_i := \frac{(K_0(R_2) - K_0(R_1))I_1(R_i) - (I_0(R_1) - I_0(R_2))K_1(R_i)}{I_0(R_1)K_0(R_2) - I_0(R_2)K_0(R_1)},$$

$$b_i := \frac{R_1^2 - R_2^2}{4\ln(R_1/R_2)} \frac{1}{R_i} - \frac{R_i}{2},$$

$$c_i := -\psi_0 \frac{K_0(R_1)I_1(R_i) + I_0(R_1)K_1(R_i)}{I_0(R_1)K_0(R_2) - I_0(R_2)K_0(R_1)} + \frac{R_1^{-1} + R_2^{-1} + \psi_0}{\ln(R_1/R_2)} \frac{1}{R_i}.$$

The system of equations (8) has a (unique) solution (A, G) with $G \neq 0$ provided that

$$a_1 b_2 - a_2 b_1 \neq 0, \qquad c_1 b_2 - c_2 b_1 \neq 0,$$
$$\text{and} \quad c_1 \neq 0 \quad \text{or} \quad c_2 \neq 0. \tag{9}$$

The computation done for the case $G = 0$ shows that c_1 and c_2 cannot be simultaneously zero when $R_2 < R_1$. For fixed $R_1 > 0$ we may see the expression $a_1 b_2 - a_2 b_1$ as a function of $R_2 \in (0, R_1)$. This function is strictly decreasing with respect to R_2 (see Fig. 2), thus $a_1 b_2 = a_2 b_1$ only when $R_1 = R_2$. Furthermore, $b_1 c_2 - b_2 c_1$ if and only if $\psi_0 = \psi_0^c$, where

$$\psi_0^c := \frac{\left(b_1/R_1 - b_2/R_2\right)^{\frac{1/R_1 + 1/R_2}{\ln(R_1/R_2)}}}{\frac{K_0(R_1)(b_1 I_1(R_2) - b_2 I_1(R_1)) + I_0(R_1)(b_1 K_1(R_2) - b_2 K_1(R_1))}{I_0(R_1)K_0(R_2) - I_0(R_2)K_0(R_1)} + \frac{R_1^2 - R_2^2}{2R_1 R_2 \ln(R_1/R_2)}}. \tag{10}$$

It is not difficult to see that the numerator of the fraction is negative, and the same holds true for the denominator, implying that $\psi_0^c > 0$. We plotted in Fig. 2 the expression on the right hand side of (10) for fixed $R_1 > 0$ in dependence of $R_2 \in (0, R_1)$. Consequently, (9) are fulfilled provided $\psi_0 \neq \psi_0^c$, meaning that $A(R_1, R_2)$ is a stationary solution of (1) if and only if ψ_0 is not the critical constant given by (10) and

$$A = \frac{a_1 c_2 - a_2 c_1}{c_1 b_2 - c_2 b_1}, \qquad G = \frac{c_1 b_2 - c_2 b_1}{a_1 b_2 - a_2 b_1}. \tag{11}$$

This proves Theorem 1.

3 The Moving Boundary Problem

This last section is dedicated entirely to the proof of our second main result, Theorem 2. In order to prove well-posedness of problem (1), in the context defined in the introduction, we transform first (2) into a problem on the fixed domain $\Omega := \Omega(0,0)$, with boundary $\Gamma_1 := R_1 \mathbb{S}$ and $\Gamma_2 := R_2 \mathbb{S}$. This transformation will allow us to introduce solution operators related to problem (2) and which will enable us to reduce system (2) into an abstract nonlinear evolution equation for the pair (ρ_1, ρ_2).

Denote therefore $0 < R_2 < R_1$, $(A, G, \psi_0) \in \mathbb{R}^3$, and $\alpha \in (0,1)$. Given $(\rho_1, \rho_2) \in \mathcal{V}^2$, we define the mapping $\Theta_{\rho_1, \rho_2} : \Omega \to \Omega(\rho_1, \rho_2)$ by the relation

$$\Theta_{\rho_1,\rho_2}(x) = \frac{(R_1 - |x|)R_2(1 + \rho_2(x/|x|)) + (|x| - R_2)R_1(1 + \rho_1(x/|x|))}{R_1 - R_2} \, \frac{x}{|x|}$$

for $x \in \Omega$. One can easily check that Θ_{ρ_1,ρ_2} is a diffeomorphism, i.e. $\Theta_{\rho_1,\rho_2} \in Diff^{4+\alpha}(\Omega, \Omega(\rho_1,\rho_2))$, which maps Γ_i onto $\Gamma(\rho_i)$, $i = 1, 2$. Using this diffeomorphism, we define the transformed operators

$$\mathscr{A}(\rho_1,\rho_2) : buc^{2+\alpha}(\Omega) \to buc^{\alpha}(\Omega), \quad \mathscr{A}(\rho_1,\rho_2)v := \Delta(v \circ \Theta_{\rho_1,\rho_2}^{-1}) \circ \Theta_{\rho_1,\rho_2},$$

which is an elliptic operator depending analytically on (ρ_1, ρ_2), i.e.

$$\mathscr{A} \in C^{\omega}(\mathcal{V}^2, \mathscr{L}(buc^{2+\alpha}(\Omega), buc^{\alpha}(\Omega))), \tag{12}$$

respectively the trace operators $\mathscr{B}_i : \mathcal{V}^2 \times (buc^{2+\alpha}(\Omega))^2 \to h^{1+\alpha}(\mathbb{S})$ by

$$\mathscr{B}_i(\rho_1,\rho_2,v,q) := \frac{1}{R_i}\mathscr{C}_i(\rho_1,\rho_2)v - \frac{1}{R_i}\mathscr{C}_i(\rho_1,\rho_2)q - \mathscr{D}_i(\rho_1,\rho_2).$$

Given $(\rho_1, \rho_2) \in \mathcal{V}^2$, the linear operators $\mathscr{C}_i(\rho_1,\rho_2) \in \mathscr{L}(buc^{2+\alpha}(\Omega), h^{1+\alpha}(\mathbb{S}))$, $i = 1, 2$, are given by

$$\mathscr{C}_i(\rho_1,\rho_2)v(y) := \langle \nabla(v \circ \Theta_{\rho_1,\rho_2}^{-1}) | \nabla N_{\rho_i} \rangle \circ \Theta_{\rho_1,\rho_2}(R_i y)$$

for $v \in buc^{2+\alpha}(\Omega)$ and $y \in \mathbb{S}$. Moreover,

$$\mathscr{D}_i(\rho_1,\rho_2) := -\frac{AG}{R_i}\langle \frac{x}{2} | \nabla N_{\rho_i} \rangle \circ \Theta_{\rho_1,\rho_2}(R_i y).$$

The operators $\mathscr{C}_i$ and $\mathscr{D}_i, i = 1, 2$, depend analytically on (ρ_1, ρ_2) too,

$$\mathscr{C}_i \in C^\omega(\mathscr{V}^2, \mathscr{L}(buc^{2+\alpha}(\Omega), h^{1+\alpha}(\mathbb{S}))) \quad \text{and} \quad \mathscr{D}_i \in C^\omega(\mathscr{V}^2, h^{1+\alpha}(\mathbb{S})). \tag{13}$$

Having defined these operators we may re-write now (2) in an equivalent form. Namely, if $(\rho_1, \rho_2, \psi, p)$ is a solution of (2), $v := \psi \circ \Theta_{\rho_1, \rho_2}$, and $q := p \circ \Theta_{\rho_1, \rho_2}$, then the tupel (ρ_1, ρ_2, v, q) solves the following system:

$$\begin{cases}
\mathscr{A}(\rho_1, \rho_2)v = v & \text{in } \Omega, \ t \geq 0, \\
\mathscr{A}(\rho_1, \rho_2)q = 0 & \text{in } \Omega, \ t \geq 0, \\
v = G & \text{on } \Gamma_1, \ t \geq 0, \\
v = G - \psi_0 & \text{on } \Gamma_2, \ t \geq 0, \\
q = \dfrac{1}{R_1}\kappa(\rho_1) - \dfrac{AGR_1^2}{4}(1 + \rho_1)^2 & \text{on } \Gamma_1, \ t \geq 0, \\
q = -\dfrac{1}{R_2}\kappa(\rho_2) - \dfrac{AGR_2^2}{4}(1 + \rho_2)^2 - \psi_0 & \text{on } \Gamma_2, \ t \geq 0, \\
\partial_t \rho_i = \mathscr{B}_i(\rho_1, \rho_2, v, q) & \text{on } \mathbb{S}, \ t > 0, \ i = 1, 2, \\
\rho_1(0) = \rho_{01}, \\
\rho_2(0) = \rho_{02},
\end{cases} \tag{14}$$

where $\kappa : \mathscr{V} \to h^{2+\alpha}(\mathbb{S})$ is defined by

$$\kappa(\rho) := \frac{(1+\rho)^2 + 2\rho'^2 - (1+\rho)\rho''}{((1+\rho)^2 + \rho'^2)^{3/2}}, \qquad \rho \in \mathscr{V},$$

and we identified functions on Γ_i with those on $\mathbb{S}$, $i = 1, 2$, via the diffeomorphisms $[\mathbb{S} \ni y \mapsto R_i y \in \Gamma_i]$.

Though the problem becomes more involved (the diffeomorphism introduces additional nonlinearities), (14) has the advantage that the sets where the differential equations and the boundary conditions are defined do not change with time. It is convenient now to introduce solution operators to Dirichlet problem closely related to system (14).

Lemma 1. *Given* $(\rho_1, \rho_2) \in \mathscr{V}^2$, *we let* $\mathscr{T}(\rho_1, \rho_2), \mathscr{S}(\rho_1, \rho_2) \in buc^{2+\alpha}(\Omega)$ *denote the unique solution of the Dirichlet problem*

$$\begin{cases}
\mathscr{A}(\rho_1, \rho_2)v = v & \text{in } \Omega, \\
v = G & \text{on } \Gamma_1, \\
v = G - \psi_0 & \text{on } \Gamma_2,
\end{cases} \tag{15}$$

and

$$\begin{cases} \mathscr{A}(\rho_1,\rho_2)q = 0 & in\ \Omega, \\[2mm] q = \dfrac{1}{R_1}\kappa(\rho_1) - \dfrac{AGR_1^2}{4}(1+\rho_1)^2 & on\ \Gamma_1, \\[3mm] q = -\dfrac{1}{R_2}\kappa(\rho_2) - \dfrac{AGR_2^2}{4}(1+\rho_2)^2 - \psi_0 & on\ \Gamma_2, \end{cases} \qquad (16)$$

respectively. The operators $\mathscr{T}$ and $\mathscr{S}$ depend analytically on (ρ_1,ρ_2).

Proof. Given $(\rho_1,\rho_2) \in \mathscr{V}^2$, the Dirichlet problems (15) and (16) are uniquely solvable, cf. [16, Theorem 6.14]. Moreover, since $\mathscr{A}$ and κ depend analytically on their variables we deduce that also $\mathscr{T}$ and $\mathscr{S}$ do that. We may take now into consideration that $\mathscr{T}$ and $\mathscr{S}$ both map the smooth functions into $BUC^\infty(\Omega)$ and conclude that their range is contained in $buc^{2+\alpha}(\Omega)$. $\qquad\square$

With this definition, (14) reduces to the following evolution equation:

$$\partial_t X = \Phi(X) \qquad X(0) = X_0, \qquad (17)$$

where $X := (\rho_1,\rho_2)$, $X_0 := (\rho_{01},\rho_{02})$, and $\Phi := (\Phi_1,\Phi_2)$. The components of the nonlocal and nonlinear operator Φ are defined as follows:

$$\Phi_i(\rho_1,\rho_2) := \mathscr{B}_i(\rho_1,\rho_2,\mathscr{T}(\rho_1,\rho_2),\mathscr{S}(\rho_1,\rho_2)), \qquad i = 1,2.$$

In order to prove well-posedness of problem (17) it suffices to show that

$$\partial\Phi(0) = \begin{bmatrix} \partial_{\rho_1}\Phi_1(0) & \partial_{\rho_2}\Phi_1(0) \\[2mm] \partial_{\rho_1}\Phi_2(0) & \partial_{\rho_2}\Phi_2(0) \end{bmatrix}$$

generates a strongly continuous and analytic semigroup. The key role is played by the operator $\mathscr{S}$ which depends on the highest order derivatives of ρ_i, $i = 1,2$. We have:

Theorem 3. *The operator Φ is analytic, i.e. $\Phi \in C^\omega(\mathscr{V}^2,(h^{1+\alpha}(\mathbb{S}))^2)$. Given $\beta \in (0,1)$, the Fréchet derivative $\partial\Phi(0)$, seen as an unbounded operator in $(h^{1+\beta}(\mathbb{S}))^2$ with domain $(h^{4+\beta}(\mathbb{S}))^2$ generates a strongly continuous and analytic semigroup in $\mathscr{L}((h^{1+\beta}(\mathbb{S}))^2)$, i.e.*

$$-\partial\Phi(0) \in \mathscr{H}((h^{4+\beta}(\mathbb{S}))^2,(h^{1+\beta}(\mathbb{S}))^2).$$

Proof. The regularity assumption follows directly from (12) and (13). Moreover, since the constant α fixed at the beginning of this section was arbitrary, we may replace α by β and all the assertions already established remain valid.

Let us now study the Fréchet derivative of Φ in 0. One can easily see that the highest order terms in (ρ_1,ρ_2) of $\partial\Phi(0,0)[(\rho_1,\rho_2)]$ are those obtained when differentiating the curvature operator.

Consider first $\partial_{\rho_1}\Phi_1(0)$. Since $\partial\kappa(0)[\rho] = -\rho'' - \rho$ for $\rho \in \rho^{4+\alpha}(\mathbb{S})$, we may decompose

$$\partial_{\rho_1}\Phi_1(0)[\rho_1] = A_{11} + B_{11},$$

where B_{11} is an operator of first order, i.e. $B_{11} \in \mathscr{L}(h^{2+\beta}(\mathbb{S}), h^{1+\beta}(\mathbb{S}))$,

$$A_{11}\rho_1 := \frac{1}{R_1^2}\mathscr{C}_1(0)(\Delta, \mathrm{tr}_1, \mathrm{tr}_2)^{-1}(0, \rho_1'', 0), \quad \forall \rho_1 \in h^{4+\beta}(\mathbb{S}),$$

and tr_i, $i = 1, 2$, is the trace operator with respect to Γ_i. We determine now a Fourier expansion for the highest order term of $\partial_{\rho_1}\Phi_1(0)[\rho_1]$. Given $\rho_1 \in h^{4+\beta}(\mathbb{S})$, the function $w := (\Delta, \mathrm{tr}_1, \mathrm{tr}_2)^{-1}(0, \rho_1'', 0)$ is the solution of the Dirichlet problem

$$\begin{cases} \Delta w = 0 & \text{in } \Omega, \\ \quad w = \rho_1'' & \text{on } \Gamma_1, \\ \quad w = 0 & \text{on } \Gamma_2. \end{cases} \tag{18}$$

If we expand

$$\rho_1(y) = \sum_m \widehat{\rho}_1(m)y^m \quad \text{and} \quad w(ry) = \sum_{m \in \mathbb{Z}} w_m(r)y^k$$

for $y \in \mathbb{S}$ and $R_2 < r < R_1$, we find out that $w_0 = 0$, and w_m solves, for $|m| \geq 1$, the problem

$$\begin{cases} w_m'' + \frac{1}{r}w_m' - \frac{m^2}{r^2}w_m = 0 & R_2 < r < R_1, \\ \qquad w_m(R_1) = -m^2\widehat{\rho}_1(m) \\ \qquad w_m(R_2) = 0. \end{cases}$$

Hence

$$w_m(r) = -\frac{R_2^m r^{-m} - R_2^{-m}r^m}{R_2^m R_1^{-m} - R_2^{-m}R_1^m}m^2\widehat{\rho}_1(m),$$

and therewith

$$A_{11}\rho_1(y) = \frac{1}{R_1^2}\mathscr{C}_1(0)w(y) = \frac{1}{R_1^2}\langle\nabla w(R_1 y)|y\rangle - \frac{1}{R_1^2}\frac{d}{dr}(w(ry))|_{r=R_1}$$

$$= -\frac{1}{R_1^3}\sum_{m \in \mathbb{Z}\setminus\{0\}}\frac{R_1^{|m|}R_2^{-|m|} + R_1^{-|m|}R_2^{|m|}}{R_1^{|m|}R_2^{-|m|} - R_1^{-|m|}R_2^{|m|}}|m|^3\widehat{\rho}_1(m)y^m.$$

We proceed similarly and write $\partial_{\rho_2}\Phi_1(0) = A_{12} + B_{12}$, where the operator $B_{12} \in \mathscr{L}(h^{2+\beta}(\mathbb{S}), h^{1+\beta}(\mathbb{S}))$ and

$$A_{12}\rho_2 := -\frac{1}{R_1 R_2}\mathscr{C}_1(0)(\Delta, \mathrm{tr}_1, \mathrm{tr}_2)^{-1}(0, 0, \rho_2'') \quad \forall \rho_2 \in h^{4+\beta}(\mathbb{S}).$$

Given $\rho_2 \in h^{4+\beta}(\mathbb{S})$, the function $w := (\Delta, \mathrm{tr}_1, \mathrm{tr}_2)^{-1}(0, 0, \rho_2'')$ is the solution of linear Dirichlet problem

$$\begin{cases} \Delta w = 0 & \text{in } \Omega, \\ w = 0 & \text{on } \Gamma_1, \\ w = \rho_2'' & \text{on } \Gamma_2. \end{cases} \tag{19}$$

A Fourier series ansatz, as we used before, yields that

$$w(ry) = -\sum_{m \in \mathbb{Z}\setminus\{0\}} \frac{R_1^m r^{-m} - R_1^{-m} r^m}{R_1^m R_2^{-m} - R_1^{-m} R_2^m} m^2 \widehat{\rho}_2(m) y^m$$

for all $y \in \mathbb{S}$ and $R_2 < r < R_1$, provided that $\rho_2 = \sum_{m \in \mathbb{Z}} \widehat{\rho}_2(m) y^m$. Hence,

$$A_{12} \sum_{m \in \mathbb{Z}} \widehat{\rho}_2(m) y^m = -\frac{1}{R_1^2 R_2} \sum_{m \in \mathbb{Z}\setminus\{0\}} \frac{2}{R_1^{|m|} R_2^{-|m|} - R_1^{-|m|} R_2^{|m|}} |m|^3 \widehat{\rho}_2(m) y^m.$$

We consider now the second component Φ_2 and continue our computation following the same scheme. The second diagonal element of the matrix $\partial \Phi(0)$ may be also written as the sum $\partial_{\rho_2} \Phi_2(0) = A_{22} + B_{22}$, with $B_{22} \in \mathscr{L}(h^{2+\beta}(\mathbb{S}), h^{1+\beta}(\mathbb{S}))$ and

$$A_{22}\rho_2 := -\frac{1}{R_2^2} \mathscr{C}_2(0)(\Delta, \mathrm{tr}_1, \mathrm{tr}_2)^{-1}(0, 0, \rho_2'') \quad \forall \rho_2 \in h^{4+\beta}(\mathbb{S}).$$

Using once more the expansion for the solution of (19), we find out that

$$A_{22} \sum_{m \in \mathbb{Z}} \widehat{\rho}_2(m) y^m = -\frac{1}{R_2^3} \sum_{m \in \mathbb{Z}\setminus\{0\}} \frac{R_1^{|m|} R_2^{-|m|} + R_1^{-|m|} R_2^{|m|}}{R_1^{|m|} R_2^{-|m|} - R_1^{-|m|} R_2^{|m|}} |m|^3 \widehat{\rho}_2(m) y^m.$$

for all $\rho_2 = \sum_{m \in \mathbb{Z}} \widehat{\rho}_2(m) y^m$ within $h^{4+\beta}(\mathbb{S})$. Finally, $\partial_{\rho_1} \Phi_2(0) = A_{21} + B_{21}$, where $B_{21} \in \mathscr{L}(h^{2+\beta}(\mathbb{S}), h^{1+\beta}(\mathbb{S}))$ and

$$A_{21}\rho_1 := \frac{1}{R_1 R_2} \mathscr{C}_2(0)(\Delta, \mathrm{tr}_1, \mathrm{tr}_2)^{-1}(0, \rho_1'', 0) \quad \forall \rho_2 \in h^{4+\beta}(\mathbb{S}).$$

Since $(\Delta, \mathrm{tr}_1, \mathrm{tr}_2)^{-1}(0, \rho_1'', 0)$ is the solution of (18), we may use the expansion found at that point of the proof and get

$$A_{21} \sum_{m \in \mathbb{Z}} \widehat{\rho}_1(m) y^m = -\frac{1}{R_1 R_2^2} \sum_{m \in \mathbb{Z}\setminus\{0\}} \frac{2}{R_1^{|m|} R_2^{-|m|} - R_1^{-|m|} R_2^{|m|}} |m|^3 \widehat{\rho}_1(m) y^m$$

for all functions $\rho_1 = \sum_{m \in \mathbb{Z}} \widehat{\rho}_1(m) y^m$ in $h^{4+\beta}(\mathbb{S})$.

Let us notice that the operators $A_{ij}, 1 \le i, j \le 2$, found above are all Fourier multipliers, since they are of the form

$$\sum_{m \in \mathbb{Z}} \widehat{\rho}(m) y^m \mapsto \sum_{m \in \mathbb{Z}} M_k \widehat{\rho}(m) y^m$$

with symbol $(M_k)_{k\in\mathbb{Z}} \subset \mathbb{C}$. Using [14, theorem 3.4], which is a theorem characterising multiplier operators acting between Hölder spaces based on some generalised Marcinkiewicz conditions for the symbol of the operator, we find out that $-A_{ii} \in \mathscr{H}(h^{4+\beta}(\mathbb{S}), h^{1+\beta}(\mathbb{S})), i = 1, 2$, and that $A_{12}, A_{21} \in \mathscr{L}(h^{2+\beta}(\mathbb{S}))$. This may be seen from the following relations:

$$\frac{2|m|^3}{R_1^{|m|}R_2^{-|m|} - R_1^{-|m|}R_2^{|m|}} \xrightarrow[|m|\to\infty]{} 0 \text{ and } \frac{R_1^{|m|}R_2^{-|m|} + R_1^{-|m|}R_2^{|m|}}{R_1^{|m|}R_2^{-|m|} - R_1^{-|m|}R_2^{|m|}} \xrightarrow[|m|\to\infty]{} 1.$$

Since $h^{2+\beta}(\mathbb{S})$ is an intermediate space between $h^{1+\beta}(\mathbb{S})$ and $h^{4+\beta}(\mathbb{S})$:

$$h^{2+\beta}(\mathbb{S}) = (h^{1+\beta}(\mathbb{S}), h^{4+\beta}(\mathbb{S}))_{1/3},$$

where $(\cdot|\cdot)$ denotes the interpolation functor introduced by Da Prato and Grisvard [11], we get by [19, proposition 2.4.1] that the elements on the diagonal of $\partial\Phi(0)$ generate analytic semigroups, that is

$$-\partial_{\rho_i}\Phi_i(0) \in \mathscr{H}(h^{4+\beta}(\mathbb{S}), h^{1+\beta}(\mathbb{S})), i = 1, 2,$$

while the elements on the secondary diagonal belong to $\mathscr{L}(h^{2+\beta}(\mathbb{S}), h^{1+\beta}(\mathbb{S}))$, and having thus lower order. We obtain then from [1, theorem 1.6.1] that the matrix $\partial\Phi(0)$ is a generator, which completes the proof. $\qquad\square$

We give now a short proof of our second main result, Theorem 2:

Proof (Proof of Theorem 2). Let $0 < \beta < \alpha$. Since Φ is analytic and $\partial\Phi(0)$ generates a strongly continuous and analytic semigroup, we find an open neighborhood $\widetilde{\mathcal{O}}$ of 0 in $h^{4+\beta}(\mathbb{S})$ such that $-\partial\Phi(\rho_1,\rho_2) \in \mathscr{H}((h^{4+\beta}(\mathbb{S}))^2, (h^{1+\beta}(\mathbb{S}))^2)$ for all $(\rho_1,\rho_2) \in \widetilde{\mathcal{O}}^2$. Letting $\mathcal{O} := \widetilde{\mathcal{O}} \cap h^{4+\alpha}(\mathbb{S}))$, we find that $-\partial\Phi(\rho_1,\rho_2) \in \mathscr{H}((h^{4+\alpha}(\mathbb{S}))^2, (h^{1+\alpha}(\mathbb{S}))^2)$ is, for all $(\rho_1,\rho_2) \in \mathcal{O}^2$, the realisation of the operator $-\partial\Phi(\rho_1,\rho_2) \in \mathscr{H}((h^{4+\beta}(\mathbb{S}))^2, (h^{1+\beta}(\mathbb{S}))^2)$. Hence, the assumptions of [19, theorem 8.4.1] are all satisfied and the desired assertion follows at once. $\qquad\square$

References

1. H. Amann. *Linear and Quasilinear Parabolic Problems*, Volume I, Birkhäuser, Basel, 1995.
2. G.B. Arfken and H.J. Weber. *Mathematical Methods for Physicists*, Elsevier Academic Press, Amsterdam, 2005.
3. A. Borisovich and A. Friedman. Symmetric-breaking bifurcation for free boundary problems, *Indiana Univ. Math. J.* 54:927–947, 2005.
4. H.M. Byrne and M.A. Chaplain. Growth of nonnecrotic tumors in the presence and absence of inhibitors, *Math. Biosci.* 130:151–181, 1995.
5. V. Cristini, J. Lowengrub, and Q. Nie. Nonlinear simulation of tumor growth, *Journal of Mathematical Biology* 46:191–224, 2003.
6. S.B. Cui. Analysis of a free boundary problem modeling tumor growth, *Acta Math. Sin. (Engl. Ser.)* 21(5):1071–1082, 2005.

7. S.B. Cui and J. Escher. Bifurcation analysis of an elliptic free boundary problem modelling the growth of avascular tumors, *SIAM J. Math. Anal.* 39(1):210–235, 2007.

8. S.B. Cui and J. Escher. Asymptotic behaviour of solutions of a multidimensional moving boundary problem modeling tumor growth, *Comm. Part. Diff. Eq.* 33(4):636–655, 2008.

9. S.B. Cui, J. Escher, and F. Zhou. Bifurcation for a free boundary problem with surface tension modelling the growth of multi-layer tumors, *J. Math. Anal. Appl.* 337(1):443–457, 2008.

10. S.B. Cui and A. Friedman. Analysis of a mathematical model of the growth of necrotic tumors, *J. Math. Anal. Appl.* 255:636–677, 2001.

11. G. Da Prato and P. Grisvard. Equations d'évolution abstraites nonlinéaires de type parabolique, *Ann. Mat. Pura Appl.* 120:329–336, 1979.

12. J. Escher and A-V. Matioc. Radially symmetric growth of nonnecrotic tumors, *Nonlinear Differ. Equ. Appl.* 17:1–20, 2010.

13. J. Escher and A-V. Matioc. Well-posedness and stability analysis for a moving boundary problem modelling the growth of nonnecrotic tumors, *Discrete and Continuous Dynamical System – B*, to appear, 2010.

14. J. Escher and B-V. Matioc. A moving boundary problem for periodic Stokesian Hele–Shaw flows, *Interfaces Free Bound.* 11:119–137, 2009.

15. A. Friedman and F. Reitich. Analysis of a mathematical model for the growth of tumors, *J. Math. Biol.* 38:262–284, 1999.

16. D. Gilbarg and T.S. Trudinger. *Elliptic Partial Differential Equations of Second Order*, Springer-Verlag, New York, 1998.

17. H.P. Greenspan. On the growth and stability of cell cultures and solid tumors, *J. Theor. Biol.* 56:229–242, 1976.

18. H.P. Greenspan. Models for the growth of a solid tumor by diffusion, *Stud. Appl. Math.* LI(4):317–340, 1972.

19. A. Lunardi. *Analytic Semigroups and Optimal Regularity in Parabolic Problems*, Birkhäuser, Basel, 1995.

Author Index

GPSR Compliance
The European Union's (EU) General Product Safety Regulation (GPSR) is a set
of rules that requires consumer products to be safe and our obligations to
ensure this.

If you have any concerns about our products, you can contact us on

ProductSafety@springernature.com

In case Publisher is established outside the EU, the EU authorized
representative is:

Springer Nature Customer Service Center GmbH
Europaplatz 3
69115 Heidelberg, Germany